Applied Probability and Statistics

BAILEY · The Elements of Stochastic Processes with Applications to the Natural Sciences

BARTHOLOMEW · Stochastic Models for Social Processes, *Second Edition*

BENNETT and FRANKLIN · Statistical Analysis in Chemistry and the Chemical Industry

BHAT · Elements of Applied Stochastic Processes

BLOOMFIELD · Fourier Analysis of Time Series: An Introduction

BOX and DRAPER · Evolutionary Operation: A Statistical Method for Process Improvement

BROWNLEE · Statistical Theory and Methodology in Science and Engineering, *Second Edition*

BURY · Statistical Models in Applied Science

CHERNOFF and MOSES · Elementary Decision Theory

CHOW · Analysis and Control of Dynamic Economic Systems

CLELLAND, deCANI, BROWN, BURSK, and MURRAY · Basic Statistics with Business Applications, *Second Edition*

COCHRAN · Sampling Techniques, *Second Edition*

COCHRAN and COX · Experimental Designs, *Second Edition*

COX · Planning of Experiments

COX and MILLER · The Theory of Stochastic Processes, *Second Edition*

DANIEL · Application of Statistics to Industrial Experimentation

DANIEL and WOOD · Fitting Equations to Data

DAVID · Order Statistics

DEMING · Sample Design in Business Research

DODGE and ROMIG · Sampling Inspection Tables. *Second Edition*

DRAPER and SMITH · Applied Regression Analysis

DUNN and CLARK · Applied Statistics: Analysis of Variance and Regression

ELANDT-JOHNSON · Probability Models and Statistical Methods in Genetics

FLEISS · Statistical Methods for Rates and Proportions

GNANADESIKAN · Methods for Statistical Data Analysis of Multivariate Observations

GOLDBERGER · Econometric Theory

GROSS and CLARK · Survival Distributions

GROSS and HARRIS · Fundamentals of Queueing Theory

GUTTMAN, WILKS and HUNTER · Introductory Engineering Statistics, *Second Edition*

HAHN and SHAPIRO · Statistical Models in Engineering

HALD · Statistical Tables and Formulas

HALD · Statistical Theory with Engineering Applications

HARTIGAN · Clustering Algorithms

HILDEBRAND, LAING and ROSENTHAL · Prediction Analysis of Cross Classifications

HOEL · Elementary Statistics, *Third Edition*

HOLLANDER and WOLFE · Nonparametric Statistical Methods

HUANG · Regression and Econometric Methods

JAGERS · Branching Processes with Biological Applications

continued on back

Applied
Regression
Analysis

Applied Regression Analysis

N. R. DRAPER

University of Wisconsin

H. SMITH

Rensselaer Polytechnic Institute

John Wiley & Sons, Inc.

New York · London · Sydney

13 14 15

LIBRARY OF CONGRESS CATALOG CARD NUMBER: 66-17641
PRINTED IN THE UNITED STATES OF AMERICA
ISBN 0 471 22170 8

Preface

In 1962 we were approached by officers of the Chemical Division of the American Society for Quality Control (A.S.Q.C.) and asked to prepare a short course on regression analysis. In order to do this effectively, we constructed a set of class notes dealing with a number of topics we felt were important to regression analysis practitioners. These notes were well received, many additions were made, and several revisions undertaken in the months that followed. This book represents the results of these labors.

We have tried to bring together in this book a number of procedures developed for regression problems in current use. Since our emphasis is on *practical* application, we have stated theoretical results without proofs in many cases. We have recognized that although much teaching of regression work is done in class, using no or comparatively primitive computing machinery (whether the class is in a university or a company), most practical applied regression work is now performed on high-speed digital computing equipment. Thus while the text can be used without any computing equipment at all (or perhaps with only a desk calculator), we have made use of computer printouts in some parts of the book. While the number of decimal places in these printouts is of little use in practice, we have included the numbers just as the computer printer would normally show them. We have also provided various exercises, some of which can be solved easily "by hand," and other more extensive ones for which use of an electronic computer would be helpful, though not absolutely essential.

This book provides a standard, basic course in multiple linear regression, but it also includes material that either has not previously appeared in a textbook or, if it has appeared, is not generally available. For example, Chapter 3 discusses the examination of residuals; Chapter 6 examines the methods employed as selection procedures in various types of regression programs; Chapter 8 discusses the planning of large regression studies; and Chapter 10 provides a basic introduction to the theory of nonlinear estimation.

Chapters 1 and 3 together provide a course in fitting a straight line without using matrix algebra at all. If parts of Chapter 2 are added, the idea of matrix representation of regression problems can be introduced as well. A one-semester regression course could consist of the material in Chapters 1 through 7, perhaps with the addition of Chapter 8 for industrial statisticians and industrial management students. It would be possible to cover the whole of the book in one semester if some previous knowledge of parts of the material is assumed. A more complete treatment would take two semesters; this would allow the instructor to add proofs of stated but unproved results, where desired, and would allow a full discussion of all exercises, some of which are extensive.

It is assumed that the reader has a basic knowledge of elementary statistics such as may be obtained from a typical first course. Tables of the F-distribution and the t-distribution are on pages 305–307.

We are grateful to a number of friends for help in various ways: Dr. Duane Meeter read and commented on an earlier version of Chapter 10. Ralph Pollard performed the calculations for Chapter 10. Bill Whiston contributed a great deal to Chapters 6 and 8. The layout for the stepwise computation algorithm is a result of his work for a Master's thesis. Ernest Gloyd helped with the manuscript, the index, and with proofreading. Proofreading was also done by W. Beggs, W. Hill, D. Khanna, R. Lochner, D. Lund, A. MacCormick, G. Minich, T. Mitchell, C. Palit, G. Phipp, D. Pierce, R. Pollard, D. Schauer, J. Sredni, D. Stoneman, and D. Wichern. Toby Mitchell and J. B. Willis each contributed an exercise. Finally, the book could never have been completed without the assistance of Miss Judy McClellan, who diligently typed and retyped the manuscript, and Mrs. Rita Eberhardt, who helped obtain solutions to all exercises.

We gratefully acknowledge permission from various authors and publishers to reproduce portions of several books and papers. Sources are given where the material appears.

N. R. DRAPER
H. SMITH

March 1966

Contents

CHAPTER 1

FITTING A STRAIGHT LINE
BY LEAST SQUARES

1.0. Introduction: The Need for Statistical Analysis

In today's industry, there is no shortage of "information." No matter how small or how straightforward a process may be, measuring instruments abound. They tell us such things as input temperature, concentration of reactant, per cent catalyst, steam temperature, consumption rate, pressure, and so on, depending on the characteristics of the process being studied. Some of these readings are available at regular intervals, every five minutes perhaps or every half hour; others are observed continuously. Still other readings are available with a little extra time and effort. Samples of the end product may be taken at intervals and, after analysis, may provide measurements of such things as purity, per cent yield, glossiness, breaking strength, color, or whatever other properties of the end product are important to the manufacturer or user. In many plants we find huge accumulations of data of these types, and many times the figures are simply collected without any real purpose or reason in mind. Or else there may have been a purpose years before, and although the purpose no longer exists, the figures are still religiously compiled hour by hour, day by day, week by week.

The purpose of this book is not, however, to explain what type of information should or should not be collected for any given process. The purpose is to explain in some detail something of the technique of extracting, from masses of data of the type just mentioned, the main features of the relationships hidden or implied in the tabulated figures.

In any system in which variable quantities change, it is of interest to examine the effects that some variables exert (or appear to exert) on others. There may in fact be a simple functional relationship between variables; in most physical processes this is the exception rather than the rule. Often there exists a functional relationship which is too complicated to grasp or

1

to describe in simple terms. In this case we may wish to approximate to this functional relationship by some simple mathematical function, such as a polynomial, which contains the appropriate variables and which graduates or approximates to the true function over some limited ranges of the variables involved. By examining such a graduating function we may be able to learn more about the underlying true relationship and to appreciate the separate and joint effects produced by changes in certain important variables.

Even where no sensible physical relationship exists between variables, we may wish to relate them by some sort of mathematical equation. While the equation might be physically meaningless, it may nevertheless be extremely valuable for predicting the values of some variables from knowledge of other variables, perhaps under certain stated restrictions.

In this book we shall use one particular method of obtaining a mathematical relationship. This involves the initial assumption that a certain type of relationship, linear in unknown parameters (except in Chapter 10, where nonlinear models are considered), holds. The unknown parameters are estimated under certain other assumptions with the help of available data, and a fitted equation is obtained. The value of the fitted equation can be gauged, and checks can be made on the underlying assumptions to see if any of these assumptions appears to be erroneous. The simplest example of this process involves the construction of a fitted straight line when pairs of observations $(X_1, Y_1), (X_2, Y_2), \ldots, (X_n, Y_n)$ are available. We shall deal with this in a simple algebraic way in this chapter. To handle problems involving large numbers of variables, matrix methods are essential. These are introduced in the context of fitting a straight line in Chapter 2, which also contains most of the basic results for more general regression problems. Some of these results are applied in Chapter 4 where the problem of relating a variable Y to two variables X_1 and X_2 by a planar equation is discussed. Chapter 5 is concerned with more complicated models, and Chapter 6 discusses certain procedures that can be used to select a "best" fitted equation. A specific problem is examined in Chapter 7, and some of the steps and problems involved in model building are given in Chapter 8. Chapter 9 discusses the regression treatment of analysis of variance problems, and Chapter 10 provides a brief introduction to nonlinear estimation. The computer printouts in the Appendix are referred to in various places throughout the book.

The reader will find that Chapters 1, 2, 3, and 4 comprise a basic course on regression analysis. We hope that readers with little or no knowledge of matrix algebra will find its step-by-step introduction in Chapter 2 to be convenient. Other readers will find it possible to omit the early sections of Chapter 2.

The reader with a very sound basic knowledge of regression analysis can regard the later sections of Chapter 2 as summary and review sections and may wish to skim quickly over the rest of Chapter 2 as well as Chapters 1, 3, and 4 (and possibly 5). We hope he will find the later chapters of interest and value.

An intermediate reader, falling between these two extremes, could profit by working steadily through the entire book, we believe.

We assume that anyone who uses this book has had a first course in statistics and understands certain basic ideas. These include the ideas of parameters, estimates, distributions, especially normal, mean and variance of a random variable, covariance between two variables, and simple hypothesis testing involving one- and two-sided t-tests and the F-test. We believe, however, that a reader whose knowledge of these topics is rusty or incomplete will nevertheless be able to make good progress.

We do not intend this as a comprehensive textbook on all aspects of regression analysis. Our intention is to provide a sound basic course plus material necessary to the solution of some practical regression problems.

We now take an early opportunity to introduce the reader to the computer printouts in the Appendices. Look, for example, on page 352 of Appendix A. Here we see observations taken at intervals from a steam plant at a large industrial concern. Ten variables are being recorded as follows:

1. Pounds of steam used monthly.
2. Pounds of real fatty acid in storage per month.
3. Pounds of crude glycerin made.
4. Average wind velocity (in mph).
5. Calendar days per month.
6. Operating days per month.
7. Days below 32°F.
8. Average atmospheric temperature (°F).
9. (Average wind velocity.)^2
10. Number of startups.

(The manner in which these figures on page 352 are read is explained on the pages preceding them.)

We can distinguish two main types of variables at this stage. We call these *independent variables* and *response variables*, or *dependent variables*. By independent variables we shall usually mean variables that can either be set to a desired value (for example, input temperature or catalyst feed rate) or else take values that can be observed but not controlled (for example, the outdoor humidity). As a result of changes that are deliberately made, or simply take place in the independent variables, an effect is transmitted to other variables, the response variables (for example, the final color or the purity of a chemical product). In general we shall be interested in finding out how changes in the independent variables affect the values of the response variables. If we can discover a simple relationship or dependence of a response variable on just one or a few independent variables we shall, of course, be pleased. The distinction between independent and response variables is not always completely clear cut and depends sometimes on our objectives. What may be considered a response variable at a midstage of a process may also be regarded as an independent variable in relation to (say) the final color of the product. In practice, however, the roles of variables are usually easily distinguished. Although the words "independent variables" are standard, they must not be too literally interpreted. In a particular body of data, two or more independent variables may vary together in some definite way due, perhaps, to the method in which an experiment is conducted. This is not usually desirable—for one thing it restricts the information on the separate roles of the variables—but it may often be unavoidable.

Returning to Appendix A, page 352, we examine the twenty-five sets of observations on the variables, one set for each of twenty-five different months. Our primary interest here is in the monthly amount of steam produced and how it changes due to variations in the other variables. Thus we shall regard variable 1 as a dependent variable in what follows and the others as independent variables.

We shall see how the method of analysis called the *method of least squares* can be used to examine data and to draw meaningful conclusions about dependency relationships that may exist. This method of analysis is often called *regression analysis* for the reasons mentioned in the next section. We shall present the least-squares method in the context of the simplest application, fitting the "best" straight line to given data in order to relate two variables X and Y, and will discuss how it can be extended to cases where more variables are involved.

1.1.   Linear Relationships between Two Variables

In much experimental work we wish to investigate how the changes in one variable affect another variable. Sometimes two variables are linked by an exact linear relationship. For example, if the resistance R of a simple circuit is kept constant the current I varies directly with the voltage V applied, for, by Ohm's law, $I = V/R$. If we were not aware of Ohm's law, we might obtain this relationship empirically by making changes in V and observing I, while keeping R fixed and then observing that the plot of I against V more or less gave a straight line through the origin. We say

regress', to tend to a mean

"more or less" because, although the relationship actually is exact, our measurements may be subject to slight errors and thus the plotted points would probably not fall exactly on the line but would vary randomly about it. For purposes of predicting I for a particular V (with R fixed), however, we should use the straight line through the origin. Sometimes a linear relationship is not exact (even apart from error) yet can be meaningful nevertheless. For example, suppose we consider the height and weight of adult males for some given population. If we plot the pair $(X_1, X_2) =$ (height, weight), a diagram something like Figure 1.1 will result. (Such a presentation is conventionally called a Scatter or Dot diagram.)

Note that for any given height there is a range of observed weights, and vice versa. This variation will be partially due to measurement errors but primarily due to variation between individuals. Thus no unique relationship between actual height and weight can be expected. But we can note that the average observed weight for a given observed height increases as height increases. This locus of average observed weight for given observed height (as height varies) is called the *regression curve* of weight on height. Let us denote it by $X_2 = f(X_1)$. There also exists a regression curve of height on weight similarly defined which we can denote by $X_1 = g(X_2)$. Let us assume that these two "curves" are both straight lines (which in general they may not be). In general, these two curves are *not* the same, as indicated by the two lines in the figure.

Suppose we now found we had recorded an individual's height but not his weight and we wished to estimate this weight; what could we do?

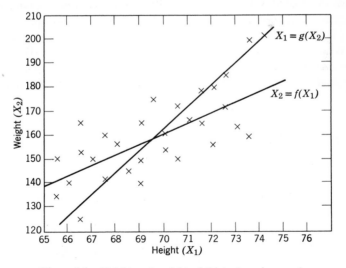

Figure 1.1   Height and weight of thirty American males.

From the regression line of weight on height we could find an average observed weight of individuals of the given height and use this as an estimate of the weight that we did not record.

[A pair of random variables such as (height, weight) follows some sort of bivariate probability distribution. When we are concerned with the dependence of a random variable Y on a quantity X which is variable but *not* a random variable, an equation that relates Y to X is usually called a *regression equation*. Although the name is, strictly speaking, incorrect, it is well established and conventional.]

We can see that whether a relationship is exactly linear or linear only insofar as mean values are concerned, knowledge of the relationship will be useful. (The relationship might, of course, be more complicated than linear but we shall consider this later.)

A linear relationship may be a valuable one even when we *know* that a linear relationship cannot be true. Consider the response relationship shown in Figure 1.2. It is obviously not linear over the range $0 \le X \le 100$. However, if we were interested primarily in the range $0 \le X \le 45$, a straight-line relationship evaluated from observations in this range might provide a perfectly adequate representation of the function *in this range*. The relationship thus fitted would, of course, not apply to values of X outside this restricted range and could not be used for predictive purposes outside this range.

(Similar remarks can be made when more than one independent variable is involved. Suppose we wish to examine the way in which a response Y depends on variables $X_1, X_2, \ldots, X_k$. We determine a regression equation from data which "cover" certain areas of the "X-space." Suppose the point $\mathbf{X}_0 = (X_{10}, X_{20}, \ldots, X_{k0})$ lies *outside* the regions covered by the original data. While we can mathematically obtain a predicted value $\hat{Y}(\mathbf{X}_0)$ for the response at the point $\mathbf{X}_0$, we must realize that reliance on such a prediction is extremely dangerous and becomes more dangerous the further $\mathbf{X}_0$ lies

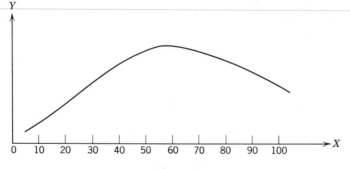

Figure 1.2

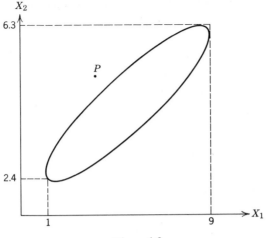

Figure 1.3

from the original regions, unless some additional knowledge is available that the regression equation is valid in a wider region of the X-space. Note that it is sometimes difficult to realize at first that a suggested point lies outside a region in a multi-dimensional space. To take a simple example consider the region defined by the ellipse in Figure 1.3. We see that there are points in the region for which $1 \leq X_1 \leq 9$ and for which $2.4 \leq X_2 \leq 6.3$. Although both coordinates of P lie within these ranges, P itself lies outside the region. When more dimensions are involved misunderstandings of this sort easily arise.)

1.2. Linear Regression: Fitting a Straight Line

We have mentioned that in many situations a straight-line relationship can be valuable in summarizing the observed dependence of one variable on another. We now show how the equation of such a straight line can be obtained by the method of least squares when data are available. Consider, in the printout on page 352, the 25 observations of variable 1 (pounds of steam used per month) and variable 8 (average atmospheric temperature in degrees Fahrenheit). The corresponding pairs of observations are given in Table 1.1 and are plotted in Figure 1.4.

Let us tentatively assume that the regression line of variable 1 which we shall denote by Y, on variable $8(X)$ has the form $\beta_0 + \beta_1 X$. Then we can write the linear, first-order model

$$Y = \beta_0 + \beta_1 X + \epsilon, \qquad (1.2.1)$$

Table 1.1   Twenty-five Observations of
Variables 1 and 8

Observation Number	Variable Number	
	1(*Y*)	8(*X*)
1	10.98	35.3
2	11.13	29.7
3	12.51	30.8
4	8.40	58.8
5	9.27	61.4
6	8.73	71.3
7	6.36	74.4
8	8.50	76.7
9	7.82	70.7
10	9.14	57.5
11	8.24	46.4
12	12.19	28.9
13	11.88	28.1
14	9.57	39.1
15	10.94	46.8
16	9.58	48.5
17	10.09	59.3
18	8.11	70.0
19	6.83	70.0
20	8.88	74.5
21	7.68	72.1
22	8.47	58.1
23	8.86	44.6
24	10.36	33.4
25	11.08	28.6

that is, for a given X, a corresponding observation Y consists of the value $\beta_0 + \beta_1 X$ plus an amount ϵ, the increment by which any individual Y may fall off the regression line. Equation (1.2.1) is the *model* of what we believe. We begin by assuming that it holds; but we shall have to inquire at a later stage if indeed it does. In many aspects of statistics it is necessary to assume a mathematical model to make progress. It might be well to emphasize that what we are usually doing is to *consider* or *tentatively entertain* our model. The model must always be critically examined somewhere along the line. It is our "opinion" of the situation at one stage of the investigation and our "opinion" must be changed if we find, at a later stage, that the facts are against it. β_0 and β_1 are called the *parameters* of the model.

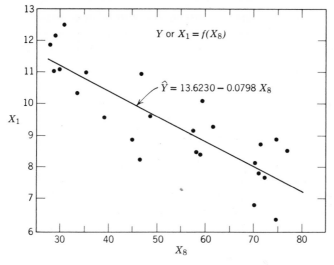

Figure 1.4

(*Note:* When we say that a model is linear or nonlinear, we are referring to linearity or nonlinearity *in the parameters*. The value of the highest power of an independent variable in the model is called the *order* of the model. For example,

$$Y = \beta_0 + \beta_1 X + \beta_{11} X^2 + \epsilon$$

is a second-order (in X) linear (in the β's) regression model. Unless a model is specifically called nonlinear it can be taken that it is linear in the parameters, and the word linear is usually omitted and understood. The order of the model could be of any size.)

Now β_0, β_1, and ϵ are unknown in Eq. (1.2.1), and in fact ϵ would be difficult to discover since it changes for each observation Y. However, β_0 and β_1 remain fixed and, although we cannot find them exactly without examining all possible occurrences of Y and X, we can use the information provided by the twenty-five observations in Table 1.1 to give us *estimates* b_0 and b_1 of β_0 and β_1; thus we can write

$$\hat{Y} = b_0 + b_1 X, \tag{1.2.2}$$

where $\hat{Y}$, read "Y hat," denotes the *predicted* value of Y for a given X, when b_0 and b_1 are determined. Equation (1.2.2) could then be used as a predictive equation; substitution for a value of X would provide a prediction of the true mean value of Y for that X.

Our estimation procedure will be that of *least squares*. Under certain assumptions to be mentioned later, this procedure has certain properties.

For the moment we state it as our chosen method of estimating the parameters without justification. Suppose we have available n sets of observations $(X_1, Y_1), (X_2, Y_2), \ldots, (X_n, Y_n)$. (In our example $n = 25$.) Then by Eq. (1.2.1) we can write

$$Y_i = \beta_0 + \beta_1 X_i + \epsilon_i, \qquad (1.2.3)$$

so that the sum of squares of deviations from the true line is

$$S = \sum_{i=1}^{n} \epsilon_i^2 = \sum_{i=1}^{n} (Y_i - \beta_0 - \beta_1 X_i)^2. \qquad (1.2.4)$$

We shall choose our estimates b_0 and b_1 to be the values which, when substituted for β_0 and β_1 in Eq. (1.2.4), produce the least possible value of S. (Note that X_i, Y_i are the fixed numbers which we have observed.) We can determine b_0 and b_1 by differentiating Eq. (1.2.4) first with respect to β_0 and then with respect to β_1 and setting the results equal to zero. Now

$$\frac{\partial S}{\partial \beta_0} = -2 \sum_{i=1}^{n} (Y_i - \beta_0 - \beta_1 X_i)$$

$$\frac{\partial S}{\partial \beta_1} = -2 \sum_{i=1}^{n} X_i(Y_i - \beta_0 - \beta_1 X_i) \qquad (1.2.5)$$

so that the estimates b_0 and b_1 are given by

$$\sum_{i=1}^{n} (Y_i - b_0 - b_1 X_i) = 0$$

$$\sum_{i=1}^{n} X_i(Y_i - b_0 - b_1 X_i) = 0 \qquad (1.2.6)$$

where we substitute (b_0, b_1) for (β_0, β_1), when we equate Eq. (1.2.5) to zero. From (1.2.6) we have

$$\sum_{i=1}^{n} Y_i - nb_0 - b_1 \sum_{i=1}^{n} X_i = 0$$

$$\sum_{i=1}^{n} X_i Y_i - b_0 \sum_{i=1}^{n} X_i - b_1 \sum_{i=1}^{n} X_i^2 = 0 \qquad (1.2.7)$$

or

$$b_0 n + b_1 \sum_{i=1}^{n} X_i = \sum_{i=1}^{n} Y_i$$

$$b_0 \sum_{i=1}^{n} X_i + b_1 \sum_{i=1}^{n} X_i^2 = \sum_{i=1}^{n} X_i Y_i \qquad (1.2.8)$$

These equations are called the *normal equations*.

The solution of Eq. (1.2.8) for b_1 is

$$b_1 = \frac{\sum X_i Y_i - [(\sum X_i)(\sum Y_i)]/n}{\sum X_i^2 - (\sum X_i)^2/n} = \frac{\sum (X_i - \bar{X})(Y_i - \bar{Y})}{\sum (X_i - \bar{X})^2} \qquad (1.2.9)$$

where all summations are from $i = 1$ to n and the two expressions for b_1 are just slightly different forms of the same quantity since, defining

$$\bar{X} = (X_1 + X_2 + \cdots + X_n)/n = \sum X_i/n,$$
$$\bar{Y} = (Y_1 + Y_2 + \cdots + Y_n)/n = \sum Y_i/n,$$

we have that

$$\sum (X_i - \bar{X})(Y_i - \bar{Y}) = \sum X_i Y_i - \bar{X} \sum Y_i - \bar{Y} \sum X_i + n\bar{X}\bar{Y}$$
$$= \sum X_i Y_i - n\bar{X}\bar{Y}$$
$$= \sum X_i Y_i - (\sum X_i)(\sum Y_i)/n.$$

The first form in Eq. (1.2.9) is normally used when actually computing the value of b_1. The solution of Eq. (1.2.8) for b_0 is

$$b_0 = \bar{Y} - b_1 \bar{X} \tag{1.2.10}$$

The quantity $\sum X_i^2$ is called the *uncorrected sum of squares of the X's,* and $(\sum X_i)^2/n$ is the *correction for the mean of the X's.* The difference is called the *corrected sum of squares of the X's.* Similarly, $\sum X_i Y_i$ is called the *uncorrected sum of products,* and $(\sum X_i)(\sum Y_i)/n$ is the *correction for the means.* The difference is called the *corrected sum of products of X and Y.* Substituting Eq. (1.2.10) into Eq. (1.2.2) gives the estimated regression equation

$$\hat{Y} = \bar{Y} + b_1(X - \bar{X}), \tag{1.2.11}$$

where b_1 is given by Eq. (1.2.9). Let us now perform these calculations on the data given as an example in Table 1.1. We find the following:

$$n = 25$$
$$\sum Y_i = 10.98 + 11.13 + \cdots + 11.08 = 235.60$$
$$\bar{Y} = 235.60/25 = 9.424$$
$$\sum X_i = 35.3 + 29.7 + \cdots + 28.6 = 1315$$
$$\bar{X} = 1315/25 = 52.60$$
$$\sum X_i Y_i = (10.98)(35.3) + (11.13)(29.7) + \cdots + (11.08)(28.6)$$
$$= 11821.4320$$
$$\sum X_i^2 = (35.3)^2 + (29.7)^2 + \cdots + (28.6)^2 = 76323.42$$
$$b_1 = \frac{\sum X_i Y_i - (\sum X_i)(\sum Y_i)/n}{\sum X_i^2 - (\sum X_i)^2/n}$$
$$b_1 = \frac{11821.4320 - (1315)(235.60)/25}{76323.42 - (1315)^2/25} = \frac{-571.1280}{7154.42}$$
$$b_1 = -0.079829.$$

Table 1.2   Fitted Values, Observations, and Residuals

Observation Number	Y_i	$\hat{Y}_i$	$Y_i - \hat{Y}_i$
1	10.98	10.81	0.17
2	11.13	11.25	−0.12
3	12.51	11.17	1.34
4	8.40	8.93	−0.53
5	9.27	8.72	0.55
6	8.73	7.93	0.80
7	6.36	7.68	−1.32
8	8.50	7.50	1.00
9	7.82	7.98	−0.16
10	9.14	9.03	0.11
11	8.24	9.92	−1.68
12	12.19	11.32	0.87
13	11.88	11.38	0.50
14	9.57	10.50	−0.93
15	10.94	9.89	1.05
16	9.58	9.75	−0.17
17	10.09	8.89	1.20
18	8.11	8.04	0.07
19	6.83	8.04	−1.21
20	8.88	7.68	1.20
21	7.68	7.87	−0.19
22	8.47	8.98	−0.51
23	8.86	10.06	−1.20
24	10.36	10.96	−0.60
25	11.08	11.34	−0.26

The fitted equation is thus

$$\hat{Y} = \bar{Y} + b_1(X - \bar{X})$$
$$\hat{Y} = 9.4240 - 0.079829(X - 52.60)$$
$$\hat{Y} = 13.623005 - 0.079829X.$$

The fitted regression line is plotted in Figure 1.4. We can tabulate for each of the 25 values X_i, at which a Y_i observation is available, the fitted value $\hat{Y}_i$ and the *residual* $Y_i - \hat{Y}_i$ as in Table 1.2. The residuals are given to the same number of places as the original data.

Note that since $\hat{Y}_i = \bar{Y} + b_1(X_i - \bar{X})$,

$$Y_i - \hat{Y}_i = (Y_i - \bar{Y}) - b_1(X_i - \bar{X}),$$

$$\sum_{i=1}^{n}(Y_i - \hat{Y}_i) = \sum_{i=1}^{n}(Y_i - \bar{Y}) - b_1\sum_{i=1}^{n}(X_i - \bar{X}) = 0.$$

Thus the sum of the residuals should be zero. In fact, it is -0.02 here—a rounding error. The sum of residuals in any regression problem is always zero when there is a β_0 term in the model as a consequence of the first normal equation. The omission of β_0 from a model implies that the response is zero when all the independent variables are zero. This is a very strong assumption which is usually unjustified. In a straight-line model $Y = \beta_0 + \beta_1 X + \epsilon$ omission of β_0 implies that the line passes through $X = 0$, $Y = 0$—that is, that the line has a zero *intercept* $\beta_0 = 0$ at $X = 0$. We note here, before the more general discussion in Section 5.4, that physical removal of β_0 from the model is always possible by "centering" the data, but that this is quite different from setting $\beta_0 = 0$. For example, if we write Eq. (1.2.1) in the form

$$Y - \bar{Y} = (\beta_0 + \beta_1 \bar{X} - \bar{Y}) + \beta_1(X - \bar{X}) + \epsilon$$

or

$$y = \beta_0' + \beta_1 x + \epsilon$$

say, where $y = Y - \bar{Y}$, $\beta_0' = \beta_0 + \beta_1 \bar{X} - \bar{Y}$, $x = X - \bar{X}$, then the least-squares estimates of β_0' and β_1 are given as follows:

$$b_1 = \frac{\sum (x_i - \bar{x})(y_i - \bar{y})}{\sum (x_i - \bar{x})^2} = \frac{\sum (X_i - \bar{X})(Y_i - \bar{Y})}{\sum (X_i - \bar{X})^2}$$

identical to Eq. (1.2.9), while

$$b_0' = \bar{y} - b_1 \bar{x} = 0, \qquad \text{since } \bar{x} = \bar{y} = 0,$$

whatever the value of b_1. Because this always happens, we can write the centered model as

$$Y - \bar{Y} = \beta_1(X - \bar{X}) + \epsilon$$

omitting the β_0' (intercept) term entirely. We have lost one parameter but there is a corresponding loss in the data since the quantities $Y_i - \bar{Y}$, $i = 1, 2, \ldots, n$ represent only $(n - 1)$ separate pieces of information due to the fact that their sum is zero, whereas $Y_1, Y_2, \ldots, Y_n$ represent n separate pieces of information. Effectively the "lost" piece of information has been used to enable the proper adjustments to be made to the model so that the intercept term can be removed.

1.3.   The Precision of the Estimated Regression

We now tackle the question of what measure of precision can be attached to our estimate of the regression line. Consider the following identity:

$$Y_i - \hat{Y}_i = Y_i - \bar{Y} - (\hat{Y}_i - \bar{Y}). \tag{1.3.1}$$

If we square both sides and sum from $i = 1$ to n, we obtain

$$\sum (Y_i - \hat{Y}_i)^2 = \sum \{(Y_i - \bar{Y}) - (\hat{Y}_i - \bar{Y})\}^2$$
$$= \sum \{(Y_i - \bar{Y})^2 + (\hat{Y}_i - \bar{Y})^2 - 2(Y_i - \bar{Y})(\hat{Y}_i - \bar{Y})\}$$
$$= \sum (Y_i - \bar{Y})^2 + \sum (\hat{Y}_i - \bar{Y})^2 - 2 \sum (Y_i - \bar{Y})(\hat{Y}_i - \bar{Y}).$$

The third term can be rewritten as

$$-2 \sum (Y_i - \bar{Y})b_1(X_i - \bar{X}) \quad \text{by (1.2.11)}$$
$$= -2b_1 \sum (Y_i - \bar{Y})(X_i - \bar{X})$$
$$= -2b_1{}^2 \sum (X_i - \bar{X})^2 \quad \text{by (1.2.9)}$$
$$= -2 \sum (\hat{Y}_i - \bar{Y})^2 \quad \text{by (1.2.11)}.$$

Thus

$$\sum (Y_i - \hat{Y}_i)^2 = \sum (Y_i - \bar{Y})^2 - \sum (\hat{Y}_i - \bar{Y})^2. \qquad (1.3.2)$$

Equation (1.3.2) can be rewritten

$$\sum (Y_i - \bar{Y})^2 = \sum (Y_i - \hat{Y}_i)^2 + \sum (\hat{Y}_i - \bar{Y})^2. \qquad (1.3.3)$$

Now $Y_i - \bar{Y}$ is the deviation of the ith observation from the overall mean and so the left-hand side of Eq. (1.3.3) is the sum of squares of deviations of the observations from the mean; this is shortened to *SS about the mean*, and is also the *corrected sum of squares of the Y's*. Since $Y_i - \hat{Y}_i$ is the deviation of the ith observation from its predicted or fitted value (the ith *residual*—see Chapter 3), and $\hat{Y}_i - \bar{Y}$ is the deviation of the predicted value of the ith observation from the mean, we can express Eq. (1.3.3) in words as follows:

$$\frac{\text{Sum of squares}}{\text{about the mean}} = \frac{\text{Sum of squares}}{\text{about regression}} + \frac{\text{Sum of squares}}{\text{due to regression}}.$$

This shows that, of the variation in the Y's about their mean, some of the variation can be ascribed to the regression line and some, $\sum (Y_i - \hat{Y}_i)^2$, to the fact that the actual observations do not all lie on the regression line —if they all did, the sum of squares about the regression would be zero! From this procedure we can see that a way of assessing how useful the regression line will be as a predictor is to see how much of the SS about the mean has fallen into the SS due to regression and how much into the SS about regression. We shall be pleased if the SS due to regression is much greater than the SS about regression, or what amounts to the same thing if the ratio $R^2 = $ (SS due to regression)/(SS about mean) is not too far from unity.

Any sum of squares has associated with it a number called its degrees of freedom. This number indicates how many independent pieces of information involving the n independent numbers $Y_1, Y_2, \ldots, Y_n$ are needed to compile the sum of squares. For example, the SS about the mean needs $(n - 1)$ independent pieces (for of the numbers $Y_1 - \bar{Y}, Y_2 - \bar{Y}, \ldots,$ $Y_n - \bar{Y}$, only $(n - 1)$ are independent since all n numbers sum to zero by definition of the mean). We can compute the SS due to regression from a single function of $Y_1, Y_2, \ldots, Y_n$, namely b_1 [since $\sum (\hat{Y}_i - \bar{Y})^2 = b_1^2 \sum (X_i - \bar{X})^2$], and so this sum of squares has one degree of freedom. By subtraction, the SS about regression has $(n - 2)$ degrees of freedom. Thus, corresponding to Eq. (1.3.3), we can show the split of degrees of freedom as

$$(n - 1) = (n - 2) + 1. \tag{1.3.4}$$

Using Eqs. (1.3.3) and (1.3.4) and employing alternative computational forms for the expressions of Eq. (1.3.3) we can construct an *analysis of variance* table in the following form:

Source	Sum of Squares	Degrees of Freedom	Mean Square
Regression	$b_1\left(\sum X_i Y_i - \dfrac{(\sum X_i)(\sum Y_i)}{n}\right)$	1	MS_R
About regression (residual)	By subtraction	$n - 2$	$s^2 = \dfrac{(SS)}{(n-2)}$
About mean (total, corrected for mean)	$\sum Y_i^2 - \dfrac{(\sum Y_i)^2}{n}$	$n - 1$	

The "Mean Square" column is obtained by dividing each sum of squares entry by its corresponding degrees of freedom.

A more general form of the analysis of variance table, which we do not need here but which is useful for comparison purposes later (see Section 2.2), is obtained by incorporating the correction factor for the mean of the Y's into the table where, for reasons explained in Section 2.2, it is called $SS(b_0)$. The table takes the form:

Source	Sum of Squares	Degrees of Freedom	Mean Square
Regression (b_0)	$SS(b_0) = \dfrac{(\sum Y_i)^2}{n}$	1	
Regression $(b_1 \mid b_0)$	$SS(b_1 \mid b_0) = b_1\left(\sum X_i Y_i - \dfrac{(\sum X_i)(\sum Y_i)}{n}\right)$	1	MS_R
Residual	By subtraction	$n - 2$	$s^2 = \dfrac{(SS)}{(n-2)}$
Total, uncorrected for mean	$\sum Y_i^2$	n	

The notation $SS(b_1 \mid b_0)$ is read "the sum of squares for b_1 after allowance has been made for b_0." The purpose of this notation is explained in Sections 2.2 and 2.7.

The mean square about regression, s^2, will provide an estimate based on $n - 2$ degrees of freedom of the variance about the regression, a quantity we shall call $\sigma^2_{Y \cdot X}$. If the regression equation were estimated from an indefinitely large number of observations, the variance about the regression would represent a measure of the error with which any observed value of Y could be predicted from a given value of X using the determined equation (see note 1 of Section 1.4).

We shall now carry out the calculations of this section for our example and then discuss a number of ways the regression equation can be examined. The SS due to regression is $b_1\{\sum X_i Y_i - (\sum X_i)(\sum Y_i)/n\}$

$$= \frac{\{\sum X_i Y_i - (\sum X_i)(\sum Y_i)/n\}^2}{\{\sum X_i^2 - (\sum X_i)^2/n\}}$$

$$= (-571.1280)^2/7154.42$$

$$= 45.59.$$

The Total (corrected) SS is $\sum Y_i^2 - (\sum Y_i)^2/n$

$$= 2284.1102 - (235.60)^2/25$$

$$= 63.82.$$

Table 1.3   The Analysis of Variance Table for the Example

Source	df	SS	MS	Calculated F Value
Total (corrected)	24	63.82		
Regression (b_1)	1	45.59	45.59	57.52
Residual	23	18.23	$s^2 = 0.7926$	

Note that the entries in this table are not in the same order as those in the corresponding theoretical table above. This makes no difference whatsoever. In computer printouts, for example, the order depends on the way in which the program is written. Careful inspection of analysis of variance tables should always be made and it should not be assumed that any particular order is standard. Our estimate of $\sigma^2_{Y \cdot X}$ is $s^2 = 0.7926$ based on 23 degrees of freedom. The F value will be explained shortly.

1.4. Examining the Regression Equation

Up to this point we have made no assumptions at all that involve probability distributions. A number of specified algebraic calculations have been made and that is all. We now make the basic assumptions that, in the model $Y_i = \beta_0 + \beta_1 X + \epsilon_i$, $i = 1, 2, \ldots, n$,

(1) ϵ_i is a random variable with mean zero and variance σ^2 (unknown), that is, $E(\epsilon_i) = 0$, $V(\epsilon_i) = \sigma^2$.

(2) ϵ_i and ϵ_j are uncorrelated, $i \neq j$, so that

$$\text{cov} (\epsilon_i, \epsilon_j) = 0.$$

Thus

$$E(Y_i) = \beta_0 + \beta_1 X_i, \qquad V(Y_i) = \sigma^2$$

and Y_i and Y_j, $i \neq j$, are uncorrelated. A further assumption, which is not immediately necessary and will be recalled when used, is that

(3) ϵ_i is a normally distributed random variable, with mean zero and variance σ^2 by (1), that is,

$$\epsilon_i \sim N(0, \sigma^2).$$

Under this additional assumption, ϵ_i, ϵ_j are not only uncorrelated but necessarily independent.

Notes

(1) σ^2 may or may not be equal to $\sigma^2_{Y \cdot X}$, the variance about the regression mentioned earlier. If the postulated model is the true model, then $\sigma^2 = \sigma^2_{Y \cdot X}$. If the postulated model is not the true model, then $\sigma^2 < \sigma^2_{Y \cdot X}$. It follows that s^2, the residual mean square which estimates $\sigma^2_{Y \cdot X}$ in any case, is an estimate of σ^2 if the model is correct but not otherwise. If $\sigma^2_{Y \cdot X} > \sigma^2$ we shall say that the postulated model is incorrect or *suffers from lack of fit*. Ways of deciding this will be discussed later.

(2) There is a tendency for errors that occur in many real situations to be normally distributed due to the Central Limit theorem. If an error term such as ϵ is a sum of errors from several sources, then no matter what the probability distribution of the separate errors may be, their sum ϵ will have a distribution that will tend more and more to the normal distribution as the number of components increases, by the Central Limit theorem. An experimental error in practice may be a composite of a meter error, an error due to a small leak in the system, an error in measuring the amount of catalyst used, and so on. Thus the assumption of normality is not unreasonable in most cases. In any case we shall later check the assumption by examining residuals (see Chapter 3).

We now use these assumptions in examining the regression equation.

Standard Error of the Slope b_1; Confidence Interval for β_1.

We know that $b_1 = \sum(X_i - \bar{X})(Y_i - \bar{Y})/\sum(X_i - \bar{X})^2$

$$= \sum(X_i - \bar{X})Y_i/\sum(X_i - \bar{X})^2$$

(since the other term removed from the numerator is $\sum(X_i - \bar{X})\bar{Y} = \bar{Y}\sum(X_i - \bar{X}) = 0$)

$$= \{(X_1 - \bar{X})Y_1 + \cdots + (X_n - \bar{X})Y_n\}/\sum(X_i - \bar{X})^2.$$

Now the variance of a function

$$F = a_1 Y_1 + a_2 Y_2 + \cdots + a_n Y_n$$

is

$$V(F) = a_1^2 V(Y_1) + a_2^2 V(Y_2) + \cdots + a_n^2 V(Y_n),$$

if the Y_i are pairwise uncorrelated and the a_i are constants; furthermore, if $V(Y_i) = \sigma^2$,

$$V(F) = (a_1^2 + a_2^2 + \cdots + a_n^2)\sigma^2$$
$$= (\sum a_i^2)\sigma^2.$$

In the expression for b_1, $a_i = (X_i - \bar{X})/\sum(X_i - \bar{X})^2$, since the X_i can be regarded as constants. Hence after reduction

$$V(b_1) = \frac{\sigma^2}{\sum(X_i - \bar{X})^2}. \tag{1.4.1}$$

(*Note:* An implication of this result is of interest. Suppose that, before any data had been collected, we wished to select the X_i values at which to take observations Y_i and wished to do it in a way that would minimize $V(b_1)$. Then the X_i chosen would have to maximize $\sum(X_i - \bar{X})^2$. The theoretical answer to this problem is that some X_i should be located at each of plus and minus infinity. The practical interpretation of this is that the X_i should be located at the extremes of the X-region in which experimental runs are possible. For example, if we wished to perform four runs, two would be placed at each extreme. This result is sensible and correct when the first-order model being tentatively entertained is *precisely the correct one*. When this is not true, and in practice it never really is true, this result may be quite wrong. In fact it has been shown by G. Box and N. Draper (*Journal of the American Statistical Association*, **54**, 622–654, 1959) that if the "region of interest" of the X's is scaled to the interval $(-R, R)$ and if we take $\bar{X} = 0$ and a straight line is to be fitted but some second-order tendency exists in the true model, then, the appropriate value for $\sum(X_i - \bar{X})^2$ is not infinity but a number slightly bigger than $NR/3$, where N is the number of X's to be chosen unless the model is nearly correct or the experimental error is very large. The general moral is that conclusions obtained by minimizing variance error only and assuming the postulated model to be correct are likely to be wrong in many practical design situations.)

The standard error of b_1 is the square root of the variance, that is,

$$\text{s.e.} (b_1) = \frac{\sigma}{\{\sum (X_i - \bar{X})^2\}^{1/2}}$$

or, if σ is unknown and we use the estimate s in its place, assuming the model is correct, the *estimated* standard error of b_1 is given by

$$\text{est. s.e.} (b_1) = \frac{s}{\{\sum (X_i - \bar{X})^2\}^{1/2}}. \qquad (1.4.2)$$

If we assume that the variations of the observations about the line are normal, that is, that the errors ϵ_i are all from the same normal distribution, $N(0, \sigma^2)$, it can be shown that we can assign $100(1 - \alpha)\%$ confidence limits for β_1 by calculating

$$b_1 \pm \frac{t(n - 2, 1 - \frac{1}{2}\alpha)s}{\{\sum (X_i - \bar{X})^2\}^{1/2}} \qquad (1.4.3)$$

where $t(n - 2, 1 - \frac{1}{2}\alpha)$ is the $(1 - \frac{1}{2}\alpha)$ percentage point of a t-distribution, with $(n - 2)$ degrees of freedom (the number of degrees of freedom on which the estimate s^2 is based).

On the other hand, if a test is appropriate, we can test the null hypothesis that β_1 is equal to β_{10}, where β_{10} is a specified value which could be zero, against the alternative that β_1 is different from β_{10} (usually stated "$H_0: \beta_1 = \beta_{10}$ versus $H_1: \beta_1 \neq \beta_{10}$") by calculating

$$t = \frac{(b_1 - \beta_{10})}{\{\text{est. s.e.} (b_1)\}}$$

$$= \frac{(b_1 - \beta_{10})\{\sum (X_i - \bar{X})^2\}^{1/2}}{s} \qquad (1.4.4)$$

and comparing $|t|$ with $t(n - 2, 1 - \frac{1}{2}\alpha)$ from a t-table with $(n - 2)$ degrees of freedom—the number on which s^2 is based. The test will be a two-sided test conducted at the $100(1 - \alpha)\%$ level in this form. Calculations for our example follow.

Example (*continued*).

$$V(b_1) = \sigma^2 / \sum (X_i - \bar{X})^2$$
$$= \sigma^2 / 7154.42$$
$$\text{est. } V(b_1) = s^2 / 7154.42$$
$$= 0.7926 / 7154.42$$
$$= 0.00011078$$
$$\text{est. s.e.} (b_1) = \sqrt{\text{est. } V(b_1)} = 0.0105$$

Suppose $\alpha = 0.05$, so that $t(23, 0.975) = 2.069$. Then 95% confidence limits for β_1 are $b_1 \pm t(23, 0.975) \cdot s/\{\sum (X_i - \bar{X})^2\}^{1/2}$,

or $\qquad\qquad\qquad -0.0798 \pm (2.069)(0.0105)$,

providing the interval $\quad -0.1015 \leq \beta_1 \leq -0.0581$.

In words, the true value β_1 lies in the interval (-0.1015 to -0.0581), and this statement is made with 95% confidence.

We shall also test the null hypothesis that the true value β_1 is zero, or that there is no relationship between atmospheric temperature and the amount of steam used. As noted above, we write (using $\beta_{10} = 0$),

$$H_0: \beta_1 = 0, \qquad H_1: \beta_1 \neq 0$$

and evaluate

$$t = b_1/\text{s.e. } (b_1)$$
$$= -0.0798/0.0105$$
$$= -7.60.$$

Since $|t| = 7.60$ exceeds the appropriate critical value of $t(23, 0.975) = 2.069$, $H_0: \beta_1 = 0$ is rejected. (Actually 7.60 also exceeds $t(23, 0.9995)$; we chose a two-sided 95% level test here however so that the confidence interval and the t-test would both make use of the same probability level. In this case we can effectively make the test by examining the confidence interval to see if it includes zero, as described below.) The data we have seen cause us to reject the idea that a linear relationship between Y and X might not exist.

If it had happened that the observed $|t|$ value had been smaller than the critical value we would have said that we *could not reject* the hypothesis. Note carefully that we do not use the word "accept," since we normally cannot accept a hypothesis. The most we can say is that on the basis of certain observed data we cannot reject it. It may well happen, however, that in another set of data we can find evidence which is contrary to our hypothesis and so reject it.

For example, if we see a man who is poorly dressed we may hypothesize, H_0: "This man is poor." If the man walks to save bus fare or avoids lunch to save lunch money, we have no reason to reject this hypothesis. Further observations of this kind may make us feel H_0 is true, but we still cannot accept it unless we know all the true facts about the man. However, a single observation against H_0, such as finding that the man owns a bank account containing $500,000 will be sufficient to reject the hypothesis.

Once we have the confidence interval for β_1 we do not actually have to compute the $|t|$ value for a particular t-test. It is simplest to examine the confidence interval for β_1 and see if it contains the value β_{10}. If it does,

predicted value of an *individual* observation will still be given by $\hat{Y}$ but will have variance

$$\sigma^2 + V(\hat{Y}_k) = \sigma^2 \left\{ 1 + \frac{1}{n} + \frac{(X_k - \bar{X})^2}{\sum (X_i - \bar{X})^2} \right\} \qquad (1.4.8)$$

with corresponding estimated value obtained by inserting s^2 for σ^2. Confidence values can be obtained in the same way as discussed; that is, we can calculate a 95% confidence interval for a new observation which will be centered on $\hat{Y}_k$ and whose length will depend on an estimate of this new variance from

$$\hat{Y}_k \pm t(v, 0.975) \left\{ 1 + \frac{1}{n} + \frac{(X_k - \bar{X})^2}{\sum (X_i - \bar{X})^2} \right\}^{\frac{1}{2}} s$$

where v is the number of degrees of freedom on which s^2 is based (and equals $n - 2$ here). A confidence interval for the mean of q new observations about $\hat{Y}_k$ is obtained similarly as follows:

Let $\bar{Y}_q$ be the mean of q future observations at X_k (where q could equal 1 to give the case above). Then

$$\bar{Y}_q \sim N(\beta_0 + \beta_1 X_k, \sigma_q^2)$$
$$\hat{Y}_k \sim N(\beta_0 + \beta_1 X_k, V(\hat{Y}_k))$$

so that

$$\bar{Y}_q - \hat{Y}_k \sim N(0, \sigma_q^2 + V(\hat{Y}_k))$$

and $[(\bar{Y}_q - \hat{Y}_k)/\text{est. s.e.}(\bar{Y}_q - \hat{Y}_k)]$ is distributed as a $t(v)$ variable where v is the number of degrees of freedom on which s^2, the estimate of σ^2, is based. Thus

$$\text{prob} \left\{ |\bar{Y}_q - \hat{Y}_k| \leq t(v, 0.975) \left[s^2 \left(\frac{1}{q} + \frac{1}{n} + \frac{(X_k - \bar{X})^2}{\sum (X_i - \bar{X})^2} \right) \right]^{\frac{1}{2}} \right\} = 0.95$$

from which we can obtain 95% confidence limits for $\bar{Y}_q$ about $\hat{Y}_k$ of

$$\hat{Y}_k \pm t(v, 0.975) \left[\frac{1}{q} + \frac{1}{n} + \frac{(X_k - \bar{X})^2}{\sum (X_i - \bar{X})^2} \right]^{\frac{1}{2}} s.$$

These limits are of course wider than those for the mean value of Y for given X_k, since these limits are the ones within which 95% of future observations at X_k (or future means of q observations at X_k as the case may be) are expected to lie.

F-Test for Significance of Regression

Since the Y_i are random variables, any function of them is also a random variable; two particular functions are MS_R, the mean square due to regression, and s^2, the mean square due to residual variation, which arise in the

then the hypothesis $\beta_1 = \beta_{10}$ cannot be rejected; if it does not, the hypothesis is rejected. This can be seen from Eq. (1.4.4), for $H_0: \beta_1 = \beta_{10}$ is rejected at the $(1 - \alpha)$ level if $|t| > t(n - 2, 1 - \frac{1}{2}\alpha)$, which implies that

$$|b_1 - \beta_{10}| > t(n - 2, 1 - \tfrac{1}{2}\alpha) \cdot s / \{\sum (X_i - \bar{X})^2\}^{\frac{1}{2}}$$

that is, that β_{10} lies outside the limits Eq. (1.4.3).

Standard Error of the Intercept; Confidence Interval for β_0

A confidence interval for β_0 and a test of whether or not β_0 is equal to some specified value can be constructed in a way similar to that just described for β_1. We can show (details in Section 2.3) that

$$\text{s.e.}(b_0) = \left\{ \frac{\sum X_i^2}{n \sum (X_i - \bar{X})^2} \right\}^{\frac{1}{2}} \sigma.$$

Thus $100(1 - \alpha)\%$ confidence limits for β_0 are given by

$$b_0 \pm t(n - 2, 1 - \tfrac{1}{2}\alpha) \left\{ \frac{\sum X_i^2}{n \sum (X_i - \bar{X})^2} \right\}^{\frac{1}{2}} s.$$

A t-test for the null hypothesis $H_0: \beta_0 = \beta_{00}$ against the alternative $H_1: \beta_0 \neq \beta_{00}$, where β_{00} is a specified value, will be rejected at the $(1 - \alpha)$ level if β_{00} falls outside the confidence interval, or not rejected if β_{00} falls inside, or may be conducted separately by finding the quantity

$$t = (b_0 - \beta_{00}) / \left\{ \frac{\sum X_i^2}{n \sum (X_i - \bar{X})^2} \right\}^{\frac{1}{2}} s$$

and comparing it with percentage points $t(n - 2, 1 - \frac{1}{2}\alpha)$ since $n - 2$ is the number of degrees of freedom on which s^2, the estimate of σ^2, is based. (*Note:* It is also possible to get a *joint confidence region* for β_0 and β_1 simultaneously by applying the formula (2.6.15).)

Standard Error of $\hat{Y}$

We have shown that the regression equation is

$$\hat{Y} = \bar{Y} + b_1(X - \bar{X})$$

where both $\bar{Y}$ and b_1 are subject to error, which will influence $\hat{Y}$. Now if a_i and c_i are constants, and

$$a = a_1 Y_1 + a_2 Y_2 + \cdots + a_n Y_n,$$
$$c = c_1 Y_1 + c_2 Y_2 + \cdots + c_n Y_n,$$

then provided that Y_i and Y_j are uncorrelated when $i \neq j$, and if $V(Y_i) = \sigma^2$, all i,

$$\text{cov}(a, c) = (a_1 c_1 + a_2 c_2 + \cdots + a_n c_n)\sigma^2. \quad (1.4.5)$$

It follows that setting $a = \bar{Y}$ implies $a_i = 1/n$, and setting $c = b_1$ implies $c_i = (X_i - \bar{X})/\sum(X_i - \bar{X})^2$, so that

$$\text{cov}(\bar{Y}, b_1) = 0,$$

that is, $\bar{Y}$ and b_1 are uncorrelated random variables. Thus the variance of the predicted mean value of Y, $\hat{Y}_k$ at a specific value X_k, of X is

$$V(\hat{Y}_k) = V(\bar{Y}) + (X_k - \bar{X})^2 V(b_1)$$

$$= \frac{\sigma^2}{n} + \frac{(X_k - \bar{X})^2 \sigma^2}{\sum(X_i - \bar{X})^2}. \quad (1.4.6)$$

Hence

$$\text{est. s.e.}(\hat{Y}_k) = s\left\{\frac{1}{n} + \frac{(X_k - \bar{X})^2}{\sum(X_i - \bar{X})^2}\right\}^{1/2}. \quad (1.4.7)$$

This is a minimum when $X_k = \bar{X}$ and increases as we move X_k away from $\bar{X}$ in either direction. In other words, the greater distance an X_k is (in either direction) from $\bar{X}$, the larger is the error we may expect to make when predicting, from the regression line, the mean value of Y at X_k. This is intuitively very reasonable. To state the matter loosely, we might expect to make our "best" predictions in the "middle" of our observed range of X and would expect our predictions to be less good away from the "middle." For values of X outside our experience—that is, outside the range observed—we should expect our predictions to be less good, becoming worse as we moved away from the range of observed X-values.

Example (*continued*).

$$n = 25, \qquad \sum(X_i - \bar{X})^2 = 7154.42$$

$$s^2 = 0.7926, \qquad \bar{X} = 52.60.$$

$$\text{est. } V(\hat{Y}_k) = s^2\left\{\frac{1}{n} + \frac{(X_k - \bar{X})^2}{\sum(X_i - \bar{X})^2}\right\}$$

$$= 0.7926\left\{\frac{1}{25} + \frac{(X_k - 52.60)^2}{7154.42}\right\}.$$

For example, if $X_k = \bar{X}$, then

$$\text{est. } V(\hat{Y}_k) = 0.7926\{\tfrac{1}{25}\} = 0.031704.$$

That is, est. s.e. $(\hat{Y}_k) = \sqrt{0.031704} = 0.1781.$

If $X_k = 28.6$,

$$\text{est. } V(\hat{Y}_k) = 0.7926\left\{\frac{1}{25} + \frac{(28.60 - 52.60)^2}{7154.42}\right\} = 0.095516.$$

That is, est. s.e.$(\hat{Y}_k) = \sqrt{0.095516} = 0.3091.$

Correspondingly, est. s.e. $(\hat{Y}_k)$ is 0.3091 when $X_k = 76.60$.

The 95% confidence limits for the true mean value of Y for a given X_k are then given by $\hat{Y}_k \pm (2.069)$ est. s.e. $(\hat{Y}_k)$. Figure 1.5 illustrates the situation; two curved lines about the regression line are the loci of the 95% confidence limits and show how the limits change as the position of X_k changes. These curves are hyperbolae.

These limits can be interpreted as follows. Suppose repeated samples of Y_i are taken of the same size and at the same fixed values of X as were used to determine the fitted line above. Then of all the 95% confidence intervals constructed for the mean value of Y for a given value of X, X_k say, 95% of these intervals will contain the true value of this mean value of Y at X_k. If only one prediction $\hat{Y}_k$ is made, say for $X = X_k$, then the probability that the calculated interval at this point will contain the true mean is 0.95.

The variance and standard error shown above apply to the predicted *mean value of Y for a given X_k*. Since the actual observed value of Y varies about the true mean value with variance σ^2 (independent of the $V(\hat{Y})$), a

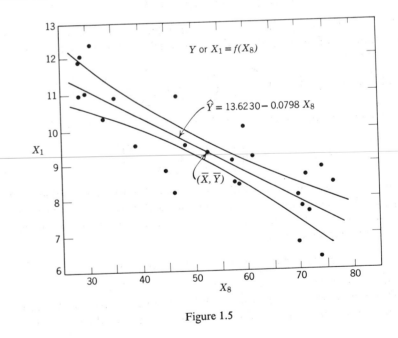

Figure 1.5

analysis of variance table shown in Section 1.3. These functions then have their own distribution, mean, variance, and moments. It can be shown that the mean values are as follows:

$$E(MS_R) = \sigma^2 + \beta_1^2 \sum (X_i - \bar{X})^2 \qquad (1.4.9)$$

$$E(s^2) = \sigma^2$$

where, if Z is a random variable, $E(Z)$ denotes its mean value or expected value. Suppose that the errors ϵ_i are independent $N(0, \sigma^2)$ variables. Then it can be shown that *if* $\beta_1 = 0$, the variable MS_R multiplied by its degrees of freedom (here, one) and divided by σ^2 follows a χ^2 distribution with the same (1) number of degrees of freedom. In addition, $(n - 2)s^2/\sigma^2$ follows a χ^2 distribution with $(n - 2)$ degrees of freedom. Also since these two variables are independent, a statistical theorem tells us that the ratio

$$F = \frac{MS_R}{s^2} \qquad (1.4.10)$$

follows an F distribution with (here) 1 and $(n - 2)$ degrees of freedom provided (recall) that $\beta_1 = 0$. This fact can thus be used as a test of $\beta_1 = 0$. We compare the ratio $F = MS_R/s^2$ with the $100(1 - \alpha)\%$ point of the tabulated $F(1, n - 2)$ distribution in order to determine whether β_1 can be considered nonzero on the basis of the data we have seen.

Example (*continued*). From Table 1.3, we see that the required ratio is $F = 45.59/0.7926 = 57.52$. If we look up percentage points of the $F(1, 23)$ distribution, we see that the 95% point $F(1, 23, 0.95) = 4.28$. Since the calculated F exceeds the critical F value in the table, that is $F = 57.52 > 4.28$, we reject the hypothesis $H_0: \beta_1 = 0$ running a risk of less than 5% of being wrong.

(*Note*. In the particular case of fitting a straight line, this F-test for "regression" is exactly the same as the t-test for $\beta_1 = 0$ given earlier. This is because the ratio

$$\frac{MS_R}{s^2} = \frac{b_1\{\sum (X_i - \bar{X})(Y_i - \bar{Y})\}}{s^2}$$

$$= \frac{b_1^2 \sum (X_i - \bar{X})^2}{s^2}$$

(by the definition of b_1)

$$= \left[\frac{b_1\{\sum (X_i - \bar{X})^2\}^{1/2}}{s} \right]^2$$

$$= t^2$$

from Eq. (1.4.4). Since the variable $F(1, n - 2)$ is the square of the $t(n - 2)$ variable, exactly the same test results. When there are more regression coefficients the overall F-test for regression, which is the extension of the one given here, does not correspond to the t-test of a coefficient. However, tests for *individual coefficients* can be made either in t or $t^2 = F$ form by a similar argument. The F form is often seen in computer programs.)

In the example we had an observed F value of 57.52 and an observed t value of -7.60. Note that $(-7.6)^2 = 57.76$. Were it not for round-off error this would be equal to the observed value of F.

Percentage Variation Explained

We define $R^2 = $ (SS due to regression)/(Total SS, corrected for mean). Then R^2 measures the "proportion of total variation about the mean $\overline{Y}$ explained by the regression." It is often expressed as a percentage by multiplying by 100.

Example (*continued*).   From Table 1.3,

$$R^2 = \frac{45.59}{63.82} \times 100 = 71.44\%.$$

Thus the regression equation obtained explains 71.44% of the total variation.

1.5.   Lack of Fit and Pure Error

We have already remarked that the fitted regression line is a calculated line based on a certain model or assumption, an assumption we should not blindly accept but should *tentatively entertain*. In certain circumstances we can check whether or not the model is correct. First, we can examine the consequences of an incorrect model. Let us recall that $e_i = Y_i - \hat{Y}_i$ is the *residual* at X_i. This is the amount by which the actual observed value Y_i exceeds the fitted value $\hat{Y}_i$. As shown in Section 1.2, $\sum e_i = 0$. The residuals contain all available information on the ways in which the fitted model fails to properly explain the observed variation in the dependent variable Y. (For examination of residuals, see Chapter 3.) Let $\eta_i = E(Y_i)$ denote the value given by the true model, whatever it is, at $X = X_i$. Then we can write

$$Y_i - \hat{Y}_i = (Y_i - \hat{Y}_i) - E(Y_i - \hat{Y}_i) + E(Y_i - \hat{Y}_i)$$

$$= \{(Y_i - \hat{Y}_i) - (\eta_i - E(\hat{Y}_i))\} + (\eta_i - E(\hat{Y}_i))$$

$$= q_i + B_i,$$

say, where

$$q_i = \{(Y_i - \hat{Y}_i) - (\eta_i - E(\hat{Y}_i))\}, \qquad B_i = \eta_i - E(\hat{Y}_i).$$

The quantity B_i is the bias error at $X = X_i$. If the model is correct then $E(\hat{Y}_i) = \eta_i$ and B_i is zero. If the model is not correct $E(\hat{Y}_i) \neq \eta_i$ and B_i is not zero but has a value that depends on the true model and the value of X_i. The quantity q_i is a random variable that has zero mean since $E(q_i) = E(Y_i - \hat{Y}_i) - (\eta_i - E(\hat{Y}_i)) = \eta_i - E(\hat{Y}_i) - (\eta_i - E(\hat{Y}_i)) = 0$, and this is true whether the model is correct or not, that is, whether $E(\hat{Y}_i) = \eta_i$ or not.

The q_i, it can be shown, are correlated, and the quantity $q_1{}^2 + q_2{}^2 + \cdots + q_n{}^2$ has expected or mean value $(n - 2)\sigma^2$ where $V(Y_i) = V(\epsilon_i) = \sigma^2$ is the error variance. From this it can be shown further that the residual mean square, value

$$\frac{1}{n - 2} \left\{ \sum_{i=1}^{n} (Y_i - \hat{Y}_i)^2 \right\}$$

has expected or mean value σ^2 if the postulated model is of the correct form, or $\sigma^2 + \sum B_i{}^2/(n - 2)$ if the model is not correct. If the model is correct, that is, $B_i = 0$, then the residuals are (correlated) random deviations q_i and the residual mean square can be used as an estimate of the error variance σ^2.

However, if the model is not correct, that is, $B_i \neq 0$, then the residuals contain both random (q_i) and systematic (B_i) components. We can refer to these as the variance error and bias error components of the residuals, respectively. Also the residual mean square will tend to be inflated and will no longer provide a satisfactory measure of the random variation present in the observations. (Since, however, the mean square *is* a random variable it may, by chance, not have a large value even when bias does exist. For some similar work on the general regression case see Section 2.12.)

In the simple case of fitting a straight line, bias error can usually be detected merely by examining a plot of the data (see Figure 1.7 for example). When the model is more complicated and/or involves more variables this may not be possible. If a prior estimate of σ^2 is available (by "prior estimate" we mean one obtained from previous experience of the variation in the situation being studied) we can see (or test by an F-test) whether or not the residual mean square is significantly greater than this prior estimate. If it is significantly greater we say that there is lack of fit and we would reconsider the model which would be inadequate in its present form. If no prior estimate of σ^2 is available, but repeat measurements of Y (i.e., two or more measurements) have been made at the same value of X, we can use these repeats to obtain an estimate of σ^2. Such an estimate is said to represent "pure error" because, if the setting of X is identical for two observations, only the random variation can influence the results and provide differences between them. Such differences will usually then provide an estimate of σ^2 which is much more reliable than we can obtain from any other source. For this reason, it is sensible when designing experiments to arrange for repeat observations.

(*Note.* It is important to understand that repeated runs must be genuine repeats and not just repetitions of the same reading. For example, suppose we were attempting to relate, by regression methods, $Y =$ Intelligence Quotient to $X =$ height of person. A genuine repeat point would be obtained if we measured the separate I.Q.'s of two people of exactly the same height. If, however, we measured *twice* the I.Q. of one person, of some specified height, this would not be a genuine repeat point in our context but merely a "reconfirmed" single point. The latter would, it is true, supply information on the variation of the testing method which is part of the variation σ^2, but it would *not* provide information on the variation in I.Q. between people of the same height, which is the σ^2 of our problem. In chemical experiments, a succession of readings made during steady state running does not provide genuine repeat points. However, if a certain set of conditions was reset anew, after intermediate runs at other X-levels had been made, and provided drifts in the response level had not occurred, genuine repeat runs would be obtained. With this in mind, repeat runs that show remarkable agreement, which is contrary to expectation, should always be regarded cautiously and subjected to additional investigation.)

The pure error estimate of σ^2 is found as follows (the same formulae also apply when there is more than one independent variable). Suppose

$Y_{11}, Y_{12}, \ldots, Y_{1n_1}$ are n_1 repeat observations at X_1.

$Y_{21}, Y_{22}, \ldots, Y_{2n_2}$ are n_2 repeat observations at X_2.

$\cdots$

$Y_{k1}, Y_{k2}, \ldots, Y_{kn_k}$ are n_k repeat observations at X_k.

The contribution to the pure error sum of squares from the X_1 readings is then

$$\sum_{u=1}^{n_1}(Y_{1u} - \overline{Y}_1)^2 = \sum_{u=1}^{n_1} Y_{1u}^2 - n_1 \overline{Y}_1^{\,2}$$

where $\overline{Y}_1 = (Y_{11} + Y_{12} + \cdots + Y_{1n_1})/n_1$, and this sum of squares has $(n_1 - 1)$ degrees of freedom. Similar quantities are evaluated for the other sets of Y's. The total SS (pure error) $= \sum_{i=1}^{k} \sum_{u=1}^{n_i} (Y_{iu} - \overline{Y}_i)^2$ with total degrees of freedom $= \sum_{i=1}^{k} (n_i - 1) = \sum_{i=1}^{k} n_i - k = n_e$, say.

then the hypothesis $\beta_1 = \beta_{10}$ cannot be rejected; if it does not, the hypothesis is rejected. This can be seen from Eq. (1.4.4), for $H_0: \beta_1 = \beta_{10}$ is rejected at the $(1 - \alpha)$ level if $|t| > t(n - 2, 1 - \tfrac{1}{2}\alpha)$, which implies that

$$|b_1 - \beta_{10}| > t(n - 2, 1 - \tfrac{1}{2}\alpha) \cdot s/\{\sum (X_i - \bar{X})^2\}^{1/2}$$

that is, that β_{10} lies outside the limits Eq. (1.4.3).

Standard Error of the Intercept; Confidence Interval for β_0

A confidence interval for β_0 and a test of whether or not β_0 is equal to some specified value can be constructed in a way similar to that just described for β_1. We can show (details in Section 2.3) that

$$\text{s.e.}(b_0) = \left\{ \frac{\sum X_i^2}{n \sum (X_i - \bar{X})^2} \right\}^{1/2} \sigma.$$

Thus $100(1 - \alpha)\%$ confidence limits for β_0 are given by

$$b_0 \pm t(n - 2, 1 - \tfrac{1}{2}\alpha)\left\{ \frac{\sum X_i^2}{n \sum (X_i - \bar{X})^2} \right\}^{1/2} s.$$

A t-test for the null hypothesis $H_0: \beta_0 = \beta_{00}$ against the alternative $H_1: \beta_0 \neq \beta_{00}$, where β_{00} is a specified value, will be rejected at the $(1 - \alpha)$ level if β_{00} falls outside the confidence interval, or not be rejected if β_{00} falls inside, or may be conducted separately by finding the quantity

$$t = (b_0 - \beta_{00}) \Big/ \left\{ \frac{\sum X_i^2}{n \sum (X_i - \bar{X})^2} \right\}^{1/2} s$$

and comparing it with percentage points $t(n - 2, 1 - \tfrac{1}{2}\alpha)$ since $n - 2$ is the number of degrees of freedom on which s^2, the estimate of σ^2, is based. (*Note:* It is also possible to get a *joint confidence region* for β_0 and β_1 simultaneously by applying the formula (2.6.15).)

Standard Error of $\hat{Y}$

We have shown that the regression equation is

$$\hat{Y} = \bar{Y} + b_1(X - \bar{X})$$

where both $\bar{Y}$ and b_1 are subject to error, which will influence $\hat{Y}$. Now if a_i and c_i are constants, and

$$a = a_1 Y_1 + a_2 Y_2 + \cdots + a_n Y_n,$$
$$c = c_1 Y_1 + c_2 Y_2 + \cdots + c_n Y_n,$$

then provided that Y_i and Y_j are uncorrelated when $i \neq j$, and if $V(Y_i) = \sigma^2$, all i,

$$\text{cov} (a, c) = (a_1 c_1 + a_2 c_2 + \cdots + a_n c_n)\sigma^2. \tag{1.4.5}$$

It follows that setting $a = \bar{Y}$ implies $a_i = 1/n$, and setting $c = b_1$ implies $c_i = (X_i - \bar{X})/\sum (X_i - \bar{X})^2$, so that

$$\text{cov} (\bar{Y}, b_1) = 0,$$

that is, $\bar{Y}$ and b_1 are uncorrelated random variables. Thus the variance of the predicted mean value of Y, $\hat{Y}_k$ at a specific value X_k, of X is

$$V(\hat{Y}_k) = V(\bar{Y}) + (X_k - \bar{X})^2 V(b_1)$$

$$= \frac{\sigma^2}{n} + \frac{(X_k - \bar{X})^2 \sigma^2}{\sum (X_i - \bar{X})^2}. \tag{1.4.6}$$

Hence

$$\text{est. s.e. } (\hat{Y}_k) = s \left\{ \frac{1}{n} + \frac{(X_k - \bar{X})^2}{\sum (X_i - \bar{X})^2} \right\}^{1/2}. \tag{1.4.7}$$

This is a minimum when $X_k = \bar{X}$ and increases as we move X_k away from $\bar{X}$ in either direction. In other words, the greater distance an X_k is (in either direction) from $\bar{X}$, the larger is the error we may expect to make when predicting, from the regression line, the mean value of Y at X_k. This is intuitively very reasonable. To state the matter loosely, we might expect to make our "best" predictions in the "middle" of our observed range of X and would expect our predictions to be less good away from the "middle." For values of X outside our experience—that is, outside the range observed—we should expect our predictions to be less good, becoming worse as we moved away from the range of observed X-values.

Example (*continued*).

$$n = 25, \qquad \sum (X_i - \bar{X})^2 = 7154.42$$

$$s^2 = 0.7926, \qquad \bar{X} = 52.60.$$

$$\text{est. } V(\hat{Y}_k) = s^2 \left\{ \frac{1}{n} + \frac{(X_k - \bar{X})^2}{\sum (X_i - \bar{X})^2} \right\}$$

$$= 0.7926 \left\{ \frac{1}{25} + \frac{(X_k - 52.60)^2}{7154.42} \right\}.$$

For example, if $X_k = \bar{X}$, then

$$\text{est. } V(\hat{Y}_k) = 0.7926\{\tfrac{1}{25}\} = 0.031704.$$

That is, est. s.e. $(\hat{Y}_k) = \sqrt{0.031704} = 0.1781.$

If $X_k = 28.6$,

$$\text{est. } V(\hat{Y}_k) = 0.7926\left\{\frac{1}{25} + \frac{(28.60 - 52.60)^2}{7154.42}\right\} = 0.095516.$$

That is, est. s.e.$(\hat{Y}_k) = \sqrt{0.095516} = 0.3091$.

Correspondingly, est. s.e. $(\hat{Y}_k)$ is 0.3091 when $X_k = 76.60$.

The 95% confidence limits for the true mean value of Y for a given X_k are then given by $\hat{Y}_k \pm (2.069)$ est. s.e. $(\hat{Y}_k)$. Figure 1.5 illustrates the situation; two curved lines about the regression line are the loci of the 95% confidence limits and show how the limits change as the position of X_k changes. These curves are hyperbolae.

These limits can be interpreted as follows. Suppose repeated samples of Y_i are taken of the same size and at the same fixed values of X as were used to determine the fitted line above. Then of all the 95% confidence intervals constructed for the mean value of Y for a given value of X, X_k say, 95% of these intervals will contain the true value of this mean value of Y at X_k. If only one prediction $\hat{Y}_k$ is made, say for $X = X_k$, then the probability that the calculated interval at this point will contain the true mean is 0.95.

The variance and standard error shown above apply to the predicted *mean value of* Y for a given X_k. Since the actual observed value of Y varies about the true mean value with variance σ^2 (independent of the $V(\hat{Y})$), a

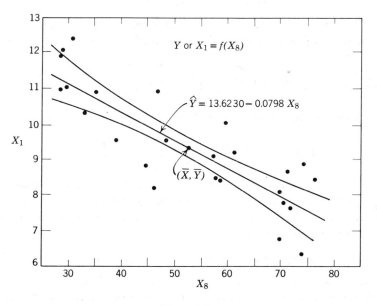

Figure 1.5

predicted value of an *individual* observation will still be given by $\hat{Y}$ but will have variance

$$\sigma^2 + V(\hat{Y}_k) = \sigma^2 \left\{ 1 + \frac{1}{n} + \frac{(X_k - \bar{X})^2}{\sum (X_i - \bar{X})^2} \right\} \qquad (1.4.8)$$

with corresponding estimated value obtained by inserting s^2 for σ^2. Confidence values can be obtained in the same way as discussed; that is, we can calculate a 95% confidence interval for a new observation which will be centered on $\hat{Y}_k$ and whose length will depend on an estimate of this new variance from

$$\hat{Y}_k \pm t(v, 0.975) \left\{ 1 + \frac{1}{n} + \frac{(X_k - \bar{X})^2}{\sum (X_i - \bar{X})^2} \right\}^{1/2} s$$

where v is the number of degrees of freedom on which s^2 is based (and equals $n - 2$ here). A confidence interval for the mean of q new observations about $\hat{Y}_k$ is obtained similarly as follows:

Let $\bar{Y}_q$ be the mean of q future observations at X_k (where q could equal 1 to give the case above). Then

$$\bar{Y}_q \sim N(\beta_0 + \beta_1 X_k, \sigma_q^2)$$
$$\hat{Y}_k \sim N(\beta_0 + \beta_1 X_k, V(\hat{Y}_k))$$

so that

$$\bar{Y}_q - \hat{Y}_k \sim N(0, \sigma_q^2 + V(\hat{Y}_k))$$

and $[(\bar{Y}_q - \hat{Y}_k)/\text{est. s.e. } (\bar{Y}_q - Y_k)]$ is distributed as a $t(v)$ variable where v is the number of degrees of freedom on which s^2, the estimate of σ^2, is based. Thus

$$\text{prob} \left\{ |\bar{Y}_q - \hat{Y}_k| \le t(v, 0.975) \left[s^2 \left(\frac{1}{q} + \frac{1}{n} + \frac{(X_k - \bar{X})^2}{\sum (X_i - \bar{X})^2} \right) \right]^{1/2} \right\} = 0.95$$

from which we can obtain 95% confidence limits for $\bar{Y}_q$ about $\hat{Y}_k$ of

$$\hat{Y}_k \pm t(v, 0.975) \left[\frac{1}{q} + \frac{1}{n} + \frac{(X_k - \bar{X})^2}{\sum (X_i - \bar{X})^2} \right]^{1/2} s.$$

These limits are of course wider than those for the mean value of Y for given X_k, since these limits are the ones within which 95% of future observations at X_k (or future means of q observations at X_k as the case may be) are expected to lie.

F-Test for Significance of Regression

Since the Y_i are random variables, any function of them is also a random variable; two particular functions are MS_R, the mean square due to regression, and s^2, the mean square due to residual variation, which arise in the analysis of variance table shown in Section 1.3. These functions then have their own distribution, mean, variance, and moments. It can be shown that the mean values are as follows:

$$E(MS_R) = \sigma^2 + \beta_1^2 \sum (X_i - \bar{X})^2 \qquad (1.4.9)$$

$$E(s^2) = \sigma^2$$

where, if Z is a random variable, $E(Z)$ denotes its mean value or expected value. Suppose that the errors ϵ_i are independent $N(0, \sigma^2)$ variables. Then it can be shown that *if $\beta_1 = 0$*, the variable MS_R multiplied by its degrees of freedom (here, one) and divided by σ^2 follows a χ^2 distribution with the same (1) number of degrees of freedom. In addition, $(n - 2)s^2/\sigma^2$ follows a χ^2 distribution with $(n - 2)$ degrees of freedom. Also since these two variables are independent, a statistical theorem tells us that the ratio

$$F = \frac{MS_R}{s^2} \qquad (1.4.10)$$

follows an F distribution with (here) 1 and $(n - 2)$ degrees of freedom provided (recall) that $\beta_1 = 0$. This fact can thus be used as a test of $\beta_1 = 0$. We compare the ratio $F = MS_R/s^2$ with the $100(1 - \alpha)\%$ point of the tabulated $F(1, n - 2)$ distribution in order to determine whether β_1 can be considered nonzero on the basis of the data we have seen.

Example (*continued*). From Table 1.3, we see that the required ratio is $F = 45.59/0.7926 = 57.52$. If we look up percentage points of the $F(1, 23)$ distribution, we see that the 95% point $F(1, 23, 0.95) = 4.28$. Since the calculated F exceeds the critical F value in the table, that is $F = 57.52 > 4.28$, we reject the hypothesis $H_0: \beta_1 = 0$ running a risk of less than 5% of being wrong.

(*Note.* In the particular case of fitting a straight line, this F-test for "regression" is exactly the same as the t-test for $\beta_1 = 0$ given earlier. This is because the ratio

$$\frac{MS_R}{s^2} = \frac{b_1\{\sum (X_i - \bar{X})(Y_i - \bar{Y})\}}{s^2}$$

$$= \frac{b_1^2 \sum (X_i - \bar{X})^2}{s^2}$$

(by the definition of b_1)

$$= \left[\frac{b_1\{\sum (X_i - \bar{X})^2\}^{1/2}}{s}\right]^2$$

$$= t^2$$

from Eq. (1.4.4). Since the variable $F(1, n - 2)$ is the square of the $t(n - 2)$ variable, exactly the same test results. When there are more regression coefficients the overall F-test for regression, which is the extension of the one given here, does not correspond to the t-test of a coefficient. However, tests for *individual coefficients* can be made either in t or $t^2 = F$ form by a similar argument. The F form is often seen in computer programs.)

In the example we had an observed F value of 57.52 and an observed t value of -7.60. Note that $(-7.6)^2 = 57.76$. Were it not for round-off error this would be equal to the observed value of F.

Percentage Variation Explained

We define $R^2 = $ (SS due to regression)/(Total SS, corrected for mean). Then R^2 measures the "proportion of total variation about the mean $\overline{Y}$ explained by the regression." It is often expressed as a percentage by multiplying by 100.

Example (*continued*).   From Table 1.3,

$$R^2 = \frac{45.59}{63.82} \times 100 = 71.44\%.$$

Thus the regression equation obtained explains 71.44% of the total variation.

1.5.   Lack of Fit and Pure Error

We have already remarked that the fitted regression line is a calculated line based on a certain model or assumption, an assumption we should not blindly accept but should *tentatively entertain*. In certain circumstances we can check whether or not the model is correct. First, we can examine the consequences of an incorrect model. Let us recall that $e_i = Y_i - \hat{Y}_i$ is the *residual* at X_i. This is the amount by which the actual observed value Y_i exceeds the fitted value $\hat{Y}_i$. As shown in Section 1.2, $\sum e_i = 0$. The residuals contain all available information on the ways in which the fitted model fails to properly explain the observed variation in the dependent variable Y. (For examination of residuals, see Chapter 3.) Let $\eta_i = E(Y_i)$ denote the value given by the true model, whatever it is, at $X = X_i$. Then we can write

$$Y_i - \hat{Y}_i = (Y_i - \hat{Y}_i) - E(Y_i - \hat{Y}_i) + E(Y_i - \hat{Y}_i)$$

$$= \{(Y_i - \hat{Y}_i) - (\eta_i - E(\hat{Y}_i))\} + (\eta_i - E(\hat{Y}_i))$$

$$= q_i + B_i,$$

say, where

$$q_i = \{(Y_i - \hat{Y}_i) - (\eta_i - E(\hat{Y}_i))\}, \qquad B_i = \eta_i - E(\hat{Y}_i).$$

The quantity B_i is the bias error at $X = X_i$. If the model is correct then $E(\hat{Y}_i) = \eta_i$ and B_i is zero. If the model is not correct $E(\hat{Y}_i) \neq \eta_i$ and B_i is not zero but has a value that depends on the true model and the value of X_i. The quantity q_i is a random variable that has zero mean since $E(q_i) = E(Y_i - \hat{Y}_i) - (\eta_i - E(\hat{Y}_i)) = \eta_i - E(\hat{Y}_i) - (\eta_i - E(\hat{Y}_i)) = 0$, and this is true whether the model is correct or not, that is, whether $E(\hat{Y}_i) = \eta_i$ or not.

The q_i, it can be shown, are correlated, and the quantity $q_1{}^2 + q_2{}^2 + \cdots + q_n{}^2$ has expected or mean value $(n - 2)\sigma^2$ where $V(Y_i) = V(\epsilon_i) = \sigma^2$ is the error variance. From this it can be shown further that the residual mean square, value

$$\frac{1}{n-2}\left\{\sum_{i=1}^{n}(Y_i - \hat{Y}_i)^2\right\}$$

has expected or mean value σ^2 if the postulated model is of the correct form, or $\sigma^2 + \sum B_i{}^2/(n - 2)$ if the model is not correct. If the model is correct, that is, $B_i = 0$, then the residuals are (correlated) random deviations q_i and the residual mean square can be used as an estimate of the error variance σ^2.

However, if the model is not correct, that is, $B_i \neq 0$, then the residuals contain both random (q_i) and systematic (B_i) components. We can refer to these as the variance error and bias error components of the residuals, respectively. Also the residual mean square will tend to be inflated and will no longer provide a satisfactory measure of the random variation present in the observations. (Since, however, the mean square *is* a random variable it may, by chance, not have a large value even when bias does exist. For some similar work on the general regression case see Section 2.12.)

In the simple case of fitting a straight line, bias error can usually be detected merely by examining a plot of the data (see Figure 1.7 for example). When the model is more complicated and/or involves more variables this may not be possible. If a prior estimate of σ^2 is available (by "prior estimate" we mean one obtained from previous experience of the variation in the situation being studied) we can see (or test by an F-test) whether or not the residual mean square is significantly greater than this prior estimate. If it is significantly greater we say that there is lack of fit and we would reconsider the model which would be inadequate in its present form. If no prior estimate of σ^2 is available, but repeat measurements of Y (i.e., two or more measurements) have been made at the same value of X, we can use these repeats to obtain an estimate of σ^2. Such an estimate is said to represent "pure error" because, if the setting of X is identical for two observations, only the random variation can influence the results and provide differences between them. Such differences will usually then provide an estimate of σ^2 which is much more reliable than we can obtain from any other source. For this reason, it is sensible when designing experiments to arrange for repeat observations.

(*Note.* It is important to understand that repeated runs must be genuine repeats and not just repetitions of the same reading. For example, suppose we were attempting to relate, by regression methods, $Y =$ Intelligence Quotient to $X =$ height of person. A genuine repeat point would be obtained if we measured the separate I.Q.'s of two people of exactly the same height. If, however, we measured *twice* the I.Q. of one person, of some specified height, this would not be a genuine repeat point in our context but merely a "reconfirmed" single point. The latter would, it is true, supply information on the variation of the testing method which is part of the variation σ^2, but it would *not* provide information on the variation in I.Q. between people of the same height, which is the σ^2 of our problem. In chemical experiments, a succession of readings made during steady state running does not provide genuine repeat points. However, if a certain set of conditions was reset anew, after intermediate runs at other X-levels had been made, and provided drifts in the response level had not occurred, genuine repeat runs would be obtained. With this in mind, repeat runs that show remarkable agreement, which is contrary to expectation, should always be regarded cautiously and subjected to additional investigation.)

The pure error estimate of σ^2 is found as follows (the same formulae also apply when there is more than one independent variable). Suppose

$Y_{11}, Y_{12}, \ldots, Y_{1n_1}$ are n_1 repeat observations at X_1.

$Y_{21}, Y_{22}, \ldots, Y_{2n_2}$ are n_2 repeat observations at X_2.

$\cdots$

$Y_{k1}, Y_{k2}, \ldots, Y_{kn_k}$ are n_k repeat observations at X_k.

The contribution to the pure error sum of squares from the X_1 readings is then

$$\sum_{u=1}^{n_1} (Y_{1u} - \overline{Y}_1)^2 = \sum_{u=1}^{n_1} Y_{1u}^2 - n_1 \overline{Y}_1^2$$

where $\overline{Y}_1 = (Y_{11} + Y_{12} + \cdots + Y_{1n_1})/n_1$, and this sum of squares has $(n_1 - 1)$ degrees of freedom. Similar quantities are evaluated for the other sets of Y's. The total SS (pure error) $= \sum_{i=1}^{k} \sum_{u=1}^{n_i} (Y_{iu} - \overline{Y}_i)^2$ with total degrees of freedom $= \sum_{i=1}^{k} (n_i - 1) = \sum_{i=1}^{k} n_i - k = n_e$, say.

Then the mean square for pure error is

$$s_e^2 = \left\{ \frac{\sum\limits_{i=1}^{k} \sum\limits_{u=1}^{n_i} (Y_{iu} - \bar{Y}_i)^2}{\sum\limits_{i=1}^{k} n_i - k} \right\}$$

and is an estimate of σ^2. In words this quantity is the total "within repeats" sum of squares divided by the total degrees of freedom.

(*Note.*   If there are only two observations Y_{i1}, Y_{i2} at the point X_i then

$$\sum\limits_{u=1}^{2} (Y_{iu} - \bar{Y}_i)^2 = \tfrac{1}{2}(Y_{i1} - Y_{i2})^2$$

and this is an easier form to compute. This SS has one degree of freedom.)

The pure error sum of squares can be introduced into the analysis of variance table as shown in Figure 1.6. The usual procedure is then to compare the ratio $F = MS_L/s_e^2$ with the $100(1 - \alpha)\%$ point of an F distribution with $(n_r - n_e)$ and n_e degrees of freedom. If the ratio is

1. Significant. This indicates that the model appears to be inadequate. Attempts would be made to discover where and how the inadequacy occurs. (See comments on the various residuals plots discussed in Chapter 3. Note, however, that the plotting of residuals is a standard technique for all regression analyses, not only those in which lack of fit can be demonstrated by this particular test.)

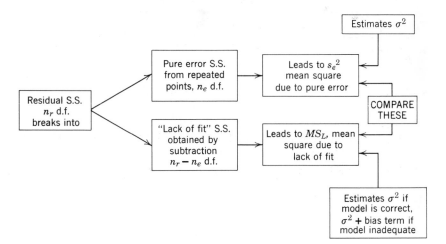

Figure 1.6

2. Not significant. This indicates that there appears to be no reason to doubt the adequacy of the model and both pure error and lack of fit mean squares can be used as estimates of σ^2. A pooled estimate of σ^2 can be obtained by recombining the pure error and lack of fit sums of squares into the residual sum of squares and dividing by the residual degrees of freedom n_r to give $s^2 = $ (Residual SS)$/n_r$. (Note that the residuals should *still* be examined—see the comments after the following example.)

Example. Since our previous example which involved data taken from Appendix A did not contain repeat observations, we shall employ a specially constructed example to illustrate the lack of fit and pure error section. A regression line $\hat{Y} = 1.436 + 0.338X$ was estimated from the following data. The analysis of variance table is shown below.

Obsn.	Y	X	Obsn.	Y	X	Obsn.	Y	X
1	2.3	1.3	9	1.7	3.7	17	3.5	5.3
2	1.8	1.3	10	2.8	4.0	18	2.8	5.3
3	2.8	2.0	11	2.8	4.0	19	2.1	5.3
4	1.5	2.0	12	2.2	4.0	20	3.4	5.7
5	2.2	2.7	13	5.4	4.7	21	3.2	6.0
6	3.8	3.3	14	3.2	4.7	22	3.0	6.0
7	1.8	3.3	15	1.9	4.7	23	3.0	6.3
8	3.7	3.7	16	1.8	5.0	24	5.9	6.7

ANOVA Table

Source	df	SS	MS	F ratio
Total (corrected)	23	27.518		
Regression	1	6.326	6.326	6.569 significant at $\alpha = $ 0.05 level
Residual	22	21.192	$0.963 = s^2$	

We shall now find the pure error, and hence the lack of fit.
1. Pure error SS from repeats at $X = 1.3$ is $\frac{1}{2}(2.3 - 1.8)^2 = 0.125$, with 1 degree of freedom.
2. Pure error SS from repeats at $X = 4.7$ is

$$(5.4)^2 + (3.2)^2 + (1.9)^2 - 3\{(5.4 + 3.2 + 1.9)/3\}^2$$
$$= 43.01 - (10.5)^2/3$$
$$= 43.01 - 36.75$$
$$= 6.26, \text{ with 2 degrees of freedom.}$$

Similar calculations provide the following quantities:

Level of X	$\sum_{u=1}^{n_i} (Y_{iu} - \bar{Y}_i)^2$	df
1.3	0.125	1
2.0	0.845	1
3.3	2.000	1
3.7	2.000	1
4.0	0.240	2
4.7	6.260	2
5.3	0.980	2
6.0	0.020	1
Totals	12.470	11

We can thus rewrite the analysis of variance as shown in Table 1.4. The F ratio $MS_L/s_e^2 = 0.699$ is not significant since it is less than unity. Thus, on the basis of this test at least, we have no reason to doubt the adequacy of our model and can use $s^2 = 0.963$ as an estimate of σ^2. (Even though this test does not make us doubt the model we should still, as a standard procedure in general, examine the residuals; see Chapter 3. Note that the fact that the model passes all hurdles does not mean that it is *the* correct model—merely that it is a plausible one which has not been found inadequate by the data.) If lack of fit had been found, a different model would have been necessary—perhaps the quadratic one $Y = \alpha + \beta X + \gamma X^2 + \epsilon$. The following diagrams (Figure 1.7) illustrate some situations that may arise when a straight line is fitted to data and the consequent action to be taken.

Table 1.4   ANOVA (Showing Lack-of-Fit)

Source	df	SS	MS	F ratio
Total	23	27.518		
Regression	1	6.326	6.326	6.569 significant at $\alpha = 0.05$
Residual	22	21.192	$0.963 = s^2$	
Lack of fit	11	8.722	$0.793 = MS_L$	0.699 (not significant)
Pure error	11	12.470	$1.134 = s_e^2$	

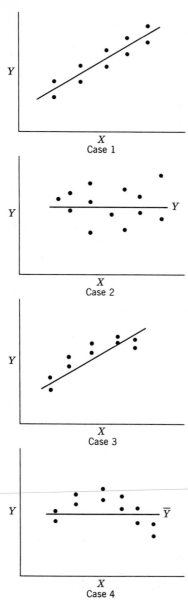

Case 1: (1) Try $Y = \beta_0 + \beta_1 X + \epsilon$. (2) No lack of fit. (3) Significant linear regression. (4) Use model $\hat{Y} = b_0 + b_1 X$.

Case 2: (1) Try $Y = \beta_0 + \beta_1 X + \epsilon$. (2) No lack of fit. (3) Linear regression not significant. (4) Use model $\hat{Y} = \bar{Y}$.

Case 3: (1) Try $Y = \beta_0 + \beta_1 X + \epsilon$. (2) Significant lack of fit. (3) Significant linear regression. (4) Try model $Y = \beta_0 + \beta_1 X + \beta_{11} X^2 + \epsilon$.

Case 4: (1) Try $Y = \beta_0 + \beta_1 X + \epsilon$. (2) Significant lack of fit. (3) Linear regression not significant. (4) Try model $Y = \beta_0 + \beta_1 X + \beta_{11} X^2 + \epsilon$. (*Note:* β_1 may become significantly different from zero when residual error term is reduced by taking out $\beta_{11} X^2$. See Chapter 6.)

Figure 1.7 Typical straight-line regression situations.

1.6. The Correlation Between X and Y

When we fit a postulated straight line model $Y = \beta_0 + \beta_1 X + \epsilon$, we are tentatively entertaining the idea that Y can be expressed, apart from error, as a first-order function of X. In such a relationship X is usually assumed to be fixed while Y is a random variable subject to error.

If X and Y were both random variables following some (unknown) bivariate distribution then we could define the *correlation coefficient between X and Y* as

$$\rho_{XY} = \frac{\text{covariance}\,(X,\,Y)}{\{V(X)\,V(Y)\}^{\frac{1}{2}}} \tag{1.6.1}$$

where if $f(X,\,Y)$ is the continuous joint probability distribution of X and Y,

$$\text{covar}\,(X,\,Y) = \int_{-\infty}^{\infty}\int_{-\infty}^{\infty} \{Y - E(Y)\}\{X - E(X)\} f(X,\,Y)\,dX\,dY$$

and

$$\tag{1.6.2}$$

$$V(Y) = \int_{-\infty}^{\infty}\int_{-\infty}^{\infty} \{Y - E(Y)\}^2 f(X,\,Y)\,dY\,dX$$

where

$$E(Y) = \int_{-\infty}^{\infty}\int_{-\infty}^{\infty} Y f(X,\,Y)\,dY\,dX,$$

and $V(X)$ and $E(X)$ are similarly defined in terms of X. (If the distributions are discrete, summations replace integrals in the usual way.)

It can be shown that $-1 \le \rho_{XY} \le 1$. The quantity ρ_{XY} is a measure of the association between the random variables X and Y. For example, if $\rho_{XY} = 1$, X and Y are perfectly positively correlated and the possible values of X and Y all lie on a straight line with positive slope in the $(X,\,Y)$ plane. If $\rho_{XY} = 0$ the variables are said to be uncorrelated, that is, linearly unassociated with each other. This does *not* mean that X and Y are statistically independent, as most elementary textbooks emphasize. If $\rho_{XY} = -1$, X and Y are perfectly negatively correlated and the possible values of X and Y again all lie on a straight line, with negative slope, in the $(X,\,Y)$ plane.

If a sample of size n, $(X_1,\,Y_1), (X_2,\,Y_2), \ldots, (X_n,\,Y_n)$ is available from the joint distribution, the quantity

$$r_{XY} = \frac{\sum_{i=1}^{n} (X_i - \bar{X})(Y_i - \bar{Y})}{\left\{\sum_{i=1}^{n} (X_i - \bar{X})^2\right\}^{\frac{1}{2}} \left\{\sum_{i=1}^{n} (Y_i - \bar{Y})^2\right\}^{\frac{1}{2}}} \tag{1.6.3}$$

called the *sample correlation coefficient between X and Y* is an estimate of ρ_{XY} and provides an empirical measure of the association between X and Y. (If factors $1/(n-1)$ are placed before all summations, then r_{XY} has the form of ρ_{XY} with the variances and the covariance replaced by sample values.) Like ρ_{XY}, r_{XY} lies between -1 and 1.

When X_i and Y_i, $i = 1, 2, \ldots, n$, are all constants, rather than sample values from some distribution, r_{XY} can still be used as a measure of association. Since the set of values (X_i, Y_i) $i = 1, 2, \ldots, n$ can be thought of as a complete finite distribution, r_{XY} is, effectively, a population rather than a sample value, that is, $r_{XY} = \rho_{XY}$ in this case.

If Y is a random variable and $X_1, X_2, \ldots, X_n$ represent the values of a finite X-distribution, the correlation coefficient ρ_{XY} can still be defined by Eq. (1.6.1) provided that all integrations with respect to X in the expression (1.6.2) are properly replaced by summations over the discrete values $X_1, X_2, \ldots, X_n$. The expression (1.6.3) can still be used to estimate ρ_{XY} by r_{XY} if a sample of observations $Y_1, Y_2, \ldots, Y_n$ at the n X-values $X_1, X_2, \ldots, X_n$, respectively, is available.

In this book we shall make use of expressions of the form of r_{XY} in Eq. (1.6.3); their actual names and roles will depend on whether the quantities which stand in the place of Y and X are to be considered sample or population values. We shall call all such quantities r_{XY} the *correlation (coefficient) between X and Y* and use them as appropriate measures of association between various quantities of interest. The distinctions made above as to whether they are actually population or sample values are not necessary for our purposes and will be ignored throughout. Note the fact that the correlation r_{XY} is nonzero implies only that there is association between the values X_i and Y_i, $i = 1, 2, \ldots, n$, and does not by itself imply any sort of causal relationship between X and Y. Such a false assumption has led to erroneous conclusions on many occasions (For some examples of such conclusions, including "Lice make a man healthy" see Chapter 8 of *How to Lie with Statistics* by Darrell Huff, W. W. Norton and Company, New York, 1954).

Correlation and Regression

Suppose that data $(X_1, Y_1), (X_2, Y_2), \ldots, (X_n, Y_n)$ are available. We can obtain $r_{XY} = r_{YX}$ by applying Eq. (1.6.3) and, if we postulate a model $Y = \beta_0 + \beta_1 X + \epsilon$ we can also obtain an estimated regression coefficient b_1 given by Eq. (1.2.9). Comparing these two expressions we see that

$$b_1 = \left\{ \frac{\sum (Y_i - \bar{Y})^2}{\sum (X_i - \bar{X})^2} \right\}^{1/2} r_{XY}$$

where summations are over $i = 1, 2, \ldots, n$. In other words b_1 is a scaled version of r_{XY} scaled by the ratio of the "spread" of the Y_i divided by the "spread" of the X_i. If we write

$$(n - 1)s_Y^2 = \sum (Y_i - \bar{Y})^2$$

$$(n - 1)s_X^2 = \sum (X_i - \bar{X})^2$$

then

$$b_1 = \frac{s_Y}{s_X} r_{XY}.$$

Thus b_1 and r_{XY} are closely related but provide different interpretations. The correlation r_{XY} measures association between X and Y, while b_1 measures the size of the change in Y, which can be predicted when a unit change is made in X. In more general regression problems, the regression coefficients are also related to correlations of the type of Eq. (1.6.3) but in a more complicated manner (see Section 5.4).

EXERCISES

A. A study was made on the effect of temperature on the yield of a chemical process. The following data (in coded form) were collected:

X	Y
-5	1
-4	5
-3	4
-2	7
-1	10
0	8
1	9
2	13
3	14
4	13
5	18

(a) Assuming a model, $Y = \beta_0 + \beta_1 X + \epsilon$, what are the least squares estimates of β_0 and β_1? What is the prediction equation?

(b) Construct the analysis of variance table and test the hypothesis, $H_0: \beta_1 = 0$ with an α risk of 0.05.

(c) What are the confidence limits ($\alpha = 0.05$) for β_1?

(d) What are the confidence limits ($\alpha = 0.05$) for the true mean value of Y when $X = 3$?

(e) What are the confidence limits ($\alpha = 0.05$) for the difference between the true mean value of Y when $X_1 = 3$ and the true mean value of Y when $X_2 = -2$?

(f) Are there any indications that a better model should be tried?

(g) Comment on the number of levels of temperature investigated with respect to the estimate of β_1 in the assumed model.

B. A test is to be run on a given process for the purpose of determining the effect of an independent variable X (such as process temperature) on a certain characteristic property of the finished product Y (such as density). Four observations are to be taken at each of five settings of the independent variable X.

1. In what order would you take the twenty observations required in this test?

2. When the test was actually run, the following results were obtained:

$$\bar{X} = 5.0 \qquad \sum (X_i - \bar{X})^2 = 160.0 \qquad \sum (X_i - \bar{X})(Y_i - \bar{Y}) = 80.0$$

$$\bar{Y} = 3.0 \qquad \sum (Y_i - \bar{Y})^2 = 83.2.$$

Assume a model of the type $Y = \beta_0 + \beta_1 X + \epsilon$.
   a. Calculate the fitted regression equation.
   b. Prepare the analysis of variance table.
   c. Determine 95% confidence limits for the true mean value of Y when

   (1) $X = 5.0$
   (2) $X = 9.0$

3. Suppose that the actual data plot is as shown in Figure 1.8 and that the sum of squares due to replication (pure error) is 42. Answer the the following questions on the basis of this additional information.

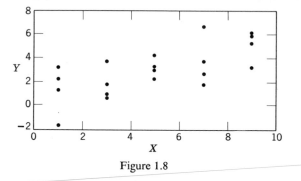

Figure 1.8

   a. Does the regression equation you derived in 2a adequately represent the data? Give reasons and include the result of a test for lack of fit.
   b. Are the confidence limits you calculated in 2c applicable? If not, state your reasons.
   c. If the model used in part 2 does not seem appropriate, suggest a possible alternate.

4. Suppose that the actual data plot is as shown in Figure 1.9 and that the sum of squares due to replication is 42.0. Answer the questions 3a, 3b, and 3c on the basis of this information.

5. Suppose that the actual data plot is as shown in Figure 1.10, and the sum of squares due to replication is 23.2. Answer the questions 3a, 3b, 3c on the basis of this information.

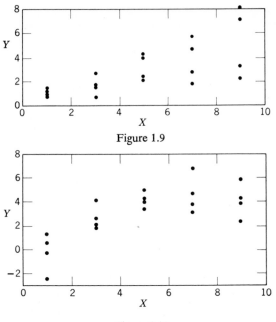

Figure 1.9

Figure 1.10

C. Thirteen specimens of 90/10 Cu–Ni alloys, each with a specific iron content, were tested in a corrosion-wheel setup. The wheel was rotated in salt sea water at 30 ft/sec for 60 days. The corrosion was measured in weight loss in milligrams/square decimeter/day, MDD. The following data were collected:

X (fe)	Y (loss in MDD)
0.01	127.6
0.48	124.0
0.71	110.8
0.95	103.9
1.19	101.5
0.01	130.1
0.48	122.0
1.44	92.3
0.71	113.1
1.96	83.7
0.01	128.0
1.44	91.4
1.96	86.2

Requirement

Determine if the effect of iron content on the corrosion resistance of 90/10 Cu–Ni alloys in sea water can be justifiably represented by a straight line model. Assume $\alpha = 0.05$.

D. (*Source:* "Application of the principle of least squares to families of straight lines" by S. Ergun, *Industrial and Engineering Chemistry*, **48**, 1956, November, 2063–2068.)

a. Show that the least squares estimates $a_1, a_2, \ldots, a_m, b$, of the parameters $\alpha_1, \alpha_2, \ldots, \alpha_m, \beta$ in the family of straight lines

$$E(Y_i) = \alpha_i + \beta X_i, \qquad i = 1, 2, \ldots, m,$$

are given by

$$b = \frac{\displaystyle\sum_{i=1}^{m} \sum_{u=1}^{n_i} (X_{iu} - \bar{X}_i)(Y_{iu} - \bar{Y}_i)}{\displaystyle\sum_{i=1}^{m} \sum_{u=1}^{n_i} (X_{iu} - \bar{X}_i)^2}$$

$$a_i = \bar{Y}_i - b\bar{X}_i$$

where

$$X_{i1}, X_{i2}, \ldots, X_{iu}, \ldots, X_{in_i}$$
$$Y_{i1}, Y_{i2}, \ldots, Y_{iu}, \ldots, Y_{in_i}$$

denote the observed values of X_i and Y_i which relate to the ith line, $i = 1, 2, \ldots, m$.

Show also that the residual sum of squares is

$$S^2 = \sum_{i=1}^{m} \sum_{u=1}^{n_i} (Y_{iu} - \bar{Y}_i)^2 - b^2 \sum_{i=1}^{m} \sum_{u=1}^{n_i} (X_{iu} - \bar{X}_i)^2$$

with $\left(\displaystyle\sum_{i=1}^{m} n_i - m - 1 \right)$ degrees of freedom, that

$$\sigma_b^2 = \frac{\sigma^2}{\displaystyle\sum_{i=1}^{m} \sum_{u=1}^{n_i} (X_{iu} - \bar{X}_i)^2}$$

and that

$$\sigma_{a_i}^2 = \frac{\sigma^2}{n_i} \left[1 + \frac{n_i \bar{X}_i^2}{\displaystyle\sum_{i=1}^{m} \sum_{u=1}^{n_i} (X_{iu} - \bar{X}_i)^2} \right]$$

b. Show that the least squares estimates $a, b_1, b_2, \ldots, b_m$, of the parameters $\alpha, \beta_1, \beta_2, \ldots, \beta_m$ in the family of straight lines

$$E(Y_i) = \alpha + \beta_i X_i, \qquad i = 1, 2, \ldots, m,$$

are given by

$$a = \frac{\displaystyle\sum_{i=1}^{m} n_i \left(\bar{Y}_i - \bar{X}_i \left[\sum_{u=1}^{n_i} X_{iu} Y_{iu} \middle/ \sum_{u=1}^{n_i} X_{iu}^2 \right] \right)}{\displaystyle\sum_{i=1}^{m} n_i \left(1 - n_i \bar{X}_i^2 \middle/ \sum_{u=1}^{n_i} X_{iu}^2 \right)}$$

$$b_i = \frac{\left\{ \displaystyle\sum_{u=1}^{n_i} X_{iu} Y_{iu} - a \sum_{u=1}^{n_i} X_{iu} \right\}}{\displaystyle\sum_{u=1}^{n_i} X_{iu}^2}$$

where X_{iu}, Y_{iu} are as above. Also show that the residual sum of squares is

$$S^2 = \sum_{i=1}^{m} \sum_{u=1}^{n_i} Y_{iu}^2 - \sum_{i=1}^{m} b_i^2 \sum_{u=1}^{n_i} X_{iu}^2 + a^2 \sum_{i=1}^{m} n_i - 2a \sum_{i=1}^{m} n_i \bar{Y}_i$$

with $\left(\sum_{i=1}^{m} n_i - m - 1\right)$ degrees of freedom, that

$$\sigma_a^2 = \sigma^2 \Big/ \left\{ \sum_{i=1}^{m} n_i \left(1 - n_i \bar{X}_i^2 \Big/ \sum_{u=1}^{n_i} X_{iu}^2\right)\right\}$$

and that

$$\sigma_{b_i}^2 = \left\{ \frac{1}{\sum_{u=1}^{n_i} X_{iu}^2} + \frac{\left(\sum_{u=1}^{n_i} X_{iu}\right)^2 \Big/ \left(\sum_{u=1}^{n_i} X_{iu}^2\right)^2}{\sum_{i=1}^{m} n_i\left(1 - n_i \bar{X}_i^2 \Big/ \sum_{u=1}^{n_i} X_{iu}^2\right)}\right\} \sigma^2.$$

E. A test for the equality of the slopes β_i of m lines represented by the first order models

$$Y_{iu} - \bar{Y}_i = \beta_i(X_{iu} - \bar{X}_i) + \epsilon_{iu}, \qquad i = 1, 2, \ldots, m,$$

can be conducted as follows. Suppose data

$$X_{i1}, X_{i2}, \ldots, X_{iu}, \ldots, X_{in_i} \quad \text{(fixed)}$$
$$Y_{i1}, Y_{i2}, \ldots, Y_{iu}, \ldots, Y_{in_i} \quad (\epsilon_{iu} \sim N(0, \sigma^2), \text{independent})$$

are available for estimation of the parameters of the ith line. The least squares estimate of β_i is

$$b_i = \left\{ \frac{\sum_{u=1}^{n_i}(X_{iu} - \bar{X}_i)(Y_{iu} - \bar{Y}_i)}{\sum_{u=1}^{n_i}(X_{iu} - \bar{X}_i)^2}\right\}$$

with sum of squares (1 df)

$$SS(b_i) = b_i^2 \left\{ \sum_{u=1}^{n_i}(X_{iu} - \bar{X}_i)^2\right\}$$

and residual sum of squares ($n_i - 2$ df)

$$S_i = \sum_{u=1}^{n_i}(Y_{iu} - \bar{Y}_i)^2 - SS(b_i).$$

If we assume $\beta_i = \beta$, all i, then the least squares estimate of β is

$$b = \left\{ \frac{\sum_{i=1}^{m} \sum_{u=1}^{n_i}(X_{iu} - \bar{X}_i)(Y_{iu} - \bar{Y}_i)}{\sum_{i=1}^{m} \sum_{u=1}^{n_i}(X_{iu} - \bar{X}_i)^2}\right\}$$

with sum of squares (1 df),

$$SS(b) = b^2 \left\{ \sum_{i=1}^{m} \sum_{u=1}^{n_i} (X_{iu} - \bar{X}_i)^2 \right\}$$

and residual sum of squares $(\sum n_i - 2m$ df$)$

$$S = \sum_{i=1}^{m} \sum_{u=1}^{n_i} (Y_{iu} - \bar{Y}_i)^2 - SS(b).$$

We can form an analysis of variance table as follows:

Source	SS	df	MS	F
b	$SS(b)$	1	M_1	$F_1 = M_1/s^2$
All $b_i \mid b$	$\sum_{i=1}^{m} SS(b_i) - SS(b)$	$m-1$	M_2	$F_2 = M_2/s^2$
Residual	by subtraction	$\sum_{i=1}^{m} n_i - 2m$	s^2 (estimates σ^2 if first order models are correct)	
Total	$\sum_{i=1}^{m} \sum_{u=1}^{n_i} (Y_{iu} - \bar{Y}_i)^2$	$\sum_{i=1}^{m} n_i - m$		

The hypothesis $H_0 : \beta_i = \beta$ is tested by comparing F_2 with an appropriate percentage point of the $F\{(m-1), (\sum_{i=1}^{m} n_i - 2m)\}$ distribution. If H_0 is not rejected, b is used as the common slope of the lines. (This is a special case of testing a linear hypothesis. A test for the equality of intercepts of two lines can also be constructed.) F_1 is used to test $H_0 : \beta = 0$.
   Apply the above procedure to the data below.

u	X_1	Y_1	X_2	Y_2	X_3	Y_3
1	3.5	24	3.2	22	3.0	32
2	4.1	32	3.9	33	4.0	36
3	4.4	37	4.9	39	5.0	47
4	5.0	40	6.1	44	6.0	49
5	5.5	43	7.0	53	6.5	55
6	6.1	51	8.1	57	7.0	59
7	6.6	62			7.3	64
8					7.4	64

F. The moisture of the wet mix of a product is considered to have an effect on the finished product density. The moisture of the mix was controlled and finished product densities measured as shown in the following data:

Mix Moisture (Coded)	Density (Coded)
X	Y
4.7	3
5.0	3
5.2	4
5.2	5
5.9	10
4.7	2
5.9	9
5.2	3
5.3	7
5.9	6
5.6	6
5.0	4

$$\sum X = 63.6 \qquad \sum Y = 62$$
$$\sum X^2 = 339.18 \qquad \sum Y^2 = 390$$
$$\sum x^2 = 2.10 \qquad \sum y^2 = 69.67$$
$$\bar{X} = 5.3 \qquad \bar{Y} = 5.17$$
$$\sum XY = 339.1$$
$$\sum xy = 10.5$$

Requirements

1. Fit the model $Y = \beta_0 + \beta_1 X + \epsilon$ to the data.
2. Place 95% confidence limits on β_1.
3. Is there any evidence in the data that a more complex model should be tried? (Use $\alpha = 0.05$.)

G. The cost of the maintenance of shipping tractors seems to increase with the age of the tractor. The following data were collected.

Age (yr)	6 Months Cost ($)
X	Y
4.5	$ 619
4.5	1049
4.5	1033
4.0	495
4.0	723
4.0	681
5.0	890
5.0	1522
5.5	987
5.0	1194
0.5	163
0.5	182
6.0	764
6.0	1373
1.0	978
1.0	466
1.0	549

Requirements

1. Determine if a straight line relationship is sensible. (Use $\alpha = 0.10$.)
2. Can a better model be selected?

H. It has been proposed in a manufacturing organization that a cup loss figure performed on the line supersede a bottle loss analysis which is a costly, time consuming laboratory procedure. The cup loss analysis would yield better control of the process because of the gain in time. If it can be shown that cup loss is a function of bottle loss, it would be a reasonable decision to make. Given the following data, what is your conclusion? (Use $\alpha = 0.05$.)

Bottle Loss (%)	Cup Loss (%)
X	Y
3.0	3.1
3.1	3.9
3.0	3.4
3.6	4.0
3.8	3.6
2.7	3.6
3.1	3.1
2.7	3.6
2.7	2.9
3.3	3.6
3.2	4.1
2.1	2.6
3.0	3.1
2.6	2.8

I. It is thought that the number of cans damaged in a boxcar shipment of cans is a function of the speed of the boxcar at impact. Thirteen boxcars selected at random were used to examine whether this appeared to be true. The data collected were as follows:

Speed of Car at Impact	No. of Cans Damaged
X	Y
4	27
3	54
5	86
8	136
4	65
3	109
3	28
4	75
3	53
5	33
7	168
3	47
8	52

What are your conclusions? (Use $\alpha = 0.05$.)

J. The effect of the temperature of the deodorizing process on the color of the finished product was determined experimentally. The data collected were as follows:

Temperature	Color
X	Y
460	0.3
450	0.3
440	0.4
430	0.4
420	0.6
410	0.5
450	0.5
440	0.6
430	0.6
420	0.6
410	0.7
400	0.6
420	0.6
410	0.6
400	0.6

1. Fit the model $Y = \beta_0 + \beta_1 X + \epsilon$.
2. Is this model sensible? (Use $\alpha = 0.05$)
3. Obtain a 95% confidence interval for the true mean value of Y at any given value of X, say X_k.

CHAPTER 2

THE MATRIX APPROACH
TO LINEAR REGRESSION

2.0. Introduction

We shall now present the example given in Chapter 1 in terms of matrix algebra. The use of matrices has many advantages, not the least of these being that once the problem is written and solved in matrix terms the solution can be applied to any regression problem no matter how many terms there are in the regression equation.

A matrix is a rectangular array of symbols or numbers and is usually denoted by a single letter in **boldface** type, for example **Q**, or **q**. There are several rules for manipulating such arrays. Quite complicated expressions or equations can often be represented very simply by just a few letters properly defined and grouped.

We shall not introduce matrices formally but will use them in the context of the example. The reader with sound knowledge of matrices can proceed directly to Section 2.5 or 2.6. The reader without any matrix background should work carefully through Sections 2.1 to 2.5 and will find some of the work in later sections of this chapter difficult at first reading; he should read these later sections again after reading Chapter 4. Mastery of the ideas of Sections 2.6 to 2.12 is essential for dealing with complicated regression problems, but it is not necessary to understand them completely to proceed to later chapters.

2.1. Fitting a Straight Line in Matrix Terms: The Estimates of β_0 and β_1

We define **Y** to be the *vector of observations* Y, **X** to be the *matrix of independent variables*, β to be the *vector of parameters to be estimated*, and ϵ to be a *vector of errors*. In terms of Chapter 1, Table 1.1, and Eq. (1.2.3) we then define for our main example:

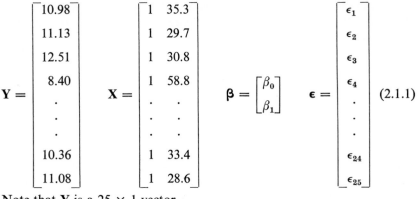

$$Y = \begin{bmatrix} 10.98 \\ 11.13 \\ 12.51 \\ 8.40 \\ \cdot \\ \cdot \\ \cdot \\ 10.36 \\ 11.08 \end{bmatrix} \qquad X = \begin{bmatrix} 1 & 35.3 \\ 1 & 29.7 \\ 1 & 30.8 \\ 1 & 58.8 \\ \cdot & \cdot \\ \cdot & \cdot \\ \cdot & \cdot \\ 1 & 33.4 \\ 1 & 28.6 \end{bmatrix} \qquad \beta = \begin{bmatrix} \beta_0 \\ \beta_1 \end{bmatrix} \qquad \epsilon = \begin{bmatrix} \epsilon_1 \\ \epsilon_2 \\ \epsilon_3 \\ \epsilon_4 \\ \cdot \\ \cdot \\ \cdot \\ \epsilon_{24} \\ \epsilon_{25} \end{bmatrix} \qquad (2.1.1)$$

Note that Y is a 25×1 vector,

$\qquad X$ is a 25×2 matrix,

$\qquad \beta$ is a 2×1 vector,

$\qquad \epsilon$ is a 25×1 vector.

(Any matrix with one column is called a *column vector;* any matrix with one row is called a *row vector*. A 1×1 "matrix" is just an ordinary number or *scalar*.)

The rules of multiplication for matrices and vectors insist that two matrices must be *conformable*. For example, if A is an $n \times p$ matrix we can

(1) *postmultiply* it by a $p \times q$ matrix to give as result an $n \times p \times p \times q = n \times q$ matrix.

(2) *premultiply* it by an $m \times n$ matrix to give as result an $m \times n \times n \times p = m \times p$ matrix.

Thus, for example, the multiplication βX is not possible since β is 2×1 and X is 25×2. But $X\beta$ is possible as follows.

$$X\beta = \begin{bmatrix} 1 & 35.3 \\ 1 & 29.7 \\ & \cdots \\ 1 & 28.6 \end{bmatrix} \begin{bmatrix} \beta_0 \\ \beta_1 \end{bmatrix} = \begin{bmatrix} \beta_0 + 35.3\beta_1 \\ \beta_0 + 29.7\beta_1 \\ \cdots \\ \beta_0 + 28.6\beta_1 \end{bmatrix} \qquad (2.1.2)$$

$$\quad 25 \times 2 \quad 2 \times 1 \qquad 25 \times 1$$

As a more general example consider the product

$$\overset{\textbf{A}}{\begin{bmatrix} 1 & 2 & 4 \\ -1 & 0 & 1 \\ 2 & 3 & 1 \end{bmatrix}} \overset{\textbf{B}}{\begin{bmatrix} 1 & -1 \\ 2 & 1 \\ 3 & 5 \end{bmatrix}} = \overset{\textbf{C}}{\begin{bmatrix} 17 & 21 \\ 2 & 6 \\ 11 & 6 \end{bmatrix}}$$

$$\quad 3 \times 3 \qquad\quad 3 \times 2 \qquad\quad 3 \times 2$$

To find the element in row i and column j of $\mathbf{C}$ we take row i of $\mathbf{A}$ and column j of $\mathbf{B}$, find the cross product of corresponding elements and add. For example,

row 2 of $\mathbf{A}$ is -1 0 1

column 1 of $\mathbf{B}$ is 1 2 3

Thus the element in row 2, column 1 of $\mathbf{C}$ is

$$-1(1) + 0(2) + 1(3) = 2.$$

Definition.   If the sum of the cross-products of corresponding elements of row i and column j is zero, then row i is said to be *orthogonal* to column j. (The same definition applies to row and row or column and column.)

The sum of two matrices or vectors is just the matrix whose elements are the sums of corresponding elements in the separate matrices or vectors. For example,

$$\mathbf{X\beta} + \mathbf{\epsilon} = \begin{bmatrix} \beta_0 + 35.3\beta_1 \\ \beta_0 + 29.7\beta_1 \\ \cdots \\ \beta_0 + 28.6\beta_1 \end{bmatrix} + \begin{bmatrix} \epsilon_1 \\ \epsilon_2 \\ \cdots \\ \epsilon_{25} \end{bmatrix} = \begin{bmatrix} \beta_0 + 35.3\beta_1 + \epsilon_1 \\ \beta_0 + 29.7\beta_1 + \epsilon_2 \\ \cdots \\ \beta_0 + 28.6\beta_1 + \epsilon_{25} \end{bmatrix} \quad (2.1.3)$$

The two matrices or vectors must have the same dimensions for this to be possible. (The difference between two matrices is similarly defined with differences instead of sums.) If two matrices or vectors are equal, corresponding elements are equal. Thus writing the matrix equation

$$\mathbf{Y} = \mathbf{X\beta} + \mathbf{\epsilon} \qquad (2.1.4)$$

implies that

$$10.98 = \beta_0 + 35.3\beta_1 + \epsilon_1 \qquad (2.1.5)$$
$$\cdots$$
$$11.08 = \beta_0 + 28.6\beta_1 + \epsilon_{25}$$

or

$$Y_i = \beta_0 + \beta_1 X_i + \epsilon_i \qquad (i = 1, \ldots, 25) \qquad (2.1.6)$$

for each of the twenty-five observations. Thus the matrix equation (2.1.4), and Eq. (2.1.6) express the same model. Equation (2.1.6) is identical to Eq. (1.2.3).

We now define the *transpose* of a matrix. It is the matrix obtained by writing all rows as columns in the order in which they occur so that the columns all become rows. The transpose of a matrix $\mathbf{M}$ is written $\mathbf{M'}$, for example,

$$\mathbf{M} = \begin{bmatrix} 3 & 2 \\ 1 & 4 \\ 7 & 0 \end{bmatrix} \qquad \mathbf{M}' = \begin{bmatrix} 3 & 1 & 7 \\ 2 & 4 & 0 \end{bmatrix}.$$

$$3 \times 2 \qquad\qquad 2 \times 3$$

Thus, for example,

$$\boldsymbol{\epsilon}' = (\epsilon_1, \epsilon_2, \ldots, \epsilon_n).$$

Note that we can then write

$$\epsilon_1^2 + \epsilon_2^2 + \cdots + \epsilon_n^2 = \boldsymbol{\epsilon}'\boldsymbol{\epsilon},$$
$$Y_1^2 + Y_2^2 + \cdots + Y_n^2 = \mathbf{Y}'\mathbf{Y}.$$

Furthermore

$$\mathbf{X}'\mathbf{X} = \begin{bmatrix} 1 & 1 & \cdots & 1 \\ 35.3 & 29.7 & \cdots & 28.6 \end{bmatrix} \begin{bmatrix} 1 & 35.3 \\ 1 & 29.7 \\ & \cdots \\ 1 & 28.6 \end{bmatrix} = \begin{bmatrix} 25 & 1315 \\ 1315 & 76323.42 \end{bmatrix}.$$

If we refer to our example of Chapter 1 we see that

$$\mathbf{X}'\mathbf{X} = \begin{bmatrix} n & \sum X_i \\ \sum X_i & \sum X_i^2 \end{bmatrix} \qquad (2.1.7)$$

In addition,

$$\mathbf{X}'\mathbf{Y} = \begin{bmatrix} 1 & 1 & \cdots & 1 \\ 35.3 & 29.7 & \cdots & 28.6 \end{bmatrix} \begin{bmatrix} 10.98 \\ 11.13 \\ \cdots \\ 11.08 \end{bmatrix} = \begin{bmatrix} 235.60 \\ 11821.4320 \end{bmatrix}$$

so that

$$\mathbf{X}'\mathbf{Y} = \begin{bmatrix} \sum Y_i \\ \sum X_i Y_i \end{bmatrix}. \qquad (2.1.8)$$

This means that the normal equations (1.2.8) can be written

$$\mathbf{X}'\mathbf{X}\mathbf{b} = \mathbf{X}'\mathbf{Y} \qquad (2.1.9)$$

where $\mathbf{b}' = (b_0, b_1)$, and these equations, when solved, provide the least squares estimates (b_0, b_1) of (β_0, β_1). How do we solve these equations in matrix form? To do so we define the *inverse* of a matrix. This exists only when a matrix is square and when the determinant of the matrix (a quantity that we shall not define here but of which we shall provide some examples) is nonzero. This latter condition is usually stated as *when the matrix is nonsingular*. This will be true in our applications unless otherwise stated. In regression work we wish to invert the $\mathbf{X'X}$ matrix. If it is singular, and so does not have an inverse, this will be reflected in the fact that some of the normal equations will be linear combinations of others; see, for example, Eq. (9.2.3). In this case there will be fewer equations than there are unknowns for which to solve. In such a case unique estimates are not possible unless some additional conditions on the parameters apply. (See Chapter 9 for additional comments on this point.)

Suppose now that $\mathbf{M}$ is a nonsingular $p \times p$ matrix. The inverse of $\mathbf{M}$ is written $\mathbf{M}^{-1}$, is $p \times p$, and is such that

$$\mathbf{M}^{-1}\mathbf{M} = \mathbf{M}\mathbf{M}^{-1} = \mathbf{I}_p$$

where $\mathbf{I}_p$ is the *unit matrix of order p* which consists of unities (i.e., ones) in every position of the main diagonal (i.e., the diagonal running from the upper left corner to the lower right corner) and zeros elsewhere; for example,

$$\mathbf{I}_4 = \begin{bmatrix} 1 & 0 & 0 & 0 \\ 0 & 1 & 0 & 0 \\ 0 & 0 & 1 & 0 \\ 0 & 0 & 0 & 1 \end{bmatrix}.$$

(When the size of the unit matrix is obvious, the subscript is often omitted.) The unit matrix plays the same role in matrix multiplication that 1 does in ordinary multiplication—it leaves the multiplicand unchanged. The inverse of a matrix is unique.

The formulae for inverting matrices of sizes two and three are as follows:

$$\mathbf{M}^{-1} = \begin{bmatrix} a & b \\ c & d \end{bmatrix}^{-1} = \begin{bmatrix} d/D & -b/D \\ -c/D & a/D \end{bmatrix} \tag{2.1.10}$$

where $D = ad - bc$ is the *determinant* of the 2×2 matrix $\mathbf{M}$.

$$\mathbf{Q}^{-1} = \begin{bmatrix} a & b & c \\ d & e & f \\ g & h & k \end{bmatrix}^{-1} = \begin{bmatrix} A & B & C \\ D & E & F \\ G & H & K \end{bmatrix} \tag{2.1.11}$$

where

$$A = (ek - fh)/Z \qquad B = -(bk - ch)/Z \qquad C = (bf - ce)/Z$$
$$D = -(dk - fg)/Z \qquad E = (ak - cg)/Z \qquad F = -(af - cd)/Z$$
$$G = (dh - eg)/Z \qquad H = -(ah - bg)/Z \qquad K = (ae - bd)/Z$$

and where
$$Z = a(ek - fh) - b(dk - fg) + c(dh - eg)$$
$$= aek + bfg + cdh - ahf - dbk - gec$$

is the determinant of $\mathbf{Q}$.

Matrices of the form $\mathbf{X'X}$ met in regression work are always symmetric, that is to say the element in the ith row and jth column is the same as the element in the jth row and ith column. Thus the transpose of a symmetric matrix is the matrix itself. This is easy to see if we apply the general rule $(\mathbf{AB})' = \mathbf{B'A'}$ for transposes of a product. Because $(\mathbf{A'})' = \mathbf{A}$ itself, we can write $(\mathbf{X'X})' = \mathbf{X'X}$. (Working with a few simple numerical cases will clarify this point.) If the matrix $\mathbf{M}$ of size two above is symmetric, $b = c$ and the inverse also becomes symmetric. If the matrix $\mathbf{Q}$ above is symmetric $b = d, c = g, f = h$. Then, relabeling the matrix $\mathbf{S}$, we obtain the symmetric inverse

$$\mathbf{S}^{-1} = \begin{bmatrix} a & b & c \\ b & e & f \\ c & f & k \end{bmatrix}^{-1} = \begin{bmatrix} A & B & C \\ B & E & F \\ C & F & K \end{bmatrix} \tag{2.1.12}$$

where $A = (ek - f^2)/Y$ $B = -(bk - cf)/Y$ $C = (bf - ce)/Y$
$$E = (ak - c^2)/Y \qquad F = -(af - bc)/Y$$
$$K = (ae - b^2)/Y$$

and where
$$Y = a(ek - f^2) - b(bk - cf) + c(bf - ce)$$
$$= aek + 2bcf - af^2 - b^2k - c^2e$$

is the determinant of $\mathbf{S}$. The inverse of any symmetric matrix is, itself, a symmetric matrix.

Matrices of sizes greater than three are usually cumbersome to invert unless they have a special form. One matrix that is easy to invert, no matter what its size, is a *diagonal matrix* which consists of nonzero elements in the main upper-left to lower-right diagonal, and zeros elsewhere. The inverse is obtained by inverting all nonzero elements where they stand. For example,

$$\begin{bmatrix} a_1 & & & \\ & a_2 & & \mathbf{0} \\ & & \cdot & \\ & & & \cdot \\ \mathbf{0} & & & \cdot \\ & & & a_n \end{bmatrix}^{-1} = \begin{bmatrix} 1/a_1 & & & \\ & 1/a_2 & & \mathbf{0} \\ & & \cdot & \\ & & & \cdot \\ \mathbf{0} & & & \cdot \\ & & & 1/a_n \end{bmatrix}. \tag{2.1.13}$$

(Note, in this special case, the use of $\mathbf{0}$ to denote a large triangular block of zeros. This is often seen.)

Another type of simplification sometimes occurs when some columns of the $\mathbf{X}$ matrix are orthogonal to *all* other columns. The $\mathbf{X'X}$ matrix then takes the *partitioned* form

$$\begin{bmatrix} \mathbf{P} & \mathbf{0} \\ \mathbf{0} & \mathbf{R} \end{bmatrix}$$

where, for example, $\mathbf{P}$ might be $p \times p$, $\mathbf{R}$ might be $r \times r$, and the symbol $\mathbf{0}$ is used to denote two differently shaped blocks of zeros, a $p \times r$ one in the top right-hand corner and an $r \times p$ one in the lower left hand corner. The inverse of this matrix is then

$$\begin{bmatrix} \mathbf{P} & \mathbf{0} \\ \mathbf{0} & \mathbf{R} \end{bmatrix}^{-1} = \begin{bmatrix} \mathbf{P}^{-1} & \mathbf{0} \\ \mathbf{0} & \mathbf{R}^{-1} \end{bmatrix}. \tag{2.1.14}$$

For example, if

$$\mathbf{P} = \begin{bmatrix} 1 & 3 \\ 2 & 8 \end{bmatrix}, \qquad \mathbf{P}^{-1} = \begin{bmatrix} 4 & -\frac{3}{2} \\ -1 & \frac{1}{2} \end{bmatrix}$$

$$\mathbf{R} = \begin{bmatrix} 1 & 0 & 1 \\ 2 & 3 & 2 \\ 4 & 1 & 1 \end{bmatrix}, \qquad \mathbf{R}^{-1} = \begin{bmatrix} -\frac{1}{9} & -\frac{1}{9} & \frac{1}{3} \\ -\frac{2}{3} & \frac{1}{3} & 0 \\ \frac{10}{9} & \frac{1}{9} & -\frac{1}{3} \end{bmatrix}$$

then

$$\begin{bmatrix} 1 & 3 & 0 & 0 & 0 \\ 2 & 8 & 0 & 0 & 0 \\ 0 & 0 & 1 & 0 & 1 \\ 0 & 0 & 2 & 3 & 2 \\ 0 & 0 & 4 & 1 & 1 \end{bmatrix}^{-1} = \begin{bmatrix} 4 & -\frac{3}{2} & 0 & 0 & 0 \\ -1 & \frac{1}{2} & 0 & 0 & 0 \\ 0 & 0 & -\frac{1}{9} & -\frac{1}{9} & \frac{1}{3} \\ 0 & 0 & -\frac{2}{3} & \frac{1}{3} & 0 \\ 0 & 0 & \frac{10}{9} & \frac{1}{9} & -\frac{1}{3} \end{bmatrix}.$$

When there are more than two nonzero blocks, the obvious extension holds. It is important to note that the blocks must be on the main diagonal, and the off-diagonal blocks must consist entirely of zeros for the extension to apply.

The inverse formula (2.1.14) also applies even when the rows and columns containing nonzero elements are intermingled, provided that the matrix can be divided in such a way that the portions, such as $\mathbf{P}$ and $\mathbf{R}$ above, are completely separated from each other by zeros. For example, using the same numbers as above, the matrix

$$\begin{bmatrix} 1 & 0 & 0 & 0 & 1 \\ 0 & 1 & 0 & 3 & 0 \\ 2 & 0 & 3 & 0 & 2 \\ 0 & 2 & 0 & 8 & 0 \\ 4 & 0 & 1 & 0 & 1 \end{bmatrix}$$

can be partitioned and the separate portions inverted separately. Note that the second and fourth rows *and* columns are completely isolated, or insulated, from the first, third, and fifth columns by zeros. Thus the nonzero elements in the second and fourth rows and columns comprise a 2×2 matrix which can be separately inverted, whereas the other nonzero elements form a completely separate 3×3 matrix which also can be separately inverted. Thus the inverse has the form

$$\begin{bmatrix} -\frac{1}{9} & 0 & -\frac{1}{9} & 0 & \frac{1}{3} \\ 0 & 4 & 0 & -\frac{3}{2} & 0 \\ -\frac{2}{3} & 0 & \frac{1}{3} & 0 & 0 \\ 0 & -1 & 0 & \frac{1}{2} & 0 \\ \frac{10}{9} & 0 & \frac{1}{9} & 0 & -\frac{1}{3} \end{bmatrix}.$$

The correctness of all these inverses can be confirmed by actually multiplying the inverse by the original, both before and behind. The result is an **I** matrix of appropriate size in every case. In practical situations, when the size of a matrix exceeds 3×3, and no simplified form is possible, finding the inverse can be a lengthy procedure. The work would usually be performed within an electronic computer. "Hand" methods are given by some authors (see Bibliography) but we shall not deal with them here.

We wish now to invert the **X'X** matrix of our example. This is of size 2×2 and of the general form of Eq. (2.1.7). Using Eq. (2.1.10) we obtain the inverse as

$$(\mathbf{X'X})^{-1} = \begin{bmatrix} \dfrac{\sum X_i^2}{n \sum (X_i - \bar{X})^2} & \dfrac{-\bar{X}}{\sum (X_i - \bar{X})^2} \\ \dfrac{-\bar{X}}{\sum (X_i - \bar{X})^2} & \dfrac{1}{\sum (X_i - \bar{X})^2} \end{bmatrix}. \qquad (2.1.15)$$

If *every* element of a matrix has a common factor it can be taken outside the matrix. (Conversely, if a matrix is multiplied by a constant C, every element of the matrix must be multiplied by C.) Thus an alternative form is

$$(\mathbf{X'X})^{-1} = \dfrac{1}{n \sum (X_i - \bar{X})^2} \begin{bmatrix} \sum X_i^2 & -\sum X_i \\ -\sum X_i & n \end{bmatrix}. \qquad (2.1.16)$$

Since **X'X** is symmetric, so is its inverse $(\mathbf{X'X})^{-1}$ as mentioned earlier. The quantity taken outside the matrix is the determinant of **X'X**, written det $(\mathbf{X'X})$ or $|\mathbf{X'X}|$. Using the form of Eq. (2.1.15) on the data of our example we find that

$$(\mathbf{X'X})^{-1} = \begin{bmatrix} 0.4267941 & -0.0073535 \\ -0.0073535 & 0.0001398 \end{bmatrix}.$$

If we premultiply Eq. (2.1.9) by $(X'X)^{-1}$, we obtain

$$(X'X)^{-1}(X'X)b = (X'X)^{-1}X'Y,$$

that is,

$$b = (X'X)^{-1}X'Y \qquad (2.1.17)$$

since $(X'X)^{-1}X'X = I$. This is an important result to remember since the solution of linear regression normal equations can *always* be written in this form, provided $X'X$ is nonsingular and the regression problem is properly expressed.

Using the data of our example we find that

$$b = \begin{bmatrix} 0.4267941 & -0.0073535 \\ -0.0073535 & 0.0001398 \end{bmatrix}\begin{bmatrix} 235.60 \\ 11,821.4320 \end{bmatrix}$$

$$= \begin{bmatrix} 13.623790 \\ -0.079848 \end{bmatrix}.$$

Note that the results are not identical, to six places of decimals, to the values obtained in Section 1.2. Such discrepancies frequently occur because of the rounding off of numbers used in the calculation, and carelessness in such matters can cause serious errors, depending on the numbers involved. Here the numerical discrepancies are slight from a practical point of view, but they emphasize the fact that in general as many figures as possible should be carried in regression calculations. Sometimes, due to the magnitudes of the numbers in the calculation, the entire significance of the results can be lost through careless rounding.

Certain ways of performing the calculations (especially when they are done "by hand," i.e., on a desk calculator) are better than others since they are less affected by roundoff error. In particular it is wise to postpone divisions to as late a stage as possible. For example, if we had employed the form of Eq. (2.1.16) instead of (2.1.15) to obtain $(X'X)^{-1}$ we would have obtained

$$(X'X)^{-1} = \frac{1}{178,860.5}\begin{bmatrix} 76,323.42 & -1315 \\ -1315 & 25 \end{bmatrix}.$$

Then we could have obtained b from

$$b = \frac{1}{178,860.5}\begin{bmatrix} 76,323.42 & -1315 \\ -1315 & 25 \end{bmatrix}\begin{bmatrix} 235.60 \\ 11,821.432 \end{bmatrix}$$

$$= \frac{1}{178,860.5}\begin{bmatrix} 2,436,614.672 \\ -14,278.2 \end{bmatrix}$$

$$= \begin{bmatrix} 13.622989 \\ -0.079829 \end{bmatrix}$$

the division being performed last of all.

Doing the calculations the three separate ways gives these answers

	Formulae (Section 1.2)	Inverse Matrix	Inverse Matrix (division last)
b_0	13.623005	13.623790	13.622989
b_1	−0.079829	−0.079848	−0.079829

As we have said, these differences are of slight consequence in this example. The third method is actually the most accurate. To see what the consequences of rounding can be, we suggest the reader make use of the inverse matrix in the second method, and round the elements in several ways—for example, rounding to 6, 5, 4, or 3 places of decimals or the same numbers of significant figures. Rounding errors provide a major share of disagreements when several people work the same regression problem using desk calculators.

When a regression computer program is written, many significant figures are automatically retained. Even so, some programs make use of double precision arithmetic to preserve accuracy, though this is not normally required. (See the discussion in Section 5.4.)

SUMMARY TO THIS POINT.   If we express the straight line model to be fitted to the data of our example in the form

$$Y = X\beta + \epsilon$$

as in Eq. (2.1.4), then the least squares estimates of (β_0, β_1), that is, of

$\beta = \begin{bmatrix} \beta_0 \\ \beta_1 \end{bmatrix}$ are given by

$$\begin{bmatrix} b_0 \\ b_1 \end{bmatrix} = b = (X'X)^{-1}X'Y.$$

This result is of great importance and should be memorized. Note that the fitted values $\hat{Y}$ are obtained by evaluating

$$\hat{Y} = Xb.$$

2.2. The Analysis of Variance in Matrix Terms

We recall from Section 1.3, page 15, that in a more general form of the analysis of variance table we wrote

$$SS(b_1 \mid b_0) = b_1 \left[\sum X_i Y_i - \frac{(\sum X_i)(\sum Y_i)}{n} \right] = b_1 [\sum X_i Y_i - n\bar{X}\bar{Y}]$$

$$SS(b_0) = \text{Correction for mean} = \frac{(\sum Y_i)^2}{n} = n\bar{Y}^2$$

Each of these sums of squares has 1 degree of freedom. Now

$$\begin{aligned}
SS(b_1 \mid b_0) + SS(b_0) &= b_1 \sum X_i Y_i - b_1 n \bar{X}\bar{Y} + n\bar{Y}^2 \\
&= b_1 \sum X_i Y_i + n\bar{Y}(\bar{Y} - b_1\bar{X}) \\
&= b_1 \sum X_i Y_i + b_0 \sum Y_i \\
&= (b_0, b_1) \begin{bmatrix} \sum Y_i \\ \sum X_i Y_i \end{bmatrix} \\
&= \mathbf{b'X'Y}
\end{aligned} \tag{2.2.1}$$

in matrix terms, with 2 degrees of freedom. Thus we can write the analysis of variance table in matrix form as follows:

Source	Sum of Squares	Degrees of Freedom	Mean Square
$\mathbf{b'} = (b_0, b_1)$	$\mathbf{b'X'Y}$	2	
Residual	$\mathbf{Y'Y} - \mathbf{b'X'Y}$	$n-2$	s^2
Total (uncorrected)	$\mathbf{Y'Y}$	n	

In this way we can split the total variation $\mathbf{Y'Y}$ into two portions, one due to the straight line we have estimated, namely $\mathbf{b'X'Y}$, and a residual which shows the remaining variation of the points about the regression line. In order to find what portion of the total variation can be attributed to the addition of the term $\beta_1 X_i$ to the simpler model $Y_i = \beta_0 + \epsilon_i$, we would just subtract the correction factor $n\bar{Y}^2$ from the sum of squares $\mathbf{b'X'Y}$ in order to obtain $SS(b_1 \mid b_0)$ as before. The quantity $n\bar{Y}^2$ would be $SS(b_0)$ if the model $Y_i = \beta_0 + \epsilon_i$ were fitted. The remainder of $\mathbf{b'X'Y}$ thus measures the *extra sum of squares removed by* b_1 when the model $Y_i = \beta_0 + \beta_1 X_i + \epsilon_i$ is used. If an estimate of pure error from repeat points is available it is subtracted from the residual sum of squares to provide the same breakup and the same tests as described in Section 1.5.

Example.   For our main example we had

$$\mathbf{b} = \begin{bmatrix} 13.62 \\ -0.0798 \end{bmatrix} \qquad \mathbf{X'Y} = \begin{bmatrix} 235.60 \\ 11{,}821.4320 \end{bmatrix}$$

Hence
$$SS(\mathbf{b}) = \mathbf{b'X'Y} = 2265.5217$$
$$SS(b_0) = (\textstyle\sum Y_i)^2/n = 2220.2944$$
$$\begin{aligned} SS(b_1 \mid b_0) &= SS(b_1, \text{ after allowance for } b_0) \\ &= \mathbf{b'X'Y} - (\textstyle\sum Y_i)^2/n = 45.2273. \end{aligned}$$

Previously we obtained 45.59 here so that a discrepancy of 0.36 has arisen. Once again this is due to the rounding off of numbers in the calculation, and it points up the fact that, even in simple regression calculations, it is wise to carry as many figures as possible.

2.3. The Variances and Covariance of b_0 and b_1 from the Matrix Calculation

We recall that $V(b_1) = \sigma^2/\sum (X_i - \bar{X})^2$. Also

$$V(b_0) = V(\bar{Y} - b_1\bar{X}) = \sigma^2\left[\frac{1}{n} + \frac{\bar{X}^2}{\sum (X_i - \bar{X})^2}\right] = \frac{\sigma^2 \sum X_i^2}{n \sum (X_i - \bar{X})^2}$$

since, as we showed earlier, $\bar{Y}$ and b_1 have zero covariance, and the X's are regarded as constants. In addition,

$$\begin{aligned}\text{cov}\,(b_0, b_1) &= \text{cov}\,[(\bar{Y} - b_1\bar{X}), b_1]\\ &= -\bar{X}V(b_1)\\ &= -\bar{X}\sigma^2/\sum (X_i - \bar{X})^2.\end{aligned}$$

Thus we can write the *variance-covariance* matrix of the vector **b** as follows.

$$\mathbf{V(b)} = \mathbf{V}\begin{pmatrix} b_0 \\ b_1 \end{pmatrix} = \begin{bmatrix} V(b_0) & \text{cov}\,(b_0, b_1) \\ \text{cov}\,(b_0, b_1) & V(b_1) \end{bmatrix}$$

$$= \begin{bmatrix} \dfrac{\sigma^2 \sum X_i^2}{n \sum (X_i - \bar{X})^2} & -\dfrac{\bar{X}\sigma^2}{\sum (X_i - \bar{X})^2} \\[3ex] -\dfrac{\bar{X}\sigma^2}{\sum (X_i - \bar{X})^2} & \dfrac{\sigma^2}{\sum (X_i - \bar{X})^2} \end{bmatrix}. \qquad (2.3.1)$$

Now if every element of a matrix has a common factor we can remove it and set it outside the matrix, so that we can remove σ^2. The matrix that remains is seen to be $(\mathbf{X'X})^{-1}$ from Eq. (2.1.15). Thus

$$\mathbf{V(b)} = (\mathbf{X'X})^{-1}\sigma^2. \qquad (2.3.2)$$

This is an important result and should be remembered. When σ^2 is unknown we use, instead, s^2 the estimate of σ^2 obtained from the analysis of variance table, if there is no lack of fit, or s_e^2, the pure error mean square if lack of fit is shown. This provides us with the *estimated variance-covariance matrix of* **b** (see also, Section 2.12, page 85).

2.4.   Variance of $\hat{Y}$ Using the Matrix Development

Let X_k be a selected value of X. The predicted mean value of Y for this value of X is

$$\hat{Y}_k = b_0 + b_1 X_k$$

Let us define the vector $\mathbf{X}_k$ as

$$\mathbf{X}_k' = (1, X_k).$$

We can then write

$$\hat{Y}_k = (1, X_k) \begin{bmatrix} b_0 \\ b_1 \end{bmatrix} = \mathbf{X}_k'\mathbf{b} = \mathbf{b}'\mathbf{X}_k.$$

Since $\hat{Y}_k$ is a linear combination of the random variables b_0 and b_1, it follows that

$$V(\hat{Y}_k) = V(b_0) + 2X_k \operatorname{cov}(b_0, b_1) + X_k^2 V(b_1).$$

As can be verified by working out the indicated matrix and vector products, the above quantity can be expressed in the alternative form

$$V(\hat{Y}_k) = [1, \ X_k] \begin{bmatrix} V(b_0) & \operatorname{cov}(b_0, b_1) \\ \operatorname{cov}(b_0, b_1) & V(b_1) \end{bmatrix} \begin{bmatrix} 1 \\ X_k \end{bmatrix}$$

$$= \mathbf{X}_k'(\mathbf{X}'\mathbf{X})^{-1}\sigma^2\mathbf{X}_k$$

$$= \mathbf{X}_k'(\mathbf{X}'\mathbf{X})^{-1}\mathbf{X}_k\sigma^2.$$

Although now given in a different form, this is identical in value to Eq. (1.4.6). This important matrix result should be remembered. With suitable redefinition of $\mathbf{X}_k$ and $\mathbf{X}$ it is applicable to the general linear regression situation. An estimated variance is obtained when σ^2 is replaced by an estimate s^2.

2.5.   Summary of Matrix Approach to Fitting a Straight Line

1. Set down the model in the form $\mathbf{Y} = \mathbf{X}\boldsymbol{\beta} + \boldsymbol{\epsilon}$.
2. Find $\mathbf{b} = (\mathbf{X}'\mathbf{X})^{-1}\mathbf{X}'\mathbf{Y}$ to obtain the least squares estimate $\mathbf{b}$ of $\boldsymbol{\beta}$ provided by the data.

3. Construct $\mathbf{b'X'Y}$ the sum of squares due to coefficients and hence obtain the basic analysis of variance as follows:

Source	Sum of Squares	Degrees of Freedom	Mean Square
Regression	$\mathbf{b'X'Y}$	2	
Residual	$\mathbf{Y'Y} - \mathbf{b'X'Y}$	$n-2$	s^2 (estimates σ^2 if the model is correct)
Total	$\mathbf{Y'Y}$	n	

Additional subdivision of the sum of squares is achieved by finding $SS(b_1 \mid b_0)$, the extra sum of squares due to b_1, and introducing pure error. The more detailed analysis of variance table will take the form

Source		Sum of Squares	Degrees of Freedom	Mean Square
SS(b)	Mean (b_0)	$n\bar{Y}^2$	1	
	SS($b_1 \mid b_0$)	$\mathbf{b'X'Y} - n\bar{Y}^2$	1	
Residual	Lack of fit	$\mathbf{Y'Y} - \mathbf{b'X'Y} - SS(\text{p.e.})$	$n-2-n_e$	MS_L $\Big\}s^2$
	Pure error	$SS(\text{p.e.})$	n_e	s_e^2
Total		$\mathbf{Y'Y}$	n	

The tests for lack of fit and for β_1 are performed as described in Chapter 1. An additional measure of the regression is provided by the ratio

$$R^2 = \frac{(\mathbf{b'X'Y} - n\bar{Y}^2)}{(\mathbf{Y'Y} - n\bar{Y}^2)}$$

4. If no lack of fit is shown, $(\mathbf{X'X})^{-1}s^2$ will provide estimates of $V(b_0)$, $V(b_1)$, and cov (b_0, b_1) and enable individual coefficients to be tested or other calculations made as in Chapter 1.

5. The following quantities can be found:

The vector of fitted values: $\hat{\mathbf{Y}} = \mathbf{Xb}$

A prediction of Y at X_k: $\hat{Y}_k = \mathbf{X}_k'\mathbf{b} = \mathbf{b'X}_k$

with variance: $V(\hat{Y}_k) = \mathbf{X}_k'(\mathbf{X'X})^{-1}\mathbf{X}_k\sigma^2$

2.6. The General Regression Situation

We have seen how the problem of fitting a straight line by least squares can be handled through the use of matrices. This approach is important for the following reason. If we wish to fit *any* model linear in parameters $\beta_0, \beta_1, \beta_2, \ldots$, by least squares, the calculations necessary are of exactly the same form (in matrix terms) as those for the straight line involving only two parameters β_0 and β_1. The mechanics of calculation, however, increase sharply with the number of parameters. Thus while the formulae are easy to remember, the use of a digital computer is an essential for nearly all problems except when

(1) the number of parameters is small—say less than five,
(2) the data arise from a designed experiment which provides an $\mathbf{X'X}$ matrix of simple, or "patterned," form.

A general statement of linear regression methods will now be given. For the theoretical derivation of these results, the reader should consult, for example, *Regression Analysis* by R. L. Plackett.

Suppose we have a model under consideration which can be written in the form

$$\mathbf{Y} = \mathbf{X\beta} + \mathbf{\epsilon} \tag{2.6.1}$$

where $\mathbf{Y}$ is an $(n \times 1)$ vector of observations,

$\mathbf{X}$ is an $(n \times p)$ matrix of known form,

$\mathbf{\beta}$ is a $(p \times 1)$ vector of parameters,

$\mathbf{\epsilon}$ is an $(n \times 1)$ vector of errors,

and where $E(\mathbf{\epsilon}) = \mathbf{0}$, $V(\mathbf{\epsilon}) = \mathbf{I}\sigma^2$, so the elements of $\mathbf{\epsilon}$ are uncorrelated.

Since $E(\mathbf{\epsilon}) = \mathbf{0}$, an alternative way of writing the model is

$$E(\mathbf{Y}) = \mathbf{X\beta}. \tag{2.6.1}$$

The error sum of squares is then

$$\begin{aligned}
\mathbf{\epsilon'\epsilon} &= (\mathbf{Y} - \mathbf{X\beta})'(\mathbf{Y} - \mathbf{X\beta}) \\
&= \mathbf{Y'Y} - \mathbf{\beta'X'Y} - \mathbf{Y'X\beta} + \mathbf{\beta'X'X\beta} \\
&= \mathbf{Y'Y} - 2\mathbf{\beta'X'Y} + \mathbf{\beta'X'X\beta}.
\end{aligned} \tag{2.6.2}$$

(This follows due to the fact that $\mathbf{\beta'X'Y}$ is a 1×1 matrix, or a scalar, whose transpose $(\mathbf{\beta'X'Y})' = \mathbf{Y'X\beta}$ must have the same value.)

The least squares estimate of $\mathbf{\beta}$ is the value $\mathbf{b}$ which, when substituted in Eq. (2.6.2), minimizes $\mathbf{\epsilon'\epsilon}$. It can be determined by differentiating Eq. (2.6.2) with respect to $\mathbf{\beta}$ and setting the resultant matrix equation equal to zero, at the same time replacing $\boldsymbol{\beta}$ by $\mathbf{b}$. (Differentiating $\boldsymbol{\epsilon}'\boldsymbol{\epsilon}$ with respect to a vector quantity $\boldsymbol{\beta}$ is equivalent to differentiating $\boldsymbol{\epsilon}'\boldsymbol{\epsilon}$ separately with respect to each element of $\boldsymbol{\beta}$ in order, writing down the resulting derivatives one below the other, and rearranging the whole into matrix form.) This provides the *normal equations*

$$(\mathbf{X}'\mathbf{X})\mathbf{b} = \mathbf{X}'\mathbf{Y}. \tag{2.6.3}$$

Two main cases arise; either Eq. (2.6.3) consists of p independent equations in p unknowns, or some equations depend on others so that there are fewer than p independent equations in the p unknowns (the p unknowns are the elements of $\mathbf{b}$). If some of the normal equations depend on others, $\mathbf{X}'\mathbf{X}$ is singular, so that $(\mathbf{X}'\mathbf{X})^{-1}$ does not exist. Then either the model should be expressed in terms of fewer parameters or else additional restrictions on the parameters must be given or assumed. Some examples of this situation are given in Chapter 9. If the p normal equations are independent, $\mathbf{X}'\mathbf{X}$ is nonsingular, and its inverse exists. In this case the solution of the normal equations can be written

$$\mathbf{b} = (\mathbf{X}'\mathbf{X})^{-1}\mathbf{X}'\mathbf{Y}. \tag{2.6.4}$$

This solution $\mathbf{b}$ has the following properties:

1. It is an estimate of $\boldsymbol{\beta}$ which minimizes the error sum of squares $\boldsymbol{\epsilon}'\boldsymbol{\epsilon}$, *irrespective* of any distribution properties of the errors.

Note. An assumption that the errors $\boldsymbol{\epsilon}$ are normally distributed is *not* required in order to obtain the estimates $\mathbf{b}$ but it *is* required later in order to make tests which depend on the assumption of normality, such as t- or F-tests, or for obtaining confidence intervals based on the t- and F-distributions.

2. The elements of $\mathbf{b}$ are linear functions of the observations $Y_1, Y_2, \ldots, Y_n$, and provide unbiased estimates of the elements of $\boldsymbol{\beta}$ which have the minimum variances (of *any* linear functions of the Y's which provide unbiased estimates), irrespective of distribution properties of the errors.

Note. Suppose we have an expression $T = l_1 Y_1 + l_2 Y_2 + \cdots + l_n Y_n$, which is a linear function of observations $Y_1, Y_2, \ldots, Y_n$, and which we use as an estimate of a parameter θ. Then T is a random variable whose probability distribution will depend on the distribution from which the Y's arise. If we repeatedly take samples of Y's and evaluate the corresponding T's we shall generate the distribution of T empirically. Whether we do this or not, the distribution of T will have some definite mean value which we can write $E(T)$ and a variance which we can write $V(T)$. If it happens that the mean of the distribution of T is equal to the parameter θ we are estimating by T—i.e., if $E(T) = \theta$—then we say that T is an unbiased estimator of θ. The word *estimator* is normally used when referring to the theoretical expression for T in terms of a sample of Y's. A specific numerical value of T would be called an unbiased *estimate* of θ. This distinction, though correct, is not always maintained in statistical writings. If we have all possible linear functions $T_1, T_2, \ldots$, say, of n observations $Y_1, Y_2, \ldots, Y_n$, and if the T's satisfy

$$\theta = E(T_1) = E(T_2) \cdots$$

that is, they are all unbiased estimators of θ, then the one with the smallest value of $V(T_j)$, $j = 1, 2, \ldots$, is the *minimum variance unbiased estimator* of θ. (The result (2) is "Gauss's Theorem.")

3. If the errors are independent and $\epsilon_i \sim N(0, \sigma^2)$, then $\mathbf{b}$ is the maximum likelihood estimate of $\boldsymbol{\beta}$. (In vector terms we can write $\boldsymbol{\epsilon} \sim N(\mathbf{0}, \mathbf{I}\sigma^2)$, meaning that $\boldsymbol{\epsilon}$ follows an n-dimensional multivariate normal distribution with $E(\boldsymbol{\epsilon}) = \mathbf{0}$ (where $\mathbf{0}$ denotes a vector consisting entirely of zeros and of the same length as $\boldsymbol{\epsilon}$) and $\mathbf{V}(\boldsymbol{\epsilon}) = \mathbf{I}\sigma^2$; that is, $\boldsymbol{\epsilon}$ has a variance-covariance matrix whose diagonal elements, $V(\epsilon_i)$, $i = 1, 2, \ldots, n$ are all σ^2 and whose off-diagonal elements, covariance (ϵ_i, ϵ_j), $i \neq j = 1, \ldots, n$ are all zero. The likelihood function for the sample $Y_1, Y_2, \ldots, Y_n$ is defined in this case as the product

$$\prod_{i=1}^{n} \frac{1}{\sigma(2\pi)^{1/2}} e^{-\epsilon_i^2/2\sigma^2} = \frac{1}{\sigma^n(2\pi)^{n/2}} e^{-\boldsymbol{\epsilon}'\boldsymbol{\epsilon}/2\sigma^2}. \tag{2.6.5}$$

Thus for a fixed value of σ, maximizing the likelihood function is equivalent to minimizing the quantity $\boldsymbol{\epsilon}'\boldsymbol{\epsilon}$. Note that this fact can be used to provide a justification for the least squares procedure (i.e., for minimizing the sum of *squares* of errors), because in many physical situations the assumption that errors are normally distributed is quite sensible. We shall, in any case, find out if this assumption appears to be violated by examining the residuals from the regression analysis. If any definite *a priori* knowledge *is* available about the error distribution, perhaps from theoretical considerations or from sound prior knowledge of the process under study, the maximum likelihood argument could be used to obtain estimates based on a criterion other than least squares. For example, suppose the errors ϵ_i, $i = 1, 2, \ldots, n$ were independent and followed the double exponential distribution:

$$f(\epsilon_i) = (2\sigma)^{-1} e^{-|\epsilon_i|/\sigma} \qquad (-\infty \leq \epsilon_i \leq \infty) \tag{2.6.6}$$

rather than the normal distribution:

$$f(\epsilon_i) = \frac{1}{\sigma(2\pi)^{1/2}} e^{-\epsilon_i^2/2\sigma^2} \tag{2.6.7}$$

as is usually assumed. The double exponential frequency function has a pointed peak of height $1/2\sigma$ at $\epsilon_i = 0$, and tails off to zero as ϵ_i goes to both plus and minus infinity. Then application of the maximum likelihood principle for estimating β, assuming σ fixed, would involve minimization of

$$\sum_{i=1}^{n} |\epsilon_i|$$

the sum of absolute errors and not the minimization of

$$\sum_{i=1}^{n} \epsilon_i^2$$

the sum of *squares* of errors. For further reading on minimizing the sum of absolute errors, see papers by Fisher, Karst, and Wagner listed in the Bibliography.

Suppose we have used the method of least squares to estimate β by b. We can proceed with the following steps whether the errors are normally distributed or not.

1. The fitted values are obtained from $\hat{\mathbf{Y}} = \mathbf{Xb}$.

2. The vector of residuals is given by $\mathbf{e} = \mathbf{Y} - \hat{\mathbf{Y}}$ (see Chapter 3 for their examination). It is true that $\sum_{i=1}^{n} e_i \hat{Y}_i = 0$, whatever the model. This can be seen by multiplying the jth normal equation by the jth β and adding the results. If there is a β_0 term in the model, it is also true that $\sum_{i=1}^{n} e_i = 0$. (The e_i and $\hat{Y}_i$, $i = 1, 2, \ldots, n$ are the ith elements of the vectors $\mathbf{e}$ and $\hat{\mathbf{Y}}$, respectively.)

3. $V(\mathbf{b}) = (\mathbf{X'X})^{-1}\sigma^2$ provides the variances (diagonal terms) and co-variances (off-diagonal terms) of the estimates. (An estimate of σ^2 is obtained as described below.)

4. Suppose $\mathbf{X_0}'$ is a specified $1 \times p$ vector whose elements are of the same form as a row of $\mathbf{X}$ so that $\hat{Y}_0 = \mathbf{X_0}'\mathbf{b} = \mathbf{b}'\mathbf{X_0}$ is *the fitted value at a specified point* $\mathbf{X_0}$. For example, if the model were $Y = \beta_0 + \beta_1 X + \beta_{11} X^2 + \epsilon$, then $\mathbf{X_0}' = (1, X_0, X_0^2)$ for a given value X_0. Then $\hat{Y}_0$ is the value *predicted at* $\mathbf{X_0}$ *by the regression equation* and has variance

$$V(\hat{Y}_0) = \mathbf{X_0}'V(\mathbf{b})\mathbf{X_0} = \mathbf{X_0}'(\mathbf{X'X})^{-1}\mathbf{X_0}\sigma^2. \tag{2.6.8}$$

If we use an estimate s_v^2 for σ^2, $1 - \alpha$ confidence limits for the mean value of Y at $\mathbf{X_0}$ are obtained from

$$\hat{Y}_0 \pm t(v, 1 - \tfrac{1}{2}\alpha) \, s\sqrt{\mathbf{X_0}'(\mathbf{X'X})^{-1}\mathbf{X_0}}. \tag{2.6.9}$$

5. A basic analysis of variance table can be constructed as follows.

Source	Sum of Squares	Degrees of Freedom	Mean Square
Regression	$b'X'Y$	p	MS_R
Residual	$Y'Y - b'X'Y$	$n - p$	MS_E
Total	$Y'Y$	n	

Correction: Eq. (2.6.9) and the two lines above it should appear in the later part of the section *after* the normality assumption.

A further subdivision of the parts of this table can be carried out as follows.

5(a). If a β_0 term is in the model we can subdivide the regression sum of squares into

$$SS(b_0) = \frac{(\sum Y_i)^2}{n} = n\,\overline{Y}^2$$

$$SS(\text{Regression} \mid b_0) = SS(R \mid b_0) = \mathbf{b'X'Y} - \frac{(\sum Y_i)^2}{n}. \quad (2.6.10)$$

These sums of squares are based on 1 and $p - 1$ degrees of freedom, respectively. (More extensive subdivision of the regression sum of squares will be discussed in Section 2.7.)

5(b). If repeat observations are available we can split the residual SS into SS(pure error) with n_e degrees of freedom, which estimates $n_e\sigma^2$ and SS(lack of fit) with $(n - p - n_e)$ degrees of freedom.

"Repeats" now must be repeats in *all* coordinates $X_1, X_2, \ldots, X_k$ of the independent variables (though approximate use of "very close" points is sometimes seen in practice). This provides an analysis of variance table as follows.

Source	Sum of Squares	Degrees of Freedom	Mean Square	
b_0	$SS(b_0)$	1		
Regression $\mid b_0$	$SS(R \mid b_0)$	$p - 1$	$MS(R \mid b_0)$	MS_R
Lack of fit	$SS(\text{l.o.f.})$	$n - p - n_e$	$MS(\text{l.o.f.})$	MS_E
Pure error	$SS(\text{p.e.})$	n_e	$MS(\text{p.e.})$	
Total	$Y'Y$	n		

Note.   The order in which terms are given in the table is not important. Most of the tables in this text are rearranged in the order often seen in computer printouts.

The ratio

$$R^2 = \frac{SS(R \mid b_0)}{Y'Y - SS(b_0)} = \frac{\sum (\hat{Y}_i - \overline{Y})^2}{\sum (Y_i - \overline{Y})^2} \quad (2.6.11)$$

is an extension of the quantity defined for the straight line regression and is the square of the *multiple correlation coefficient*. Another name for R^2 is the *coefficient of multiple determination*. (The quantity R^2 must not be confused with the R in the expressions $SS(R \mid b_0)$ and MS_R, where R is a label denoting the regression contribution.) If $\hat{Y}_i = Y_i$, that is, if the prediction is perfect, then $R^2 = 1$. If $\hat{Y}_i = \overline{Y}$, that is, $b_1 = b_2 = \cdots = b_{p-1} = 0$ (or a model $Y = \beta_0 + \epsilon$ alone has been fitted), then $R^2 = 0$. Thus R^2 is a measure of the usefulness of the terms, other than β_0, in the model. It is important to realize that R^2 can be made unity simply by employing n properly selected coefficients in the model, including β_0, since a model can then be chosen which fits the data exactly. (For example, if we have an observation of Y at four different values of X, a cubic polynomial

$$Y = \beta_0 + \beta_1 X + \beta_2 X^2 + \beta_3 X^3$$

passes exactly through all four points.) Since R^2 is often used as a convenient measure of the success of the regression equation in explaining the variation in the data, we must be sure that an improvement in R^2 due to adding a new term to the model has some real significance and is not due to the fact that the number of parameters in the model is getting close to saturation point—that is, the number of observations. This is an *especial* danger when there are *repeat* observations. For example, if we have one hundred observations which occur in five groups each of twenty repeats, we have effectively five pieces of information, represented by five mean values, and ninety-five error degrees of freedom for pure error, nineteen at each repeat point. Thus a five-parameter model will provide a perfect fit to the five means and may give a very large value of R^2, especially if the experimental error is small compared with the spread of the five means. In this case the fact that one hundred observations can be well predicted by a model with only five parameters is not surprising since there are really only five distinct data points and not one hundred as it first seemed. When there are no exact repeats, but the points in the X-space (at which observations Y are available) *are* close together, this type of situation can occur and yet be well concealed within the data. Plots of the data, and the residuals (see Chapter 3), will usually reveal such "clusters" of points.

The analysis of variance breakup is an algebraic equality (or a geometric one, depending on one's viewpoint—see Section 10.6) only and does not depend on distributive properties of the errors. However, if we assume additionally that $\epsilon_i \sim N(0, \sigma^2)$ and that the ϵ_i are independent—that is, that $\epsilon \sim N(0, I\sigma^2)$, we can do the following:

1. Test lack of fit by treating the ratio

$$\left[\frac{\text{SS(lack of fit)}/(n - p - n_e)}{\text{SS(pure error)}/n_e} \right] \qquad (2.6.12)$$

as an $F[(n - p - n_e), n_e]$ variate, and comparing its value with $F[(n - p - n_e), n_e, 1 - \alpha]$. If there is no lack of fit, $\text{SS(residual)}/(n - p) = \text{MS}_E$ usually called s^2 is an unbiased estimate of σ^2. If lack of fit cannot be tested, use of s^2 as an estimate of σ^2 *implies* an assumption that the model is correct. (If it is not, s^2 will usually be too large since it is a random variable with a mean *greater* than σ^2. Note carefully, however, that due to sampling fluctuation—since it *is* a random variable—it could also be too small.)

2. Test the overall regression equation (more specifically, test H_0: $\beta_1 = \beta_2 = \cdots = \beta_{p-1} = 0$ against H_1: not all $\beta_i = 0$) by treating the mean square ratio

$$\frac{[SS(R \mid b_0)/(p - 1)]}{s^2} \tag{2.6.13}$$

as an $F(p - 1, v)$ variate where $v = n - p$

Suppose we decide on a specified risk level α. The fact that the observed mean-square ratio exceeds $F(p - 1, v, 1 - \alpha)$ means that a "statistically significant" regression has been obtained; in other words, the proportion of the variation observed in the data, which has been accounted for by the equation, is greater than would be expected by chance in $100(1 - \alpha)\%$ similar sets of data with the same values of n and $\mathbf{X}$. This does not necessarily mean that the equation is useful for predictive purposes. Unless the range of values predicted by the fitted equation is considerably greater than the size of the random error, prediction will often be of no value even though a "significant" F-value has been obtained, since the equation will be "fitted to the errors" only.

Work by J. M. Wetz (in a 1964 Ph.D. thesis, "Criteria for judging adequacy of estimation by an approximating response function," written under the direction of Dr. G. E. P. Box at the University of Wisconsin) suggests that in order that an equation should be regarded as a satisfactory predictor (in the sense that the range of response values predicted by the equation is substantial compared with the standard error of the response), the observed F-ratio of (regression mean square)/(residual mean square) should exceed not merely the selected percentage point of the F-distribution, but about *four times* the selected percentage point. For example, if $p = 11$, $v = 20$, $\alpha = 0.05$, $F(10, 20, 0.95) = 2.35$. Thus the observed F-ratio would have to exceed about 9.4 for the fitted equation to be rated as a satisfactory prediction tool. Since (at the time of writing) work on this topic is not complete, the "four times" rule is given here as a current expedient for assessment of regression equations. It is subject to later confirmation.

3. State that

$$\mathbf{b} \sim N(\boldsymbol{\beta}, (\mathbf{X}'\mathbf{X})^{-1}\sigma^2). \tag{2.6.14}$$

4. Obtain a joint $100(1 - \alpha)\%$ confidence region for *all* the parameters $\boldsymbol{\beta}$ from the equation

$$(\boldsymbol{\beta} - \mathbf{b})'\mathbf{X}'\mathbf{X}(\boldsymbol{\beta} - \mathbf{b}) \leq ps^2 F(p, v, 1 - \alpha) \tag{2.6.15}$$

where $F(p, v, 1 - \alpha)$ is the $1 - \alpha$ point ("upper α-point") of the $F(p, v)$ distribution and where s^2 has the same meaning as in (1) above and the model is assumed correct. In general this will be useful only when p is small, say 2, 3, or 4, unless care is taken to present the information in a form in which it can be readily understood. The inequality above provides the equation of an "elliptically shaped" contour in a space which has as many dimensions, p, as there are parameters in $\boldsymbol{\beta}$. We can obtain individual confidence intervals for the various parameters separately from the formula

$$b_i \pm t(v, 1 - \tfrac{1}{2}\alpha) \quad \text{(estimated s.e. } (b_i))$$

where the "estimated s.e.(b_i)" is the square root of the ith diagonal term of the matrix $(\mathbf{X'X})^{-1}s^2$. (For a calculation of this type when there are two parameters β_0 and β_1, see Eq. (2.3.1), and after replacement of σ^2 by s^2, see pp. 19 and 21.) Separate confidence intervals of this type appear in our printouts and are often useful. We de-emphasize them, however, for the following reason. Figure 2.1 illustrates a possible situation that may arise when two parameters are considered. The joint 95% confidence region for the true parameters, β_1 and β_2, is shown as a long thin ellipse and encloses values (β_1, β_2) which the data regard as *jointly* reasonable for the parameters. It takes into account the correlation between the estimates b_1 and b_2. The individual 95% confidence intervals for β_1 and β_2 separately are appropriate for specifying ranges for the individual parameters irrespective of the value of the other parameter. If an attempt is made to interpret these intervals simultaneously—that is (wrongly), regard the rectangle which they define as a joint confidence region—then, for example, it may be thought that the coordinates of the point E provide

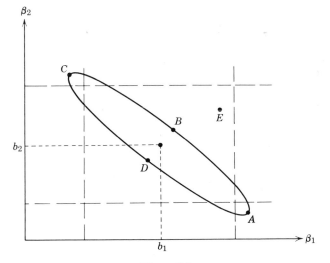

Figure 2.1

reasonable values for (β_1, β_2). The joint confidence region, however, clearly indicates that such a point is not reasonable. When only two parameters are involved, construction of the confidence ellipse is not difficult. When more parameters are involved the calculations are not difficult to handle in a computer but interpretation is difficult. One possible solution is to find the coordinates of the points at the ends of the major axes of the region. (In Figure 2.1 these would be the points A, B, C, and D.) This would involve obtaining the confidence contour and reducing it to canonical form. This is not difficult and also can be handled easily in a computer but is beyond the present scope of this book. We can however point out the moral that the "joint" message of individual confidence intervals should be regarded with caution, and attention should be paid both to the relative sizes of the $V(b_i)$ and to the sizes of the covariances of b_i and b_j. When b_i and b_j have variances of different sizes and the correlation between b_i and b_j, namely

$$\rho_{ij} = \frac{\text{cov}(b_i, b_j)}{[V(b_i)V(b_j)]^{1/2}}$$

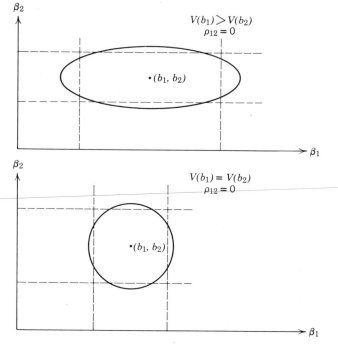

Figure 2.2

is not small, the situation illustrated in Figure 2.1 occurs. If ρ_{ij} is close to zero then the rectangular region defined by individual confidence intervals will approximate to the correct joint confidence region, though the joint region is correct. The elongation of the region will depend on the relative sizes of $V(b_i)$ and $V(b_j)$. Some examples are shown in Figure 2.2.

Note. If the model is written originally, and fitted, in the alternative form

$$E(Y - \bar{Y}) = \beta_1(X_1 - \bar{X}_1) + \beta_2(X_2 - \bar{X}_2) + \cdots + \beta_k(X_k - \bar{X}_k)$$

where $\bar{Y}$, $\bar{X}_1$, $\bar{X}_2$, ..., $\bar{X}_k$ are the observed means of the actual data, then joint confidence intervals can be obtained that do not involve β_0, which sometimes is of little interest.

2.7. The "Extra Sum of Squares" Principle

In regression work, the question often arises as to whether or not it was worthwhile to include certain terms in the model. This question can be investigated by considering the extra portion of the regression sum of squares which arises due to the fact that the terms under consideration *were* in the model. The mean square derived from this extra sum of squares can then be compared with the estimate, s^2, of σ^2 to see if it appears significantly large. If it does, the terms should have been included; if it does not, the terms would be judged unnecessary and could be removed.

We have already seen an example of this in the case of fitting a straight line where $SS(b_1 \mid b_0)$ represented the extra sum of squares due to including the term $\beta_1 X$ in the model. We now state the procedure more generally. Suppose the functions Z_1, Z_2, ..., Z_p are known functions of the basic variables X_1, X_2, ..., and suppose that values of the X's and the corresponding response Y are available. Consider the two models below.

1. $Y = \beta_0 + \beta_1 Z_1 + \beta_2 Z_2 + \cdots + \beta_p Z_p + \epsilon$.

Suppose we obtain the following least squares estimates: $b_0(1)$, $b_1(1)$, $b_2(1)$, ..., $b_p(1)$ and suppose that $SS(b_0(1), b_1(1), b_2(1), ..., b_p(1)) = S_1$, and there is no lack of fit. Let the estimate of σ^2 be s^2, obtained from the residual of model (1).

2. $Y = \beta_0 + \beta_1 Z_1 + \beta_2 Z_2 + \cdots + \beta_q Z_q + \epsilon$ $(q < p)$.

The Z's in this model (2) are the same functions as in model (1) when subscripts are the same. There are, however, fewer terms in this second model.

Suppose we now obtain the following least squares estimates: $b_0(2)$, $b_1(2)$, $b_2(2)$, ..., $b_q(2)$.

Note.   These may or may not be the same as $b_0(1), b_1(1), \ldots, b_q(1)$ above. If they are identical then $b_i(1)$ and $b_j(1)$ are orthogonal linear functions for $1 \le i \le q$, $q + 1 \le j \le p$. This happens when, in model (1), the first q columns of the $\mathbf{X}$ matrix are all orthogonal to the last $p - q$ columns. We shall see examples of this in later chapters.

Suppose that $SS(b_0(2), b_1(2), b_2(2), \ldots, b_q(2)) = S_2$, for this model. Then $S_1 - S_2$ is *the extra sum of squares* due to the inclusion of the terms $\beta_{q+1}Z_{q+1} + \cdots + \beta_p Z_p$ in model (1). Since S_1 has $(p + 1)$ degrees of freedom and S_2 has $(q + 1)$ degrees of freedom, $S_1 - S_2$ has $(p - q)$ degrees of freedom. It can be shown that, if $\beta_{q+1} = \beta_{q+2} = \cdots = \beta_p = 0$, then $E\{(S_1 - S_2)/(p - q)\} = \sigma^2$. In addition, if the errors are normally distributed, $(S_1 - S_2)$ will then be distributed as $\sigma^2 \chi^2_{p-q}$ independently of s^2. This means we can compare $(S_1 - S_2)/(p - q)$ with s^2 by an $F(p - q, v)$ test, where v is the number of degrees of freedom on which s^2 is based, to test the hypothesis H_0: $\beta_{q+1} = \beta_{q+2} = \cdots = \beta_p = 0$.

We can write $S_1 - S_2$ conveniently as $SS(b_{q+1}, \ldots, b_p \mid b_0, b_1, \ldots, b_q)$ where we must keep in mind that two models are actually involved since the notation does not show it. This is read as *the sum of squares of* $b_{q+1}, \ldots, b_p$ *given* $b_0, b_1, \ldots, b_q$. By continued application of this principle we can obtain, successively, for any regression model, $SS(b_0)$, $SS(b_1 \mid b_0)$, $SS(b_2 \mid b_0, b_1), \ldots, SS(b_p \mid b_0, b_1, \ldots, b_{p-1})$, if we wish. All these sums of squares are distributed independently of s^2 and equal their mean squares since each has one degree of freedom. The mean squares can be compared with s^2 by a series of F-tests. This is useful when the terms of the model have a logical "order of entry," as would be the case, for example, if $Z_j = X^j$. A judgment can then be made about how many terms should be in the model.

When the terms in the model occur in natural groupings, such as happens, for example, in polynomial models with (a) β_0, (b) first-order terms, (c) second-order terms, we can construct alternative extra sums of squares, for example, $SS(b_0)$, $SS(\text{first order } b\text{'s} \mid b_0)$, $SS(\text{second order } b\text{'s} \mid b_0, \text{first order } b\text{'s})$ and compare *these* with s^2. The extra sum of squares principle can be used in many ways therefore to achieve whatever breakup of the regression sum of squares seems reasonable for the problem at hand. The number of degrees of freedom for each sum of squares will be the number of parameters before the vertical division line (except when the estimates are linearly dependent; this happens when $\mathbf{X}'\mathbf{X}$ is singular and the normal equations are linearly dependent and will not usually concern us. The number of degrees of freedom is then the maximum number of linearly independent estimates in the set being considered). These extra SS's are distributed independently of s^2. The corresponding mean squares equal (sum of squares)/(degrees of freedom) can be divided by s^2 to provide an F-ratio for testing the hypothesis that the true values of the coefficients whose estimates gave rise to the extra sum of squares are zero.

The extra sum of squares principle is actually a special case of testing a general linear hypothesis. In the more general treatment the extra sum of squares is calculated from the residual sums of squares and not the regression sums of squares. Since the total sum of squares $Y'Y$ is the same for both regression calculations we would obtain the same result numerically whether we used the difference of regression or residual sums of squares. Before we discuss hypothesis testing in Section 2.10 we shall discuss an important special case of the extra sum of squares principle.

2.8. Orthogonal Columns in the X-Matrix

Suppose we have a regression problem involving parameters β_0, β_1, and β_2. Using the extra sum of squares principle we can calculate a number of quantities such as

$SS(b_2)$	from the model	$Y = \beta_2 X_2 + \epsilon$
$SS(b_2 \mid b_0)$	from the model	$Y = \beta_0 + \beta_2 X_2 + \epsilon$
$SS(b_2 \mid b_0, b_1)$	from the model	$Y = \beta_0 + \beta_1 X_1 + \beta_2 X_2 + \epsilon$

These will usually have completely different numerical values except when the "β_2" column of the X matrix is orthogonal to the "β_0" and the "β_1" columns. When this happens we can unambiguously talk about "$SS(b_2)$." We now examine this situation in more detail.

Suppose in the model $Y = X\beta + \epsilon$ we divide the matrix X up into t sets of columns denoted in matrix form by

$$X = \{X_1, X_2, \ldots, X_t\}.$$

A corresponding division can be made in β so that

$$\beta = \begin{bmatrix} \beta_1 \\ \beta_2 \\ \cdot \\ \cdot \\ \cdot \\ \beta_t \end{bmatrix}$$

where the number of columns in X_i is equal to the number of rows in β_i, $i = 1, 2, \ldots, t$. The model can then be written

$$E(Y) = X\beta = X_1\beta_1 + X_2\beta_2 + \cdots + X_t\beta_t.$$

Suppose that

$$\mathbf{b} = \begin{bmatrix} \mathbf{b}_1 \\ \mathbf{b}_2 \\ \cdot \\ \cdot \\ \cdot \\ \mathbf{b}_t \end{bmatrix}$$

is the vector estimate of $\boldsymbol{\beta}$ for this model (and given data) obtained from the normal equations

$$\mathbf{X'Xb} = \mathbf{X'Y}.$$

RESULT.   If the columns of $\mathbf{X}_i$ are orthogonal to the columns of $\mathbf{X}_j$ for all $i, j = 1, 2, \ldots, t$ $(i \neq j)$, that is, if $\mathbf{X}_i'\mathbf{X}_j = \mathbf{0}$, it is true that

$$\text{SS}(\mathbf{b}) = \text{SS}(\mathbf{b}_1) + \text{SS}(\mathbf{b}_2) + \cdots + \text{SS}(\mathbf{b}_t)$$
$$= \mathbf{b}_1'\mathbf{X}_1'\mathbf{Y} + \mathbf{b}_2'\mathbf{X}_2'\mathbf{Y} + \cdots + \mathbf{b}_t'\mathbf{X}_t'\mathbf{Y}$$

and $\mathbf{b}_i$ is the least square estimate of $\boldsymbol{\beta}_i$, and $\text{SS}(\mathbf{b}_i) = \mathbf{b}_i'\mathbf{X}_i'\mathbf{Y}$ *whether any of the other terms are in the model or not.* Thus

$$\text{SS}(\mathbf{b}_i) = \text{SS}(\mathbf{b}_i \mid \text{any set of } \mathbf{b}_j\text{'s } j \neq i).$$

(Note that it is *not* necessary for the columns of $\mathbf{X}_i$ to be orthogonal to *each other*—only for the $\mathbf{X}_i$ columns all to be orthogonal to all other columns of $\mathbf{X}$.)

We consider the case $t = 2$. Here,

$$\mathbf{X} = (\mathbf{X}_1, \mathbf{X}_2)$$

where $\mathbf{X}_1'\mathbf{X}_2 = \mathbf{X}_2'\mathbf{X}_1 = \mathbf{0}$. (This means that all the columns in $\mathbf{X}_1$ are orthogonal to all the columns in $\mathbf{X}_2$.) We can write the model as

$$Y = \mathbf{X}\boldsymbol{\beta} + \boldsymbol{\epsilon} = \mathbf{X}_1\boldsymbol{\beta}_1 + \mathbf{X}_2\boldsymbol{\beta}_2 + \boldsymbol{\epsilon}$$

where $\boldsymbol{\beta}' = (\boldsymbol{\beta}_1', \boldsymbol{\beta}_2')$ is split into the two sets of coefficients which correspond to the $\mathbf{X}_1$ and $\mathbf{X}_2$ sets of columns. The normal equations are $\mathbf{X'Xb} = \mathbf{X'Y}$; that is,

$$\begin{bmatrix} \mathbf{X}_1'\mathbf{X}_1 & \mathbf{X}_1'\mathbf{X}_2 \\ \mathbf{X}_2'\mathbf{X}_1 & \mathbf{X}_2'\mathbf{X}_2 \end{bmatrix} \begin{bmatrix} \mathbf{b}_1 \\ \mathbf{b}_2 \end{bmatrix} = \begin{bmatrix} \mathbf{X}_1'\mathbf{Y} \\ \mathbf{X}_2'\mathbf{Y} \end{bmatrix}$$

where a split in $\mathbf{b}$ corresponding to that in $\boldsymbol{\beta}$ has been made. Since the off-diagonal terms $\mathbf{X}_1'\mathbf{X}_2 = \mathbf{0}$, $\mathbf{X}_2'\mathbf{X}_1 = \mathbf{0}$, the normal equations can be split into the two sets of equations

$$\mathbf{X}_1'\mathbf{X}_1\mathbf{b}_1 = \mathbf{X}_1'\mathbf{Y}; \qquad \mathbf{X}_2'\mathbf{X}_2\mathbf{b}_2 = \mathbf{X}_2'\mathbf{Y}$$

with solutions

$$\mathbf{b}_1 = (\mathbf{X}_1'\mathbf{X}_1)^{-1}\mathbf{X}_1'\mathbf{Y}; \qquad \mathbf{b}_2 = (\mathbf{X}_2'\mathbf{X}_2)^{-1}\mathbf{X}_2'\mathbf{Y}$$

assuming that the matrices shown inverted are nonsingular. Thus $\mathbf{b}_1$ is the least squares estimate of $\boldsymbol{\beta}_1$ whether $\boldsymbol{\beta}_2$ is in the model or not, and vice versa. Now

$$SS(\mathbf{b}_1) = \mathbf{b}_1'\mathbf{X}_1'\mathbf{Y} \quad \text{and} \quad SS(\mathbf{b}_2) = \mathbf{b}_2'\mathbf{X}_2'\mathbf{Y}.$$

Thus

$$SS(\mathbf{b}_1, \mathbf{b}_2) = \mathbf{b}'\mathbf{X}'\mathbf{Y}$$
$$= (\mathbf{b}_1', \mathbf{b}_2')(\mathbf{X}_1, \mathbf{X}_2)'\mathbf{Y}$$
$$= (\mathbf{b}_1', \mathbf{b}_2')\begin{pmatrix}\mathbf{X}_1'\mathbf{Y}\\\mathbf{X}_2'\mathbf{Y}\end{pmatrix}$$
$$= \mathbf{b}_1'\mathbf{X}_1'\mathbf{Y} + \mathbf{b}_2'\mathbf{X}_2'\mathbf{Y}$$
$$= SS(\mathbf{b}_1) + SS(\mathbf{b}_2).$$

It follows that

$$SS(\mathbf{b}_1 \mid \mathbf{b}_2) = SS(\mathbf{b}_1, \mathbf{b}_2) - SS(\mathbf{b}_2) = SS(\mathbf{b}_1).$$

Similarly

$$SS(\mathbf{b}_2 \mid \mathbf{b}_1) = SS(\mathbf{b}_2)$$

and this depends only on the orthogonality of $\mathbf{X}_1$ and $\mathbf{X}_2$. The extension to cases where $t > 2$ is immediate.

2.9. Partial *F*-Tests and Sequential *F*-Tests

We have seen how to obtain extra sums of squares for one or more estimated coefficients given other coefficients by considering two models, one of which includes the coefficients in question and one of which does not.

If we have several terms in a regression model we can think of them as "entering" the equation in any desired sequence. If we find

$$SS(b_i \mid b_0, b_1, \ldots, b_{i-1}, b_{i+1}, \ldots, b_k) \qquad i = 1, 2, \ldots, k$$

we shall have a one degree of freedom sum of squares which measures the contribution to the regression sum of squares of each coefficient b_i given that all the terms which did not involve β_i were already in the model. In other words, we shall have a measure of the value of *adding a β_i term to the model* which originally did not include such a term. Another way of saying this is that we have a measure of the value of β_i *as though it were added to the model last*. The corresponding mean square, equal to the sum of squares since it has one degree of freedom, can be compared by an *F*-test to s^2 as described. This particular type of *F*-test is called a *partial*

F-test for β_i. If the extra term under consideration is $\beta_t X_t$, say, we can talk (loosely) about a partial *F*-test on the variable X_t, even though we are aware that the test actually is on the coefficient β_t.

When a suitable model is being "built" the partial *F*-test is a useful criterion for adding or removing terms from the model. The effect of an *X*-variable (X_q say) in determining a response may be large when the regression equation includes only X_q. However, when the same variable is entered into the equation after other variables, it may affect the response very little, due to the fact that X_q is highly correlated with variables already in the regression equation. The partial *F*-test can be made for all regression coefficients as though each corresponding variable were the last to enter the equation—to see the relative effects of each variable in excess of the others. This information can be combined with other information if a choice of variables need be made. Suppose, for example, either X_1 or X_2 alone could be used to provide a regression equation for a response *Y*. Suppose use of X_1 provided smaller predictive errors than use of X_2. Then if predictive accuracy were desired, X_1 would probably be used in future work. If, however, X_2 were a variable through which the response level could be controlled (whereas X_1 was a measured but non-controlling variable) and if control were important rather than prediction, then it might be preferable to use X_2 rather than X_1 as an independent variable for future work.

When variables are added one by one in stages to a regression equation, we can talk about a *sequential F-test*. This is just a name for the partial *F*-test on the variable which entered the regression at that stage.

Note. Some writers dislike (and condemn as misleading) the use of the phrases *partial F-test* and *sequential F-test*. We emphasize that these are merely convenient, short names for particular, theoretically correct *F*-tests (see Section 2.7).

2.10. Testing a General Linear Hypothesis in Regression Situations

When doing regression work it is often necessary to make a statistical test of a hypothesis H_0 which involves linear functions of the true regression coefficients $\beta_1, \beta_2, \ldots, \beta_k$.

Example 1.   Model: $E(Y) = \beta_0 + \beta_1 X_1 + \beta_2 X_2$.

$$H_0: \beta_1 = 0,$$

$$\beta_2 = 0, \quad \text{(two linear functions, independent).}$$

Example 2.   Model: $E(Y) = \beta_0 + \beta_1 X_1 + \beta_2 X_2 + \cdots + \beta_k X_k$.

$$H_0: \beta_1 = 0,$$
$$\beta_2 = 0,$$
$$\cdots$$
$$\beta_k = 0. \quad (k \text{ linear functions, all independent}),$$

Example 3.   Model: $E(Y) = \beta_0 + \beta_1 X_1 + \beta_2 X_2 + \cdots + \beta_k X_k$.

$$H_0: \beta_1 - \beta_2 = 0,$$
$$\beta_2 - \beta_3 = 0,$$
$$\cdots$$
$$\beta_{k-1} - \beta_k = 0. \quad (k-1 \text{ linear functions, independent})$$

Note that this expresses the hypothesis

$$H_0: \beta_1 = \beta_2 = \cdots = \beta_k = \beta, \text{ say}.$$

Example 4 (*General Case*).

$$\text{Model: } E(Y) = \beta_0 + \beta_1 X_1 + \beta_2 X_2 + \cdots + \beta_k X_k.$$
$$H_0: c_{11}\beta_1 + c_{12}\beta_2 + \cdots + c_{1k}\beta_k = 0,$$
$$c_{21}\beta_1 + c_{22}\beta_2 + \cdots + c_{2k}\beta_k = 0,$$
$$\cdots$$
$$c_{m1}\beta_1 + c_{m2}\beta_2 + \cdots + c_{mk}\beta_k = 0.$$

In this hypothesis there are m linear functions of $\beta_1, \beta_2, \ldots, \beta_k$, all of which may not be independent. H_0 can be expressed in matrix form as

$$H_0: \mathbf{C\beta} = \mathbf{0}$$

where

$$\mathbf{C} = \begin{bmatrix} c_{11} & c_{12} & \cdots & c_{1k} \\ c_{21} & c_{22} & \cdots & c_{2k} \\ & \cdots & & \\ c_{m1} & c_{m2} & \cdots & c_{mk} \end{bmatrix} \qquad \mathbf{\beta} = \begin{bmatrix} \beta_1 \\ \beta_2 \\ \cdot \\ \cdot \\ \cdot \\ \beta_k \end{bmatrix}$$

We shall suppose in what follows that the m functions are *dependent* and that the last $(m - q)$ of them depend upon the first q; that is, if we had these first q independent functions, we could take linear combinations of them to form the other $(m - q)$ linear functions.

We have seen earlier how it is possible to test hypotheses of the forms in Examples 1 and 2. We now explain how more general hypotheses can be tested.

Testing a General Linear Hypothesis $C\beta = 0$

Suppose that the model under consideration, assumed correct, is

$$E(\mathbf{Y}) = \mathbf{X}\boldsymbol{\beta},$$

where $\mathbf{Y}$ is $(n \times 1)$, $\mathbf{X}$ is $(n \times p)$, and $\boldsymbol{\beta}$ is $(p \times 1)$. If $\mathbf{X'X}$ is nonsingular we can estimate $\boldsymbol{\beta}$ as

$$\mathbf{b} = (\mathbf{X'X})^{-1}\mathbf{X'Y}.$$

The residual sum of squares for this analysis is given, as we have seen, by

$$\text{SSR} = \mathbf{Y'Y} - \mathbf{b'X'Y}.$$

This sum of squares has $n - p$ degrees of freedom. The linear hypothesis to be tested, H_0: $\mathbf{C}\boldsymbol{\beta} = \mathbf{0}$, provides q independent conditions on the parameters $\beta_0, \beta_1, \ldots, \beta_k$, on the assumptions (mentioned above) that $\mathbf{C}\boldsymbol{\beta} = \mathbf{0}$ represents m equations, of which only q are independent. We can use the q independent equations to solve for q of the β's in terms of the other $p - q$ of them. Substituting these solutions back into the original model provides us with a reduced model of, say,

$$E(\mathbf{Y}) = \mathbf{Z}\boldsymbol{\alpha}$$

where $\boldsymbol{\alpha}$ is a vector of parameters to be estimated. There will be $p - q$ of these parameters. The right-hand side $\mathbf{Z}\boldsymbol{\alpha}$ where $\mathbf{Z}$ is $n \times (p - q)$ and $\boldsymbol{\alpha}$ is $(p - q) \times 1$ represents the result of substituting into $\mathbf{X}\boldsymbol{\beta}$ for the dependent β's.

We can now estimate the parameter vector $\boldsymbol{\alpha}$ in the new model by

$$\mathbf{a} = (\mathbf{Z'Z})^{-1}\mathbf{Z'Y}$$

if $\mathbf{Z'Z}$ is nonsingular and can obtain a new residual sum of squares for this regression of

$$\text{SSW} = \mathbf{Y'Y} - \mathbf{a'Z'Y}.$$

This sum of squares has $n - p + q$ degrees of freedom.

Since fewer parameters are involved in this second analysis, SSW will always be larger than SSR. The difference $\text{SSW} - \text{SSR}$ is called the *sum of squares due to the hypothesis* $\mathbf{C}\boldsymbol{\beta} = \mathbf{0}$ and has $(n - p + q) - (n - p) = q$ degrees of freedom. A test of the hypothesis H_0: $\mathbf{C}\boldsymbol{\beta} = \mathbf{0}$ is now made by considering the ratio

$$\frac{\dfrac{\text{SSW} - \text{SSR}}{q}}{\dfrac{\text{SSR}}{n - p}}$$

and referring it to the $F(q, n - p)$ distribution in the usual manner. If the errors are normally distributed and independent, this is an exact test.

The appropriate test for Examples 1 and 2 (already given as Eq. (2.6.13) where $k = p - 1$) is a special case of this. The reduced model in both cases consists of

$$E(\mathbf{Y}) = \mathbf{j}\beta_0$$

where $\mathbf{j}' = (1, 1, \ldots, 1)$ is a vector of all ones. Another way of writing this model is

$$E(Y_i) = \beta_0, \qquad i = 1, 2, \ldots, n.$$

Since $b_0 = \bar{Y}$, $\mathrm{SSW} = \mathbf{Y}'\mathbf{Y} - n\bar{Y}^2$ with $(n - 1)$ degrees of freedom, whereas $\mathrm{SSR} = \mathbf{Y}'\mathbf{Y} - \mathbf{b}'\mathbf{X}'\mathbf{Y}$ with $(n - k - 1)$ degrees of freedom. So the ratio for the test $\beta_1 = \beta_2 = \cdots = \beta_k = 0$ (for Example 2; when $k = 2$, we have Example 1) is simply

$$\frac{\mathbf{b}'\mathbf{X}'\mathbf{Y} - n\bar{Y}^2}{k} \bigg/ \frac{\mathbf{Y}'\mathbf{Y} - \mathbf{b}'\mathbf{X}'\mathbf{Y}}{n - k - 1}$$

and this is referred to the $F(k, n - k - 1)$ distribution. This is exactly the procedure of Eq. (2.6.13) with $k = p - 1$, $v = n - k - 1$, and $s^2 = \mathrm{MS}_E = \mathrm{SSR}/v$.

We shall now illustrate the use of the procedure in a simple but not so typical case.

WORKED EXAMPLE. Given the model $E(\mathbf{Y}) = \mathbf{X}\boldsymbol{\beta}$, test the hypothesis $H_0 \colon \mathbf{C}\boldsymbol{\beta} = \mathbf{0}$, where

$$\mathbf{Y}' = (1, 4, 8, 9, 3, 8, 9),$$

$$\boldsymbol{\beta}' = (\beta_0, \beta_1, \beta_2, \beta_{11}),$$

$$
\begin{array}{cccc}
1 & X_1 & X_2 & X_1^2
\end{array}
$$

$$
\mathbf{X} =
\begin{bmatrix}
1 & -1 & -1 & 1 \\
1 & 1 & -1 & 1 \\
1 & -1 & 1 & 1 \\
1 & 1 & 1 & 1 \\
1 & 0 & 0 & 0 \\
1 & 0 & 1 & 0 \\
1 & 0 & 2 & 0
\end{bmatrix},
$$

and

$$\mathbf{C} = \begin{bmatrix} 0 & 0 & 0 & 1 \\ 0 & 1 & -1 & 0 \\ 0 & 1 & -1 & 1 \\ 0 & 2 & -2 & 93 \end{bmatrix}$$

Solution. We first find the residual sum of squares SSR when the original model, of form $E(Y) = \beta_0 + \beta_1 X_1 + \beta_2 X_2 + \beta_{11} X_1^2$, is fitted. We find

$$(\mathbf{X'X})^{-1} = \begin{bmatrix} 7 & 0 & 3 & 4 \\ 0 & 4 & 0 & 0 \\ 3 & 0 & 9 & 0 \\ 4 & 0 & 0 & 4 \end{bmatrix}^{-1} = \begin{bmatrix} \frac{1}{2} & 0 & -\frac{1}{6} & -\frac{1}{2} \\ 0 & \frac{1}{4} & 0 & 0 \\ -\frac{1}{6} & 0 & \frac{1}{6} & \frac{1}{6} \\ -\frac{1}{2} & 0 & \frac{1}{6} & \frac{3}{4} \end{bmatrix}$$

$$\mathbf{X'Y} = \begin{bmatrix} 42 \\ 4 \\ 38 \\ 22 \end{bmatrix}, \quad \mathbf{b} = (\mathbf{X'X})^{-1}\mathbf{X'Y} = \begin{bmatrix} \frac{11}{3} \\ 1 \\ 3 \\ \frac{11}{6} \end{bmatrix}, \quad \begin{array}{l} \mathbf{b'X'Y} = 312.33 \\ \mathbf{Y'Y} = 316 \end{array}$$

SSR $= 316 - 312.33 = 3.67$.

The equations for the null hypothesis $H_0 : \mathbf{C}\boldsymbol{\beta} = \mathbf{0}$ are

$$\beta_{11} = 0$$
$$\beta_1 - \beta_2 = 0$$
$$\beta_1 - \beta_2 + \beta_{11} = 0$$
$$2\beta_1 - 2\beta_2 + 93\beta_{11} = 0$$

The hypothesis can be more simply expressed as $H_0 : \beta_{11} = 0, \beta_1 = \beta_2 = \beta$, say, since the third and fourth equations are linear combinations of the first and second equations.

Substituting these conditions in the model gives a reduced model

$$E(Y) = \beta_0 + \beta(X_1 + X_2) = \alpha_0 + \alpha Z$$

where

$$\alpha_0 = \beta_0, \quad \alpha = \beta, \quad Z = X_1 + X_2.$$

Thus

$$Z = \begin{bmatrix} 1 & (-1-1) \\ 1 & (1-1) \\ 1 & (-1+1) \\ 1 & (1+1) \\ 1 & (0+0) \\ 1 & (0+1) \\ 1 & (0+2) \end{bmatrix} = \begin{bmatrix} 1 & -2 \\ 1 & 0 \\ 1 & 0 \\ 1 & 2 \\ 1 & 0 \\ 1 & 1 \\ 1 & 2 \end{bmatrix}$$

$$\mathbf{Z'Y} = \begin{bmatrix} 42 \\ 42 \end{bmatrix}, \qquad (\mathbf{Z'Z})^{-1} = \begin{bmatrix} 7 & 3 \\ 3 & 13 \end{bmatrix}^{-1} = \frac{1}{82}\begin{bmatrix} 13 & -3 \\ -3 & 7 \end{bmatrix}$$

$$\mathbf{a} = (\mathbf{Z'Z})^{-1}\mathbf{Z'Y} = \frac{21}{41}\begin{bmatrix} 10 \\ 4 \end{bmatrix}, \qquad \mathbf{a'Z'Y} = 301.17$$

$$SSW = 316 - 301.17 = 14.83.$$

Now $p = 4$, $n = 7$, $q = 2$, $n - p = 3$, and

$$SSW - SSR = 14.83 - 3.67 = 11.16 = SS \text{ due to the hypothesis.}$$

The appropriate test statistic for H_0 is thus $(11.16/2) \div 3.67/3 = 4.56$. Since $F(2, 3, 0.95) = 9.55$, we *do not* reject H_0. Since the original model was $E(Y) = \beta_0 + \beta_1 X_1 + \beta_2 X_2 + \beta_{11}X_1^2$ and the hypothesis *not* rejected implies $\beta_{11} = 0$, $\beta_1 = \beta_2 = \beta$, a more plausible model would be $E(Y) = \beta_0 + \beta(X_1 + X_2)$.

2.11. Weighted Least Squares

It sometimes happens that some of the observations used in a regression analysis are "less reliable" than others. What this usually means is that the variances of the observations are not all equal; in other words the matrix $\mathbf{V(\epsilon)}$ is not of the form $\mathbf{I}\sigma^2$ but is diagonal with unequal diagonal elements. It may also happen, in some problems, that the off-diagonal elements of $\mathbf{V(\epsilon)}$ are not zero, that is, that the observations are correlated.

When either or both of these events occur, the ordinary least squares estimation formula $\mathbf{b} = (\mathbf{X'X})^{-1}\mathbf{X'Y}$ does not apply and it is necessary to amend the procedures for obtaining estimates. The basic idea is to transform the observations $\mathbf{Y}$ to other variables $\mathbf{Z}$ which *do* appear to satisfy the usual tentative assumptions [that $\mathbf{Z} = \mathbf{Q\beta} + \mathbf{f}$, $E(\mathbf{f}) = \mathbf{0}$, $V(\mathbf{f}) = \mathbf{I}\sigma^2$, and, for F-tests and confidence intervals to be valid, that $\mathbf{f} \sim N(\mathbf{0}, \mathbf{I}\sigma^2)$] and to then apply the usual (unweighted) analysis to the variables so obtained. The estimates can then be re-expressed in terms of the original variables $\mathbf{Y}$. We shall describe how the usual regression procedures are changed. Suppose the model under consideration is

$$\mathbf{Y} = \mathbf{X\beta} + \mathbf{\epsilon} \tag{2.11.1}$$

where

$$E(\mathbf{\epsilon}) = \mathbf{0}, \qquad V(\mathbf{\epsilon}) = \mathbf{V}\sigma^2, \qquad \text{and} \quad \mathbf{\epsilon} \sim N(\mathbf{0}, \mathbf{V}\sigma^2). \tag{2.11.2}$$

It can be shown that it is possible to find a unique nonsingular symmetric matrix $\mathbf{P}$ such that

$$\mathbf{P'P} = \mathbf{PP} = \mathbf{P}^2 = \mathbf{V}. \tag{2.11.3}$$

Let us write

$$\mathbf{f} = \mathbf{P}^{-1}\mathbf{\epsilon}, \qquad \text{so that } E(\mathbf{f}) = \mathbf{0}. \tag{2.11.4}$$

Now it is a fact that, if $\mathbf{f}$ is a vector random variable such that $E(\mathbf{f}) = \mathbf{0}$, then $E(\mathbf{ff'}) = V(\mathbf{f})$ where the expectation is taken separately for every term in the square $n \times n$ matrix $\mathbf{ff'}$. Thus

$$V(\mathbf{f}) = E(\mathbf{ff'}) = E(\mathbf{P}^{-1}\mathbf{\epsilon\epsilon'}\mathbf{P}^{-1}), \qquad \text{since } (\mathbf{P}^{-1})' = \mathbf{P}^{-1}$$

$$= \mathbf{P}^{-1}E(\mathbf{\epsilon\epsilon'})\mathbf{P}^{-1}$$

$$= \mathbf{P}^{-1}\mathbf{PPP}^{-1}\sigma^2$$

$$= \mathbf{I}\sigma^2. \tag{2.11.5}$$

It is also true that $\mathbf{f} \sim N(\mathbf{0}, \mathbf{I}\sigma^2)$, that is, $\mathbf{f}$ is normally distributed, since the elements of $\mathbf{f}$ consist of linear combinations of the elements of $\mathbf{\epsilon}$ which is itself normally distributed.

Thus if we premultiply Eq. (2.11.1) by $\mathbf{P}^{-1}$ we obtain a new model

$$\mathbf{P}^{-1}\mathbf{Y} = \mathbf{P}^{-1}\mathbf{X\beta} + \mathbf{P}^{-1}\mathbf{\epsilon} \tag{2.11.6}$$

or

$$\mathbf{Z} = \mathbf{Q\beta} + \mathbf{f} \tag{2.11.7}$$

with an obvious notation. It is now clear that we can apply the basic least squares theory to Eq. (2.11.7) since $E(\mathbf{f}) = \mathbf{0}$ and $V(\mathbf{f}) = \mathbf{I}\sigma^2$. The residual sum of squares is

$$\mathbf{f'f} = \mathbf{\epsilon'V}^{-1}\mathbf{\epsilon} = (\mathbf{Y} - \mathbf{X\beta})'\mathbf{V}^{-1}(\mathbf{Y} - \mathbf{X\beta}). \tag{2.11.8}$$

The normal equations $\mathbf{Q'Qb} = \mathbf{Q'Z}$ become

$$\mathbf{X'V^{-1}Xb} = \mathbf{X'V^{-1}Y} \tag{2.11.9}$$

with solution

$$\mathbf{b} = (\mathbf{X'V^{-1}X})^{-1}\mathbf{X'V^{-1}Y} \tag{2.11.10}$$

when the matrix just inverted is nonsingular. The regression sum of squares is

$$\mathbf{b'Q'Z} = \mathbf{Y'V^{-1}X(X'V^{-1}X)^{-1}X'V^{-1}Y} \tag{2.11.11}$$

and the total sum of squares is

$$\mathbf{Z'Z} = \mathbf{Y'V^{-1}Y}. \tag{2.11.12}$$

The difference between Eqs. (2.11.12) and (2.11.11) provides the residual sum of squares. The sum of squares due to the mean is $(\sum Z_i)^2/n$, where Z_i are the n elements of the vector $\mathbf{Z}$. The variance-covariance matrix of $\mathbf{b}$ is

$$\mathbf{V(b)} = (\mathbf{Q'Q})^{-1}\sigma^2 = (\mathbf{X'V^{-1}X})^{-1}\sigma^2. \tag{2.11.13}$$

A joint confidence region for all the parameters can be obtained from

$$(\mathbf{b} - \boldsymbol{\beta})\mathbf{Q'Q}(\mathbf{b} - \boldsymbol{\beta}) = \left[\frac{p}{(n - p)}\right](\mathbf{Z'Z} - \mathbf{b'Q'Z})\, F(p, n - p, 1 - \alpha) \tag{2.11.14}$$

after substituting from Eqs. (2.11.11) and (2.11.12) and setting $\mathbf{Q} = \mathbf{P^{-1}X}$, if so desired.

The simplest application of weighted least squares occurs when the observations are independent but have different variances so that

$$\mathbf{V}\sigma^2 = \begin{bmatrix} \sigma_1^2 & & & \\ & \sigma_2^2 & & \mathbf{0} \\ & & \ddots & \\ & & & \ddots \\ \mathbf{0} & & & \ddots \\ & & & & \sigma_n^2 \end{bmatrix}$$

where some of the σ_i^2 may be equal.

In practical problems it is often difficult to obtain specific information on the form of $\mathbf{V}$ at first. For this reason it is sometimes necessary to make the (known to be erroneous) assumption $\mathbf{V} = \mathbf{I}$ and then attempt to discover something about the form of $\mathbf{V}$ by examining the residuals from the regression analysis (see Chapter 3).

If a weighted least squares analysis were called for but an ordinary least squares analysis were performed, the estimates obtained would still be unbiased but would not have minimum variance, since the minimum variance estimates are obtained from the correct weighted least squares analysis.

If standard least squares is used, then the estimates are obtained from $\mathbf{b_0} = (\mathbf{X'X})^{-1}\mathbf{X'Y}$ and

$$E(\mathbf{b_0}) = (\mathbf{X'X})^{-1}\mathbf{X'X}\boldsymbol{\beta} = \boldsymbol{\beta}$$

but

$$V(\mathbf{b_0}) = (\mathbf{X'X})^{-1}\mathbf{X'}[V(\mathbf{Y})]\mathbf{X}(\mathbf{X'X})^{-1}$$

$$= (\mathbf{X'X})^{-1}\mathbf{X'VX}(\mathbf{X'X})^{-1}\sigma^2.$$

We recall from Eq. (2.11.13) that if the correct analysis is performed,

$$V(\mathbf{b}) = (\mathbf{X'V^{-1}X})^{-1}\sigma^2$$

and, in general, elements of this matrix would provide smaller variances both for individual coefficients and for linear functions of the coefficients.

An Example of Weighted Least Squares

This is an extremely simple example but an interesting one. Suppose we wish to fit the model

$$E(y) = \beta x.$$

Let us suppose that

$$\mathbf{V}\sigma^2 = \mathbf{V(y)} = \begin{bmatrix} 1/w_1 & & & \\ & 1/w_2 & & \mathbf{0} \\ & & \cdot & \\ & & & \cdot \\ \mathbf{0} & & & \cdot \\ & & & & 1/w_n \end{bmatrix}\sigma^2$$

where the w's are *weights* to be specified. This means that

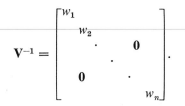

$$\mathbf{V^{-1}} = \begin{bmatrix} w_1 & & & \\ & w_2 & & \\ & & \cdot & \mathbf{0} \\ & & & \cdot \\ \mathbf{0} & & & \cdot \\ & & & & w_n \end{bmatrix}$$

Applying the general results above we find, after reduction,

$$b = \frac{\sum w_i x_i y_i}{\sum w_i x_i^2}$$

where all summations are from $i = 1, 2, \ldots, n$.

CASE 1.   Suppose $\sigma_i^2 = V(y_i) = kx_i$; that is, the variance of y_i is proportional to the size of the corresponding x_i. Then $w_i = \sigma^2/kx_i$. Hence

$$b = \frac{\sum y_i}{\sum x_i} = \frac{\bar{y}}{\bar{x}}.$$

Thus if the variance of y_i is proportional to x_i, the best estimate of the regression coefficient is the mean of the y_i divided by the mean of the x_i. In addition,

$$V(b) = \frac{\sigma^2}{\sum w_i x_i^2} = \frac{k}{\sum x_i}.$$

CASE 2.   Suppose $\sigma_i^2 = V(y_i) = kx_i^2$; that is, the variance of y_i is proportional to the square of the corresponding x_i. Then $w_i = \sigma^2/kx_i^2$. Hence

$$b = \frac{\sum(y_i/x_i)}{\sum 1}$$

$$= \frac{\sum(y_i/x_i)}{n}.$$

Thus if the variance of the y_i is proportional to x_i^2, the best estimate of the regression coefficient is the average of the n slopes obtained one from each pair of observations y_i/x_i. Also,

$$V(b) = \frac{\sigma^2}{\sum w_i x_i^2} = \frac{k}{n}.$$

2.12.   Bias in Regression Estimates

We said earlier (Section 2.6) that the least squares estimate $\mathbf{b} = (\mathbf{X'X})^{-1}\mathbf{X'Y}$ of $\boldsymbol{\beta}$ in the model $E(\mathbf{Y}) = \mathbf{X}\boldsymbol{\beta}$ is an unbiased estimate. This means that

$$E(\mathbf{b}) = \boldsymbol{\beta}.$$

That is, if we consider the distribution of $\mathbf{b}$ (obtained by taking repeated samples from the same Y-population keeping $\mathbf{X}$ fixed and estimating $\boldsymbol{\beta}$ for each sample), then the mean value of this distribution is $\boldsymbol{\beta}$.

We now emphasize that this is true *only if the postulated model is the correct model to consider.* If it is *not* the correct model, then the estimates are *biased;* that is, $E(\mathbf{b}) \neq \boldsymbol{\beta}$. The extent of the bias depends, as we shall show, not only on the postulated and the true models but also on the values of the X-variables which enter the regression calculations. When a designed experiment is used, then, the bias depends on the experimental design, as well as the models.

We shall deal with the general nonsingular regression model from the beginning, since once we have the necessary formulae in matrix terms, they can be applied universally. Special cases can be reworked in their algebraic detail as exercises if desired. Suppose we postulate the model

$$E(\mathbf{Y}) = \mathbf{X}_1\boldsymbol{\beta}_1. \tag{2.12.1}$$

This leads to the least squares estimates:

$$\mathbf{b}_1 = (\mathbf{X}_1'\mathbf{X}_1)^{-1}\mathbf{X}_1'\mathbf{Y}. \tag{2.12.2}$$

If the postulated model is correct, then, since $\mathbf{X}_1$ is a matrix of constants unaffected by expectation, and $\mathbf{b}_1$ and $\mathbf{Y}$ are the random variables,

$$E(\mathbf{b}_1) = (\mathbf{X}_1'\mathbf{X}_1)^{-1}\mathbf{X}_1'E(\mathbf{Y}) = (\mathbf{X}_1'\mathbf{X}_1)^{-1}\mathbf{X}_1'\mathbf{X}_1\boldsymbol{\beta}_1 = \boldsymbol{\beta}_1. \tag{2.12.3}$$

Thus $\mathbf{b}_1$ is an unbiased estimate of $\boldsymbol{\beta}_1$.

Now suppose we once again postulate the model (2.12.1) so that $\mathbf{b}_1$, as defined in Eq. (2.12.2), is still the vector of estimated regression coefficients. Suppose *now*, however, that the true response relationship is in fact not Eq. (2.12.1) but

$$E(\mathbf{Y}) = \mathbf{X}_1\boldsymbol{\beta}_1 + \mathbf{X}_2\boldsymbol{\beta}_2. \tag{2.12.4}$$

That is, there are terms $\mathbf{X}_2\boldsymbol{\beta}_2$ which we did not allow for in our estimation procedure. It now follows that

$$
\begin{aligned}
E(\mathbf{b}_1) = (\mathbf{X}_1'\mathbf{X}_1)^{-1}\mathbf{X}_1'E(\mathbf{Y}) &= (\mathbf{X}_1'\mathbf{X}_1)^{-1}\mathbf{X}_1'(\mathbf{X}_1\boldsymbol{\beta}_1 + \mathbf{X}_2\boldsymbol{\beta}_2) \\
&= (\mathbf{X}_1'\mathbf{X}_1)^{-1}\mathbf{X}_1'\mathbf{X}_1\boldsymbol{\beta}_1 + (\mathbf{X}_1'\mathbf{X}_1)^{-1}\mathbf{X}_1'\mathbf{X}_2\boldsymbol{\beta}_2 \\
&= \boldsymbol{\beta}_1 + \mathbf{A}\boldsymbol{\beta}_2 \tag{2.12.5}
\end{aligned}
$$

where

$$\mathbf{A} = (\mathbf{X}_1'\mathbf{X}_1)^{-1}\mathbf{X}_1'\mathbf{X}_2 \tag{2.12.6}$$

is called the *alias* matrix. Note that the bias terms $\mathbf{A}\boldsymbol{\beta}_2$ depend not only on the postulated and the true models but also on the experimental design through the matrices $\mathbf{X}_1$ and $\mathbf{X}_2$. Thus a good choice of design may cause estimates to be less biased than they would otherwise be, even if the wrong model has been postulated and fitted. We now illustrate the application of Eq. (2.12.5) to some simple numerical cases.

Example 1.   Suppose we postulate the model

$$E(Y) = \beta_0 + \beta_1 X$$

but the model

$$E(Y) = \beta_0 + \beta_1 X + \beta_{11} X^2$$

is actually the true response function, unknown to us. If we use observations of Y at $X = -1, 0$, and 1 to estimate β_0 and β_1 in the postulated model, what bias will be introduced? That is, what will the estimates b_0 and b_1 actually estimate? The true model, in terms of the observations is

$$E(\mathbf{Y}) = E\begin{bmatrix} Y_1 \\ Y_2 \\ Y_3 \end{bmatrix} = \begin{array}{ccc} 1 & X & X^2 \\ \begin{bmatrix} 1 & -1 & 1 \\ 1 & 0 & 0 \\ 1 & 1 & 1 \end{bmatrix} \end{array} \begin{bmatrix} \beta_0 \\ \beta_1 \\ \beta_{11} \end{bmatrix}$$

$$= \begin{array}{cc} 1 & X \\ \begin{bmatrix} 1 & -1 \\ 1 & 0 \\ 1 & 1 \end{bmatrix} \end{array} \begin{bmatrix} \beta_0 \\ \beta_1 \end{bmatrix} + \begin{array}{c} X^2 \\ \begin{bmatrix} 1 \\ 0 \\ 1 \end{bmatrix} \end{array} \beta_{11}$$

$$= \mathbf{X}_1\boldsymbol{\beta}_1 + \mathbf{X}_2\boldsymbol{\beta}_2$$

to achieve the form of Eq. (2.12.4) with Eq. (2.12.1) as the postulated model. It follows that

$$(\mathbf{X}_1'\mathbf{X}_1)^{-1} = \begin{bmatrix} 3 & 0 \\ 0 & 2 \end{bmatrix}^{-1} = \begin{bmatrix} \tfrac{1}{3} & 0 \\ 0 & \tfrac{1}{2} \end{bmatrix}$$

$$\mathbf{X}_1'\mathbf{X}_2 = \begin{bmatrix} 1 & 1 & 1 \\ -1 & 0 & 1 \end{bmatrix} \begin{bmatrix} 1 \\ 0 \\ 1 \end{bmatrix} = \begin{bmatrix} 2 \\ 0 \end{bmatrix}.$$

Thus

$$\mathbf{A} = \begin{bmatrix} \tfrac{1}{3} & 0 \\ 0 & \tfrac{1}{2} \end{bmatrix} \begin{bmatrix} 2 \\ 0 \end{bmatrix} = \begin{bmatrix} \tfrac{2}{3} \\ 0 \end{bmatrix}.$$

Applying Eq. (2.12.5) we obtain

$$E\begin{bmatrix} b_0 \\ b_1 \end{bmatrix} = \begin{bmatrix} \beta_0 \\ \beta_1 \end{bmatrix} + \begin{bmatrix} \tfrac{2}{3} \\ 0 \end{bmatrix} \beta_{11} = \begin{bmatrix} \beta_0 + \tfrac{2}{3}\beta_{11} \\ \beta_1 \end{bmatrix}$$

or

$$E(b_0) = \beta_0 + \tfrac{2}{3}\beta_{11}, \qquad E(b_1) = \beta_1.$$

Thus b_0 is *biased* by $\tfrac{2}{3}\beta_{11}$, and b_1 is unbiased.

Example 2. Suppose the postulated model is

$$E(Y) = \beta_0 + \beta_1 X$$

but the true model is actually

$$E(Y) = \beta_0 + \beta_1 X + \beta_{11} X^2 + \beta_{111} X^3.$$

What biases are induced by taking observations at

$$X = -3, -2, -1, 0, 1, 2, 3?$$

We find

$$\mathbf{X}_1 = \begin{bmatrix} 1 & -3 \\ 1 & -2 \\ 1 & -1 \\ 1 & 0 \\ 1 & 1 \\ 1 & 2 \\ 1 & 3 \end{bmatrix} \qquad \mathbf{X}_2 = \begin{bmatrix} 9 & -27 \\ 4 & -8 \\ 1 & -1 \\ 0 & 0 \\ 1 & 1 \\ 4 & 8 \\ 9 & 27 \end{bmatrix}$$

$$(\mathbf{X}_1'\mathbf{X}_1)^{-1} = \begin{bmatrix} 7 & 0 \\ 0 & 28 \end{bmatrix}^{-1} = \begin{bmatrix} \frac{1}{7} & 0 \\ 0 & \frac{1}{28} \end{bmatrix}$$

$$\mathbf{X}_1'\mathbf{X}_2 = \begin{bmatrix} 28 & 0 \\ 0 & 196 \end{bmatrix}$$

$$\mathbf{A} = (\mathbf{X}_1'\mathbf{X}_1)^{-1}\mathbf{X}_1'\mathbf{X}_2 = \begin{bmatrix} 4 & 0 \\ 0 & 7 \end{bmatrix}.$$

Thus

$$E(b_0) = \beta_0 + 4\beta_{11}$$
$$E(b_1) = \beta_1 + 7\beta_{111}.$$

By using the general formula (2.12.5) we can find the bias in any regression estimates once the postulated model, the true model, and the design are established. This enables us to find, in specific situations, what effect will be transmitted to our estimates if a particular departure from the assumed model occurs. A sensible procedure in many situations where a polynomial model is postulated is to work on the basis that the postulated model may be wrong because it does not contain terms of one degree higher than those present.

The Effect of Bias on the Least Squares Analysis

Note. In this part of the section *only* we shall write $\mathbf{X}$ for the matrix previously called $\mathbf{X}_1$ and $\boldsymbol{\beta}$ for the vector previously called $\boldsymbol{\beta}_1$. The notation

$X_2\beta_2$ will still denote the extra terms of the true model, however. We now summarize the effect bias has on the usual least squares analysis.

Suppose that

1. the postulated model $E(Y) = X\beta$ contains p parameters; $V(Y) = I\sigma^2$.
2. the true model is $E(Y) = X\beta + X_2\beta_2$, where β_2 may be 0 in which case the postulated model is correct.
3. the total number of observations taken is n and there are f degrees of freedom available for lack of fit and e degrees of freedom for pure error, so that $n = p + f + e$. (This means there are $p + f$ distinct points in the design.)
4. the estimates $\mathbf{b} = (X'X)^{-1}X'Y$ and the fitted values $\hat{Y} = Xb$ are obtained as usual.
5. $A = (X'X)^{-1}X'X_2$.

Then a number of results are true as given below.

1. The matrix $(X'X)^{-1}\sigma^2$ is always the correct variance covariance matrix, $V(\mathbf{b})$, of estimated coefficients $\mathbf{b}$.
2. $E(\mathbf{b}) = \beta + A\beta_2$.
3. $E(\hat{Y}) = X\beta + XA\beta_2$.
4. The analysis of variance table takes the form below.

Source	Sum of Squares	Degrees of Freedom	Expected Value of Mean Square
Estimates	$b'X'Y$	p	$\sigma^2 + (\beta + A\beta_2)'X'X(\beta + A\beta_2)/p$
Lack of fit	By difference	f	$\sigma^2 + \beta_2'(X_2 - XA)'(X_2 - XA)\beta_2/f$
Pure error	es_e^2	e	σ^2
Total	$Y'Y$	n	

5. When $\beta_2 = 0$, that is, when the postulated model is correct, the results above reduce to the following:

$$E(\mathbf{b}) = \beta, \qquad E(\hat{Y}) = X\beta$$

$$E(\text{mean square due to estimates}) = \sigma^2 + \beta'X'X\beta/p.$$

(This is why the mean square due to estimates is compared with an estimate of σ^2 to test $H_0: \beta = 0$. It has expectation σ^2 if $\beta = 0$.)

$$E(\text{lack of fit mean square}) = \sigma^2.$$

(This is why the model must be correct if a residual mean square is to provide a valid estimate of σ^2. If lack of fit exists the lack of fit mean square has an expectation greater than σ^2, as shown in the analysis of variance table. This is why the lack of fit mean square is compared with pure error to test lack of fit.)

CHAPTER 3

THE EXAMINATION
OF RESIDUALS

3.0. Introduction

Note. The work of this chapter is useful and valid not only for linear regression models, but also for nonlinear regression models and analysis of variance models. In fact, this chapter applies to *any* situation where a model is fitted and measures of unexplained variation (in the form of a set of residuals) is available for examination.

The residuals are defined as the n differences $e_i = Y_i - \hat{Y}_i$, $i = 1$, $2, \ldots, n$ where Y_i is an observation and $\hat{Y}_i$ is the corresponding fitted value obtained by use of the fitted regression equation.

We can see from this definition that the residuals e_i are the differences between what is actually observed, and what is predicted by the regression equation—that is, the amount which the regression equation has not been able to *explain*. Thus we can think of the e_i as the *observed errors if the model is correct*. (There are, however, restrictions on the e_i—see Section 3.7.) Now in performing the regression analysis we have made certain assumptions about the errors; the usual assumptions are that the errors are independent, have zero mean, a constant variance, σ^2, and follow a normal distribution. The last assumption is required for making F-tests. Thus if our fitted model is correct, the residuals should exhibit tendencies that tend to confirm the assumptions we have made, or at least, should not exhibit a denial of the assumptions. This latter idea is the one that should be kept in mind when examining the residuals; we should ask, "Do the residuals make it appear that our assumptions are wrong?" After we have examined the residuals we shall be able to conclude either

1. the assumptions appear to be violated (in a way that can be specified), or
2. the assumptions do not appear to be violated.

Note that (2) does not mean that we are concluding that the assumptions are correct; it means merely that on the basis of the data we have seen, we have no reason to say that they are incorrect. The same spirit occurs in making tests of hypotheses when we either *reject* or *do not reject* (rather than *accept*). We now give ways of examining the residuals in order to check the model. These ways are all graphical, are easy to do, and are usually very revealing when the assumptions are violated. The principal ways of plotting the residuals e_i are

1. Overall.
2. In time sequence, if the order is known.
3. Against the fitted values $\hat{Y}_i$.
4. Against the independent variables X_{ji}, for $j = 1, 2, \ldots, k$.

In addition to these basic plots, the residuals should also be plotted

5. In any way that is sensible for the particular problem under considera- tion.

We now explain these plots in more detail. The following simple example will be used for illustrative purposes.

Example. A regression analysis provides eleven residuals $e_1, e_2, \ldots, e_{11}$, with values 5, -2, -4, 4, 0, -6, 9, -2, -5, 3, -2.

3.1. Overall Plot

When the residuals of the example are plotted overall we obtain the diagram shown in Figure 3.1. If our model is correct these residuals should resemble eleven observations from a normal distribution with zero mean. Does our overall plot contradict this idea? First we note that the mean of the residuals is zero, but this is necessarily the case for any regression model with a constant term β_0 in it. This is easily seen from the first normal equa- tion obtained by differentiating the error sum of squares with respect to β_0. If the model fitted is $E(Y) = \beta_0 + \beta_1 X_1 + \cdots + \beta_k X_k$, the equation can be written

$$- 2 \sum (Y_i - b_0 - b_1 X_{1i} - \cdots - b_k X_{ki}) = 0$$

where the summation is taken over $i = 1, 2, \ldots, n$. This reduces to

$$\sum (Y_i - \hat{Y}_i) = 0.$$

Thus

$$\sum e_i = \frac{\sum e_i}{n} = 0.$$

While the plot exhibits slight irregularity it does not appear abnormal for a sample of eleven observations from a normal distribution. How can we tell? To help set up standards for judging plots like this, one can make use of a table of random normal deviates. (An extensive table is published

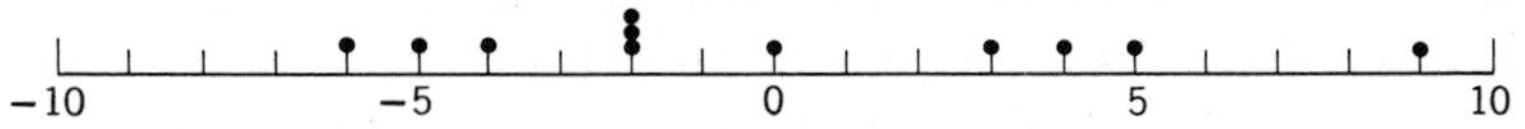

Figure 3.1

by the Rand Corporation but shorter tables are given in some statistical textbooks.) A number of samples of selected size (here eleven) can be taken from the table and plotted in the way illustrated above. A little practice of this sort will provide an excellent "feel" of how abnormal a plot should look before it can be said to appear to contradict the normality assumption.

An alternative procedure is to construct either a *normal plot* or a *half normal* plot of the residuals on standard probability paper. (See "Use of half-normal plots in interpreting factorial two level experiments," by C. Daniel, *Technometrics*, **1** (1959), 311–341.) The points should fall, approximately, on a straight line. Once again, however, standards of judgment are required for assessing the plot. Thus there is no particular advantage to be gained by using these alternatives, although they may be preferred by some readers.

The "Unit Normal Deviate" Form of the Residuals

We usually assume that $\epsilon_i \sim N(0, \sigma^2)$, so that $\epsilon_i/\sigma \sim N(0, 1)$. Now, if the model is correct, the residual mean square

$$s^2 = \frac{\sum_{i=1}^{n}(e_i - \bar{e})^2}{(n - p)} = \frac{\sum_{i=1}^{n}e_i^2}{(n - p)}$$

estimates σ^2. (*Note:* If we ignore rounding error $\bar{e} = \sum e_i/n = 0$.) The quantity e_i/s is often called the *unit normal deviate form* of the residual e_i. The e_i/s, $i = 1, 2, \ldots, n$, can be examined in an overall plot to see if they make it appear that the assumption $\epsilon_i/\sigma \sim N(0, 1)$ is wrong. Since 95% of an $N(0, 1)$ distribution lies between the limits $(-1.96, 1.96)$, we might expect that roughly 95% of the e_i/s were between the limits $(-2, 2)$. It is sometimes convenient to examine the residuals in this alternative manner. for example, to check for outliers (see page 94). If $n - p$ is small, 95% limits of the $t(n - p)$ distribution can be used instead.

3.2. Time Sequence Plot

Let us assume that the residuals in the example above occurred in the time order as given. The plot would then be as shown in Figure 3.2. We see that if we "step back" from this diagram we obtain the impression of a horizontal "band" of residuals which can be represented by Figure 3.3. This is indicative that a long-term time effect is not influencing the data (or if it is, has somehow been accounted for by an X-variable also subject to a time effect). However, if our "step back" view of the residuals resembled any of those shown in Figure 3.4 we should conclude that we had not taken into account a time effect, as follows in the cases shown.

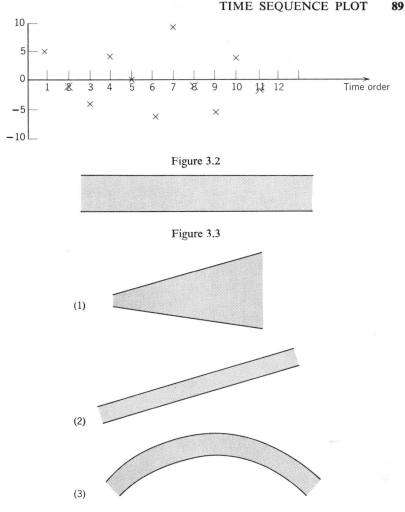

Figure 3.2

Figure 3.3

Figure 3.4

(1) The variance is not constant but increases with time, implying that a weighted least squares analysis should have been used.
(2) A linear term in time should have been included in the model.
(3) Linear and quadratic terms in time should have been included in the model.

Combinations and variations (e.g., (2) sloped in the opposite direction, and so on) of these defects can, of course, occur.

We said above that the residuals reveal no apparent long-term time trend. A closer examination of the time plot shows that the comment cannot be extended to short-term time trends. If we take the residuals in groups of three, namely (1, 2, 3), (4, 5, 6), (7, 8, 9), (10, 11) we can observe a downward linear trend in each group, revealing some sort of "seasonal" disturbance which must be introduced into a revised model. If we tentatively assume a common slope for the linear trends we might, for example, add a term of form $\delta\{(t - 1) \bmod 3\}$ to the model where δ denotes a regression coefficient to be estimated and $(t - 1) \bmod 3$ is the variable obtained by finding the remainder after dividing the value of $t - 1$ by 3. The values of this new variable are taken as follows:

Old variable t	1	2	3	4	5	6	7	8	9	10	11
$(t - 1) \bmod 3$	0	1	2	0	1	2	0	1	2	0	1

(Another alternative is to use $\{(t - 1) \bmod 3 - 1\}$ as a variable. This provides levels $-1, 0, 1$ instead of 0, 1, 2 and may sometimes be convenient.)

While use of a dummy variable in this way will take care of the variation, efforts should also be made to find an assignable cause for it. It is easy to imagine patterns in residual plots, and dummy variables should not be used without some real justification.

3.3. Plot Against $\hat{Y}_i$

Assume that the $\hat{Y}_i$ which correspond respectively to the e_i given above were 44, 8, 10, 62, 22, 48, 56, 30, 24, 16, 34. Then we should plot as shown in Figure 3.5. The "horizontal band" indicates no abnormality, and our least squares analysis would not appear to be invalidated.

Abnormality would be indicated by plots of the form shown as (1), (2), and (3) in Figure 3.4. Here the plots would indicate:

(1) Variance not constant, as assumed; need for weighted least squares or a transformation on the observations Y_i before making a regression analysis.

(2) Error in analysis; the departure from fitted equation is systematic

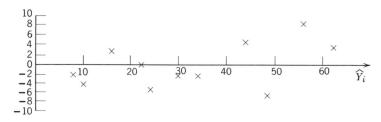

Figure 3.5

negative residuals correspond to low $\hat{Y}$'s, positive residuals to high $\hat{Y}$'s). The effect can also be caused by wrongly omitting a β_0 term in the model.

(3) Model inadequate—need for extra terms in the model (e.g., square or cross-product terms), or need for a transformation on the observations Y_i before analysis.

3.4. Plot Against the Independent Variables $X_{ji}, i = 1, 2, \ldots, n$

The form of these plots is the same as that against the $\hat{Y}_i$, except that we use (instead of the values of the corresponding $\hat{Y}_i$) the values of the corresponding X_{ji}, namely $X_{j1}, X_{j2}, \ldots, X_{jn}$. Once again an overall impression of a horizontal band of residuals is regarded as satisfactory. The anomalies illustrated in Figure 3.4 indicate here:

(1) Variance not constant; need for weighted least squares or a preliminary transformation on the Y's.

(2) Error in calculations; linear effect of X_j not removed.

(3) Need for extra terms, for example, a quadratic term, in X_j in the model or a transformation on the Y's.

In small regression problems which involve two or three X's only, it is possible to draw a diagram of the two or three dimensional space in which the data points occur. In such a case, the points at which the observations were taken can be plotted and the residuals written in next to the points. Where this is possible, it often provides a good visual grasp of the situation. When more than three X-variables occur it is possible to make such diagrams for subsets of the variables and this is sometimes appropriate.

3.5. Other Residual Plots

Specialist knowledge of the problem under study often suggests that other types of residual plots should be examined. For example, suppose it were known that the eleven observations which led to the eleven residuals given above came from three separate machines A, B, and C, so that the residuals when grouped by machines were

$$A: -2, -4, -6,$$
$$B: -2, -5, -2,$$
$$C: 5, 4, 0, 9, 3,$$

Figure 3.6 shows a plot against machines. This would suggest that there is a basic difference in level of response Y of machine C compared with A

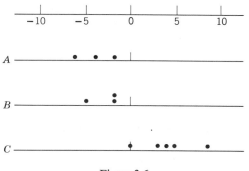

Figure 3.6

and B. Such a difference could be incorporated into the model by the introduction of a dummy variable; this is discussed in Chapter 5.

Another example of "other residual plots" occurs when a possible new variable comes into consideration. Suppose it is suspected that the ambient temperature is affecting the contents of a large vessel. Although vessel temperature has been recorded at a selected, protected, measuring point, the temperature at the other side of the vessel may possibly be affected by exposure to the outside air. If the ambient temperatures are recorded for the period during which data were collected, the residuals could now be plotted against the temperatures observed, to see if any dependency of response on ambient temperature is revealed. If it is, new terms of appropriate kinds can be added to the model to take account of the dependency.

These are two examples of what "other residual plots" might be used. In general residuals should be plotted in *any* reasonable way that occurs to the experimenter or statistician, based on specialist knowledge of the problem under study. The plots of Sections 3.1 to 3.4 are, however, the basic ones, and should always be performed for a full analysis.

3.6. Statistics for Examination of Residuals

The plots that have been recommended in earlier sections are visual techniques for checking some of the basic regression assumptions. Certain statistics have been suggested which provide a numerical measure for some of the discrepancies previously described. We shall mention these only briefly and obliquely. This lack of emphasis is deliberate since in practical regression situations a detailed examination of the corresponding residual plots is usually far more informative, and the plots will almost certainly reveal any violations of assumptions serious enough to require corrective action.

Consider the plot of e_i against $\hat{Y}_i$ described in Section 3.3. Three particular types of discrepancies were mentioned and related to the diagrams of Figure 3.4. We can measure each of these defects by appropriate statistics as follows. Define

$$T_{pq} = \sum_{i=1}^{n} e_i^p \hat{Y}_i^q. \tag{3.6.1}$$

Then

1. $T_{21} = \sum_{i=1}^{n} e_i^2 \hat{Y}_i$ provides a measure for the type of defect shown in Figure 3.4(1) of Section 3.2 (but in the context of Section 3.3). It is related to a more general statistic given by F. J. Anscombe in his "Examination of residuals" paper mentioned below.
2. $T_{11} = \sum_{i=1}^{n} e_i \hat{Y}_i$. This should always be zero. This provides a measure for the defect shown in Figure 3.4(2) of Section 3.2 (but in the sense of Section 3.3). Evaluation of this statistic could be done as a routine check, if desired.
3. $T_{12} = \sum_{i=1}^{n} e_i \hat{Y}_i^2$ provides a measure for the type of defect shown in Figure 3.4(3) of Section 3.2 (but in the context of Section 3.3). It is related to Tukey's "one degree of freedom for nonadditivity" statistic.

Other types of statistics are also available. Readers who would like to learn more about these should consult the following papers: F. J. Anscombe, "Examination of residuals," *Proceedings of the Fourth Berkeley Symposium on Mathematical Statistics and Probability*, **1** (1961), 1–36; and F. J. Anscombe and J. W. Tukey, "The examination and analysis of residuals," *Technometrics*, **5** (1963), 141–160.

3.7. Correlations among the Residuals

In a general regression situation, when p parameters are estimated from n observations, the n residuals are associated with only $n - p$ degrees of freedom. Thus the residuals cannot be independent and correlations exist among them. If the postulated model is $\mathbf{Y} = \mathbf{X}\boldsymbol{\beta} + \boldsymbol{\epsilon}$, and if $\mathbf{X}'\mathbf{X}$ is nonsingular, then the residuals can be written in matrix form as

$$\begin{aligned} \mathbf{e} = \mathbf{Y} - \hat{\mathbf{Y}} &= \mathbf{Y} - \mathbf{Xb} \\ &= \mathbf{Y} - \mathbf{X}(\mathbf{X}'\mathbf{X})^{-1}\mathbf{X}'\mathbf{Y} \\ &= (\mathbf{I} - \mathbf{X}(\mathbf{X}'\mathbf{X})^{-1}\mathbf{X}')\,\mathbf{Y} \\ &= (\mathbf{I} - \mathbf{R})\mathbf{Y} \end{aligned}$$

say, where $\mathbf{R} = \mathbf{X}(\mathbf{X}'\mathbf{X})^{-1}\mathbf{X}'$. (The matrix $\mathbf{R}$ is an important matrix which occurs repeatedly in regression work, incidentally.) Since $E(\mathbf{Y}) = \mathbf{X}\boldsymbol{\beta}$, it follows that

$$\mathbf{e} - E(\mathbf{e}) = (\mathbf{I} - \mathbf{R})(\mathbf{Y} - \mathbf{X}\boldsymbol{\beta}) = (\mathbf{I} - \mathbf{R})\boldsymbol{\epsilon}$$

and the variance-covariance matrix of $\mathbf{e}$ is defined as

$$V(\mathbf{e}) = E\{[\mathbf{e} - E(\mathbf{e})]\,[\mathbf{e} - E(\mathbf{e})]'\} = (\mathbf{I} - \mathbf{R})E(\boldsymbol{\epsilon}\boldsymbol{\epsilon}')(\mathbf{I} - \mathbf{R})'.$$

Now $E(\boldsymbol{\epsilon}\boldsymbol{\epsilon}') = \mathbf{V}(\boldsymbol{\epsilon}) = \mathbf{I}\sigma^2$ if $E(\boldsymbol{\epsilon}) = \mathbf{0}$, as we usually assume, and when unweighted least squares are used. Furthermore, $(\mathbf{I} - \mathbf{R})' = (\mathbf{I}' - \mathbf{R}') = \mathbf{I} - [\mathbf{X}(\mathbf{X}'\mathbf{X})^{-1}\mathbf{X}']' = \mathbf{I} - \mathbf{X}(\mathbf{X}'\mathbf{X})^{-1}\mathbf{X}' = \mathbf{I} - \mathbf{R}$. Thus the matrix $\mathbf{I} - \mathbf{R}$ is symmetric, and

$$\begin{aligned} \mathbf{V}(\mathbf{e}) &= (\mathbf{I} - \mathbf{R})\mathbf{I}\sigma^2(\mathbf{I} - \mathbf{R})' \\ &= (\mathbf{I} - \mathbf{R})(\mathbf{I} - \mathbf{R})\sigma^2 \\ &= (\mathbf{I} - \mathbf{R} - \mathbf{R} + \mathbf{R}\mathbf{R})\sigma^2 \\ &= (\mathbf{I} - \mathbf{R})\sigma^2 \end{aligned}$$

since $\mathbf{R}\mathbf{R} = \mathbf{R}^2 = \mathbf{R}$ as is easily confirmed[1]. Thus $V(e_i)$ is given by the ith diagonal element, and covar (e_i, e_j) is given by the (i, j)th element of the matrix $(\mathbf{I} - \mathbf{R})\sigma^2$. The correlation between e_i and e_j is given by

$$\rho_{ij} = \frac{\text{covar } (e_i, e_j)}{\{V(e_i) \cdot V(e_j)\}^{\frac{1}{2}}} .$$

The values of these correlations thus depend entirely on the elements of the matrix $\mathbf{X}$, since σ^2 cancels.

Do these correlations invalidate the residuals plots? Anscombe remarks on this point in "The examination and analysis of residuals," *Technometrics*, **5** page 144. In discussing the two-way analysis of variance (where there are several constraints on the residuals) he remarks that, although correlations and constraints affect distributions of functions of the residuals, the ". . . corresponding effects on the graphical procedures . . . can usually be neglected. This is mainly because of the way in which graphical appearances arise from residuals, though in part because of the absence of precisely defined significance levels. (This is also true for most other balanced designs.)" In a later sentence Anscombe states that in a two-way table with four or more rows and four or more columns, ". . . the effect of correlation upon graphical procedures is usually negligible . . . ." It would appear that in general regression situations the effect of correlation between residuals need not be considered when plots are made, except when the ratio $(n - p)/n$ equals (number of degrees of freedom in residuals)/(number of residuals) is quite small.

3.8.   Outliers

An outlier among residuals is one that is far greater than the rest in absolute value and perhaps lies three or four standard deviations or further from the mean of the residuals. The outlier is a peculiarity and indicates a data point which is not at all typical of the rest of the data. It follows that

---

[1] When $\mathbf{R}\mathbf{R} = \mathbf{R}$, $\mathbf{R}$ is said to be idempotent. Both $\mathbf{R}$ and $\mathbf{I} - \mathbf{R}$ are symmetric and idempotent.

an outlier should be submitted to particularly careful examination to see if the reason for its peculiarity can be determined.

Rules have been proposed for rejecting outliers (i.e., for deciding to remove the corresponding observation(s) from the data, after which the data is reanalyzed without these observations). Automatic rejection of outliers is not always a very wise procedure. Sometimes the outlier is providing information which other data points cannot due to the fact that it arises from an unusual combination of circumstances which may be of vital interest and requires further investigation rather than rejection. (See, e.g., Section 4.1.) As a general rule, outliers should be rejected out of hand only if they can be traced to causes such as errors in recording the observations or in setting up the apparatus. Otherwise careful investigation is in order. (A paper by F. J. Anscombe on "Rejection of outliers" appears in *Technometrics*, **2** (1960), 123–147.)

3.9.   Examining Runs in the Time Sequence Plot of Residuals

When the time sequence of a set of residuals is known, it is sometimes noticeable that groups of positive or negative residuals occur in what might be an unusual pattern. To take an extreme case, if thirty residuals in time sequence consisted of ten negative followed by twenty positive residuals, we might suspect that an unconsidered variable had changed levels between the tenth and eleventh run. We would then search for an assignable cause for this behavior. When there is a sequence of such runs it is useful to have a method which will enable a decision to be made on the abnormality of the run pattern.

Suppose we have a sequence of signs such as

$$+ \; + \; - \; + \; - \; - \; - \; - \; + \; + \; - \; + \; + \; +.$$

These may be the signs of residuals in time sequence (which will be our application) but equally well the "plus and minus" signs might denote "male and female," "heads and tails," "better and worse," "treatment A and treatment B," or the two levels of any other dichotomous classification. Suppose there are n signs in all, n_1 plus signs and n_2 minus signs, and there are u runs. In the above example $n_1 = 8$, $n_2 = 6$, and there are $u = 7$ runs indicated by the parentheses below

$$(+ \; +)(-)(+)(- \; - \; - \; -)(+ \; +)(-)(+ \; + \; +).$$

We can ask if the particular arrangement of signs we observe is an "extreme" arrangement or not. For example, if there are six signs, two of which are plus, the following arrangements are possible.

Arrangement	Number of runs, u
+ + − − − −	2
+ − + − − −	4
+ − − + − −	4
+ − − − + −	4
+ − − − − +	3
− + + − − −	3
− + − + − −	5
− + − − + −	5
− + − − − +	4
− − + + − −	3
− − + − + −	5
− − + − − +	4
− − − + + −	3
− − − + − +	4
− − − − + +	2

The distribution of runs is as follows.

$$u = 2 \quad\quad 3 \quad\quad 4 \quad\quad 5$$

Frequency $= 2 \quad\quad 4 \quad\quad 6 \quad\quad 3 \quad\quad$ (Total $= 15$)

Cumulative Probability $= 0.133 \quad 0.400 \quad 0.800 \quad 1.000$

Thus five runs would occur in $\frac{3}{15}$ or 20% of the possible cases, that is, with a probability of 0.2. Alternatively two runs would occur in $\frac{2}{15}$ or 13.3% of the possible cases, that is with a probability of 0.133. In examining residuals, only low numbers of runs would normally be of interest since we shall usually not be concerned whether or not there are "too many" runs. If we observed only $u = 2$ runs in a set of six residuals of which two were positive we would have observed an event which occurs with a probability of 0.133. For any given sequence of signs we can find the probability that the observed value of u (or a lesser value) will occur. (*Example:* when $n_1 = 2$, $n_2 = 4$, and the observed number of runs is 3, prob $(u \leq 3) = (2 + 4)/15 = 0.4$, a not unusual event occurring in 40% of cases.) On the basis of such a probability level we can decide whether or not we believe that a random arrangement of signs has occurred. (We might, for example, compare the probability with a preassigned value, say $\alpha = 0.05$, and reject the idea of a random arrangement if prob $(u \leq$ observed $u) \leq 0.05$). Table 3.1 shows the cumulative distributions for the cases $n_1 \leq 10$ and $n_1 \leq n_2 \leq 10$. (When $n_1 > n_2$, interchange n_1 and n_2.) These distributions were given originally by Frieda S. Swed, and C. Eisenhart in a paper "Tables for testing randomness of grouping in a sequence of alternatives," *Annals of Mathematical Statistics*, **14,** 1943, 66–87. In that paper more decimal places are given and the arrangement of cases is different.

When $n_1 > 10$ and $n_2 > 10$ exact values are not needed since a normal approximation to the actual distribution provides satisfactory accuracy. Let

$$\mu = \frac{2n_1 n_2}{n_1 + n_2} + 1, \qquad (3.9.1)$$

$$\sigma^2 = \frac{2n_1 n_2 (2n_1 n_2 - n_1 - n_2)}{(n_1 + n_2)^2 (n_1 + n_2 - 1)}. \qquad (3.9.2)$$

It can be shown that these are the actual mean and variance of the discrete distribution of u. Then approximately

$$z = \frac{(u - \mu + \frac{1}{2})}{\sigma} \qquad (3.9.3)$$

is a unit normal deviate where the half is the usual *continuity correction* which helps compensate for the fact that a continuous distribution is being used to approximate to a discrete distribution.

Example.   Examination of a set of twenty-seven residuals, fifteen of which were of one sign and twelve of which were the opposite sign, arranged in time sequence, revealed $u = 7$ runs. Does the arrangement of signs appear to be random?

Here $n_1 = 15$, $n_2 = 12$, $u = 7$. From Eqs. (3.9.1) and (3.9.2), $\mu = \frac{43}{3}$, $\sigma^2 = \frac{740}{117}$. Thus the observed value of z from Eq. (3.9.3) is

$$z = \frac{(7 - \frac{43}{3} + \frac{1}{2})}{(\frac{740}{117})^{1/2}} = -2.713.$$

The probability of obtaining a unit normal deviate of value -2.713 or smaller is 0.0033 (or 0.33%) so that an unusually low number of runs appears to have occurred. We should reject the idea that the arrangement of signs is random. The model would be suspect and we would now search for an assignable cause for the pattern of residuals.

Note. Strictly speaking, the test for runs is applicable only when the occurrences which produce the pattern of runs are independent. In a time sequence of residuals this is not true due to the correlations that exist among the residuals (see Section 3.7), and the probability level obtained from the procedure will be affected in a way which depends on the particular structure of the data. In most practical regression situations, unless the ratio $(n - p)/n$, that is (number of degrees of freedom in residuals)/(number of residuals), is quite small, the effect can be ignored.

Table 3.1 Cumulative Distribution of the Total Number of Runs u in Samples of Sizes (n_1, n_2)*

$(n_1, n_2)\backslash u$	2	3	4	5	6	7	8	9	10
(2, 3)	0.200	0.500	0.900	1.000					
(2, 4)	0.133	0.400	0.800	1.000					
(2, 5)	0.095	0.333	0.714	1.000					
(2, 6)	0.071	0.286	0.643	1.000					
(2, 7)	0.056	0.250	0.583	1.000					
(2, 8)	0.044	0.222	0.533	1.000					
(2, 9)	0.036	0.200	0.491	1.000					
(2, 10)	0.030	0.182	0.455	1.000					
(3, 3)	0.100	0.300	0.700	0.900	1.000				
(3, 4)	0.057	0.200	0.543	0.800	0.971	1.000			
(3, 5)	0.036	0.143	0.429	0.714	0.929	1.000			
(3, 6)	0.024	0.107	0.345	0.643	0.881	1.000			
(3, 7)	0.017	0.083	0.283	0.583	0.833	1.000			
(3, 8)	0.012	0.067	0.236	0.533	0.788	1.000			
(3, 9)	0.009	0.055	0.200	0.491	0.745	1.000			
(3, 10)	0.007	0.045	0.171	0.455	0.706	1.000			
(4, 4)	0.029	0.114	0.371	0.629	0.886	0.971	1.000		
(4, 5)	0.016	0.071	0.262	0.500	0.786	0.929	0.992	1.000	
(4, 6)	0.010	0.048	0.190	0.405	0.690	0.881	0.976	1.000	
(4, 7)	0.006	0.033	0.142	0.333	0.606	0.833	0.954	1.000	
(4, 8)	0.004	0.024	0.109	0.279	0.533	0.788	0.929	1.000	
(4, 9)	0.003	0.018	0.085	0.236	0.471	0.745	0.902	1.000	
(4, 10)	0.002	0.014	0.068	0.203	0.419	0.706	0.874	1.000	
(5, 5)	0.008	0.040	0.167	0.357	0.643	0.833	0.960	0.992	1.000
(5, 6)	0.004	0.024	0.110	0.262	0.522	0.738	0.911	0.976	0.998
(5, 7)	0.003	0.015	0.076	0.197	0.424	0.652	0.854	0.955	0.992
(5, 8)	0.002	0.010	0.054	0.152	0.347	0.576	0.793	0.929	0.984
(5, 9)	0.001	0.007	0.039	0.119	0.287	0.510	0.734	0.902	0.972
(5, 10)	0.001	0.005	0.029	0.095	0.239	0.455	0.678	0.874	0.958
(6, 6)	0.002	0.013	0.067	0.175	0.392	0.608	0.825	0.933	0.987
(6, 7)	0.001	0.008	0.043	0.121	0.296	0.500	0.733	0.879	0.966
(6, 8)	0.001	0.005	0.028	0.086	0.226	0.413	0.646	0.821	0.937
(6, 9)	0.000	0.003	0.019	0.063	0.175	0.343	0.566	0.762	0.902
(6, 10)	0.000	0.002	0.013	0.047	0.137	0.288	0.497	0.706	0.864
(7, 7)	0.001	0.004	0.025	0.078	0.209	0.383	0.617	0.791	0.922
(7, 8)	0.000	0.002	0.015	0.051	0.149	0.296	0.514	0.704	0.867
(7, 9)	0.000	0.001	0.010	0.035	0.108	0.231	0.427	0.622	0.806
(7, 10)	0.000	0.001	0.006	0.024	0.080	0.182	0.355	0.549	0.743
(8, 8)	0.000	0.001	0.009	0.032	0.100	0.214	0.405	0.595	0.786
(8, 9)	0.000	0.001	0.005	0.020	0.069	0.157	0.319	0.500	0.702
(8, 10)	0.000	0.000	0.003	0.013	0.048	0.117	0.251	0.419	0.621
(9, 9)	0.000	0.000	0.003	0.012	0.044	0.109	0.238	0.399	0.601
(9, 10)	0.000	0.000	0.002	0.008	0.029	0.077	0.179	0.319	0.510
(10, 10)	0.000	0.000	0.001	0.004	0.019	0.051	0.128	0.242	0.414

* Adapted from Swed, Frieda S., and Eisenhart, C. (1943), "Tables for testing randomness of grouping in a sequence of alternatives," *Ann. Math. Statist.*, **14**, 66–87.

11	12	13	14	15	16	17	18	19	20
1.000									
1.000									
1.000									
1.000									
1.000									
0.998	1.000								
0.992	0.999	1.000							
0.984	0.998	1.000							
0.972	0.994	1.000							
0.958	0.990	1.000							
0.975	0.996	0.999	1.000						
0.949	0.988	0.998	1.000	1.000					
0.916	0.975	0.994	0.999	1.000					
0.879	0.957	0.990	0.998	1.000					
0.900	0.968	0.991	0.999	1.000	1.000				
0.843	0.939	0.980	0.996	0.999	1.000	1.000			
0.782	0.903	0.964	0.990	0.998	1.000	1.000			
0.762	0.891	0.956	0.988	0.997	1.000	1.000	1.000		
0.681	0.834	0.923	0.974	0.992	0.999	1.000	1.000	1.000	
0.586	0.758	0.872	0.949	0.981	0.996	0.999	1.000	1.000	1.000

EXERCISES

A. A model of the form $Y = \beta_0 + \beta_1 X_1 + \beta_2 X_2 + \beta_3 X_3 + \epsilon$ was used to describe a particular process. Twenty-five data points were obtained and the resultant analysis yielded the following:

1. $n = 25$
2. $R^2 = 88.5\%$
3. $s = 0.5915$
4. $\bar{Y} = 9.42$
5. The analysis of variance:

Source of Variation	d.f.	SS	MS	F
Total (corrected)	24	63.8352		
Regression	3	56.4875	18.8292	53.81
Residual	21	7.3477	0.3499	

6. The prediction equation was:

$$\hat{Y} = -2.952790 - 0.073932 X_1 + 0.198999 X_2 + 0.401528 X_3$$

7. The residuals were:

Obs. No.	Observed Y	Pred. $\hat{Y}$	Residual
1	10.98	10.86	0.12
2	11.13	10.48	0.65
3	12.51	11.79	0.72
4	8.40	8.73	−0.33
5	9.27	9.13	0.14
6	8.73	8.20	0.53
7	6.36	6.18	0.18
8	8.50	8.40	0.10
9	7.82	8.05	−0.23
10	9.14	9.22	−0.08
11	8.24	9.64	−1.40
12	12.19	11.54	0.65
13	11.88	11.60	0.28
14	9.57	9.18	0.39
15	10.94	10.61	0.33
16	9.58	9.49	0.09
17	10.09	9.49	0.60
18	8.11	8.30	−0.19
19	6.83	6.51	0.32
20	8.88	8.56	0.32
21	7.68	7.74	−0.06
22	8.47	9.38	−0.91
23	8.86	9.78	−0.92
24	10.36	11.01	−0.65
25	11.08	11.76	−0.68

Requirements

1. Construct the following residual plots:
   a. overall,
   b. against observation number,
   c. against $\hat{Y}$.
2. Interpret the plots and draw conclusions concerning violations of the assumptions usually made in applying least squares multiple regression techniques.

B. The height of soap suds in the dishpan is of importance to soap manufacturers. An experiment was performed by varying the amount of soap and measuring the height of the suds in a standard dishpan after a given amount of agitation. The data are as follows:

Grams of Product	Suds Height
X	Y
4.0	33
4.5	42
5.0	45
5.5	51
6.0	53
6.5	61
7.0	62

Assume that a model of the form $Y = \beta_0 + \beta_1 X_1 + \epsilon$ is reasonable.

Requirements

1. Determine the best-fitting equation.
2. Test it for statistical significance.
3. Calculate the residuals and see if there is any evidence which suggests that a more complicated model would be more suitable.

C. In a similar experiment to that given in Exercise B, the experimenter stated that the model $Y = \beta_0 + \beta_1 X_1 + \epsilon$ was "a ridiculous model unless $\beta_0 = 0$, for anyone knows that if you don't put any soap in the dishpan there will be no suds." Thus, he insists on using the model $Y = \beta_1 X + \epsilon$. His data are shown as follows:

Grams of Product	Suds Height
X	Y
3.5	24.4
4.0	32.1
4.5	37.1
5.0	40.4
5.5	43.3
6.0	51.4
6.5	61.9
7.0	66.1
7.5	77.2
8.0	79.2

Requirements

1. Accepting the experimenter's model, determine the best estimate of β_1.
2. Using this equation, estimate $\hat{Y}$ for each X.
3. Examine the residuals.
4. Draw conclusions and make recommendations to the experimenter.

D. The following data indicate the relationship between the amount of β-erythroidine in an aqueous solution and the colorimeter reading of the turbidity.

X Concentration (mg/ml)	Y Colorimeter Reading
40	69
50	175
60	272
70	335
80	490
90	415
40	72
60	265
80	492
50	180

Requirement

Fit the equation $Y = \beta_0 + \beta_1 X + \epsilon$, obtain the residuals, examine them, and comment on the adequacy of the model.

E. A synthetic fiber, which because of its hairlike appearance has been found suitable in the manufacture of wigs, must necessarily be preshrunk prior to manufacture. This is accomplished in two steps:

1. The fiber is soaked in a dilute solution of chemical A, which is necessary to preserve the luster of the fiber during step 2.
2. The fiber is baked in large ovens at a very high temperature for one hour.

It is suggested that the temperature at which the fiber is baked may influence the effectiveness of the preshrinking process. An experiment is performed in which the baking temperature T is varied for various batches of fiber. The finished fiber is then soaked in rain water for a suitable length of time and put out in the sun to dry. The amount of further shrinkage Y (in per cent) resulting from the rain-water test is recorded along with the value of T for each batch:

Batch No.	T	Y
1	280	2.1
2	250	3.0
3	300	3.2
4	320	1.4
5	310	2.6
6	280	3.9
7	320	1.3
8	300	3.4
9	320	2.8

1. Fit a regression line $\hat{Y} = b_0 + b_1 T$ to the data by least squares.

Note. Coding the variable T may simplify the calculations, but remember in the end to express the fitted equation in terms of the original variable T.

2. Perform an analysis of variance and test:
   a. The lack of fit.
   b. The significance of the regression.
   What is the percentage variation explained by the regression equation?
3. What is the estimated standard deviation of b_1? Give a 95% confidence interval for the true regression coefficient β_1.
4. Give the fitted value $\hat{Y}_i$ and the residual $Y_i - \hat{Y}_i$ corresponding to each run (batch).
5. For $T_k = 315$, find an interval about the predicted value $\hat{Y}_k$ within which a single future observation Y will fall with probability 0.95.
6. Could we use the fitted equation to predict a value of Y at $T = 360$? Give reasons for your answer.

F. For the experimental situation described in Exercise E, suppose that data for the concentration of Chemical A had been recorded for each run. It is suggested that the variation found in this factor might be causing the large variation in response discovered previously. The readings of the concentration factor C (in per cent) are

Batch	1	2	3	4	5	6	7	8	9
C	6	6	8	7	9	8	5	9	11

where the batch numbers are the same as in Exercise E.

1. Plot the residuals obtained in Exercise E versus C. Notice anything?
2.* Fit the model: $Y = \beta_0 + \beta_1 X_1 + \beta_2 X_2 + \epsilon$ to the data, where $X_1 = (T - 300)/10$ and $X_2 = C - 8$.
3. Perform an analysis of variance and test:
   a. The lack of fit.
   b. The significance of the effect of including β_1 and β_2 in the model, rather than only β_0.
   c. The significance of including β_2 in the model, rather than just β_0 and β_1.
4. What percentage of the total variation (corrected) has been "taken up" by including the concentration effect in our model?
5. What is the estimated standard deviation of $\tilde{b}_1$? Of $\tilde{b}_2$? ($\tilde{b}_1$ and $\tilde{b}_2$ are the regression coefficients in the expression for $\hat{Y}$ in terms of the original variables T and C.)
6. Write the fitted value and the residual for each batch. Notice anything?
7. What is the estimated variance of the predicted value $\hat{Y}$ at the point $T = 315$, $C = 8$?

* If you employ the coding system: $X_1 = (T - 300)/10$, $X_2 = C - 8$, you may use the following hint:

$$\begin{bmatrix} 9 & -2 & -3 \\ -2 & 46 & 13 \\ -3 & 13 & 29 \end{bmatrix}^{-1} = \frac{1}{10111} \begin{bmatrix} 1165 & 19 & 112 \\ 19 & 252 & -111 \\ 112 & -111 & 410 \end{bmatrix}.$$

CHAPTER 4

TWO INDEPENDENT VARIABLES

4.0. Introduction

Up to this point we have considered, in detail, the first-order linear regression model in one variable X,

$$Y = \beta_0 + \beta_1 X + \epsilon,$$

and shown how the straightforward analysis can be expressed neatly in matrix terms. Usually more complex linear models are needed in practical situations. There are many problems in which a knowledge of more than one independent (or "predictor") variable is necessary in order to obtain better understanding and/or better prediction of a particular response. The matrix approach given at the end of Chapter 2 provides us with a general procedure for extending Chapter 1 results to more complicated linear models. In this chapter, we shall apply the matrix analysis to the first-order linear model:

$$Y = \beta_0 + \beta_1 X_1 + \beta_2 X_2 + \epsilon.$$

We shall continue with the example used in Chapter 1 (the data for which are given in Appendix A) and will now add variable number 6 to the problem. So that we are clear about which variables are being considered in the model we shall use the original variable subscripts. Thus our model will be written

$$Y = \beta_0 X_0 + \beta_8 X_8 + \beta_6 X_6 + \epsilon \qquad (4.0.1)$$

where Y = response or number of pounds of steam used per month,

X_0 = dummy variable, whose value is always unity,

X_8 = average atmospheric temperature in the month (in °F), and

X_6 = number of operating days in the month.

The following matrices can then be constructed. (The complete figures for the vector $\mathbf{Y}$ and the second and third columns of matrix $\mathbf{X}$ appear in Appendix A and are also given on page 116.)

$$Y = \begin{bmatrix} 10.98 \\ 11.13 \\ 12.51 \\ 8.4 \\ \cdot \\ \cdot \\ \cdot \\ 10.36 \\ 11.08 \end{bmatrix} \qquad X = \begin{matrix} X_0 & X_8 & X_6 \\ \begin{bmatrix} 1 & 35.3 & 20 \\ 1 & 29.7 & 20 \\ 1 & 30.8 & 23 \\ 1 & 58.8 & 20 \\ \cdot & \cdot & \cdot \\ \cdot & \cdot & \cdot \\ \cdot & \cdot & \cdot \\ 1 & 33.4 & 20 \\ 1 & 28.6 & 22 \end{bmatrix} \end{matrix} \qquad \beta = \begin{bmatrix} \beta_0 \\ \beta_8 \\ \beta_6 \end{bmatrix} \qquad \epsilon = \begin{bmatrix} \epsilon_1 \\ \epsilon_2 \\ \epsilon_3 \\ \epsilon_4 \\ \cdot \\ \cdot \\ \cdot \\ \epsilon_{24} \\ \epsilon_{25} \end{bmatrix}$$

where Y is a (25×1) vector,
$\quad X$ is a (25×3) matrix,
$\quad \beta$ is a (3×1) vector, and
$\quad \epsilon$ is a (25×1) vector.

Using the results of Chapter 2, the least-squares estimates of β_0, β_8, and β_6 are given by

$$b = (X'X)^{-1}X'Y$$

where b is the vector of estimates of the elements of β, provided that $X'X$ is nonsingular. Thus

$$b = \begin{bmatrix} b_0 \\ b_8 \\ b_6 \end{bmatrix} = \left\{ \begin{bmatrix} 1 & 1 & 1 & \cdots & 1 \\ 35.3 & 29.7 & 30.8 & \cdots & 28.6 \\ 20 & 20 & 23 & \cdots & 22 \end{bmatrix} \begin{bmatrix} 1 & 35.3 & 20 \\ 1 & 29.7 & 20 \\ 1 & 30.8 & 23 \\ \cdot & \cdot & \cdot \\ \cdot & \cdot & \cdot \\ \cdot & \cdot & \cdot \\ 1 & 28.6 & 22 \end{bmatrix} \right\}^{-1}$$

$$\times \begin{bmatrix} 1 & 1 & 1 & \cdots & 1 \\ 35.3 & 29.7 & 30.8 & \cdots & 28.6 \\ 20 & 20 & 23 & \cdots & 22 \end{bmatrix} \begin{bmatrix} 10.98 \\ 11.13 \\ 12.51 \\ \cdot \\ \cdot \\ \cdot \\ 11.08 \end{bmatrix}$$

Note the sizes of the matrices in the above statement:

$$[3 \times 1] = \{[3 \times 25][25 \times 3]\}^{-1}[3 \times 25][25 \times 1].$$

Multiplying the matrices within the large braces, we have

$$
\begin{array}{cc}
[3 \times 1] & [3 \times 3]^{-1} \\
\begin{bmatrix} b_0 \\ b_8 \\ b_6 \end{bmatrix} = &
\begin{bmatrix}
25.00 & 1315.00 & 506.00 \\
1315.00 & 76323.42 & 26353.30 \\
506.00 & 26353.30 & 10460.00
\end{bmatrix}^{-1}
\end{array}
$$

$$
\begin{array}{cc}
& [25 \times 1] \\
[3 \times 25] & \begin{bmatrix} 10.98 \\ 11.13 \\ 12.51 \\ \cdot \\ \cdot \\ \cdot \\ 11.08 \end{bmatrix} \\
\times \begin{bmatrix}
1 & 1 & \cdots & 1 \\
35.3 & 29.7 & \cdots & 28.6 \\
20 & 20 & \cdots & 22
\end{bmatrix} &
\end{array}
$$

Then,

$$
\begin{array}{ccc}
[3 \times 1] & [3 \times 3]^{-1} & [3 \times 1] \\
\begin{bmatrix} b_0 \\ b_8 \\ b_6 \end{bmatrix} = &
\begin{bmatrix}
25.00 & 1315.00 & 506.00 \\
1315.00 & 76323.42 & 26353.30 \\
506.00 & 26353.30 & 10460.00
\end{bmatrix}^{-1} &
\begin{bmatrix} 235.6000 \\ 11821.4320 \\ 4831.8600 \end{bmatrix}
\end{array}
$$

Next, the inverse of the $[3 \times 3]$ matrix is obtained to give

$$
\begin{array}{cc}
[3 \times 1] & [3 \times 3] \\
\begin{bmatrix} b_0 \\ b_8 \\ b_6 \end{bmatrix} = &
\begin{bmatrix}
2.778747 & -0.011242 & -0.106098 \\
& 0.146207 \times 10^{-3} & 0.175467 \times 10^{-3} \\
\text{(Symmetric)} & & 0.478599 \times 10^{-2}
\end{bmatrix}
\end{array}
$$

$$
\begin{array}{c}
[3 \times 1] \\
\times \begin{bmatrix} 235.6000 \\ 11821.4320 \\ 4831.8600 \end{bmatrix}
\end{array}
$$

The inverse calculation can be checked by multiplying $(\mathbf{X'X})^{-1}$ by the original $(\mathbf{X'X})$ to give a 3×3 unit matrix. Notice that, since the inverse (like the original matrix) is symmetric, only an upper triangular portion of it is recorded. Performing the matrix multiplication gives

$$
\begin{array}{cc}
[3 \times 1] & [3 \times 1]
\end{array}
$$

$$
\begin{bmatrix} b_0 \\ b_8 \\ b_6 \end{bmatrix} = \begin{bmatrix} 9.1266 \\ -0.0724 \\ 0.2029 \end{bmatrix}
$$

Thus, the fitted least squares equation is

$$
\hat{Y} = 9.1266 - 0.0724 X_8 + 0.2029 X_6.
$$

Actually, when these matrix calculations are performed by a computer routine, they are not carried through in precisely this way. One reason for this is that large rounding errors may occur when this sequence is followed. This point will be discussed in Section 5.4.

For the record, the algebraic form of the normal equations for the case of two independent variables is as follows:

$$
b_0 n + b_1 \sum_{i=1}^{n} X_{1i} + b_2 \sum_{i=1}^{n} X_{2i} = \sum_{i=1}^{n} Y_i
$$

$$
b_0 \sum_{i=1}^{n} X_{1i} + b_1 \sum_{i=1}^{n} X_{1i}^2 + b_2 \sum_{i=1}^{n} X_{1i} X_{2i} = \sum_{i=1}^{n} X_{1i} Y_i
$$

$$
b_0 \sum_{i=1}^{n} X_{2i} + b_1 \sum_{i=1}^{n} X_{2i} X_{1i} + b_2 \sum_{i=1}^{n} X_{2i}^2 = \sum_{i=1}^{n} X_{2i} Y_i.
$$

We obtained the fitted equation above by a single regression calculation. Actually it is possible to obtain the same equation through a series of simple straight-line regressions. Although this is not the best practical way of obtaining the final equation, it is instructive to consider how it is done. Thus, before we examine the fitted equation in Section 4.2, we shall discuss this alternative procedure.

4.1. Multiple Regression with Two Independent Variables as a Sequence of Straight-Line Regressions

In the previous section, we used least squares to determine the fitted equation

$$
\hat{Y} = 9.1266 - 0.0724 X_8 + 0.2029 X_6.
$$

Another way of obtaining this solution is as follows:

1. Plot Y (amount of steam) against X_8 (average atmospheric temperature). This plot is shown in Figure 1.4 in Chapter 1. Note the downward trend; this is reasonable, since as the temperature rises, the need for steam should diminish.

2. Regress Y on X_8. This straight line regression was performed in Chapter 1, and the resulting equation was

$$\hat{Y} = 13.6215 - 0.0798X_8$$

This fitted equation does not predict Y exactly (Table 1.2). Adding a new variable, say X_6 (the number of operating days), to the prediction equation might improve the prediction significantly.

In order to accomplish this, we desire to relate the number of operating days to the amount of unexplained variation in the data after the atmospheric temperature effect has been removed. However, if the atmospheric temperature variations are in any way related to the variability shown in the number of operating days, we must correct for this first. Thus, what we need to do is to determine the relationship between the unexplained variation in the amount of steam used after the effect of atmospheric temperature has been removed, and the remaining variation in the number of operating days after the effect of atmospheric temperature has been removed from it.

3. Regress X_6 on X_8; calculate residuals $X_{6i} - \hat{X}_{6i}$, $i = 1, 2, \ldots, n$. A plot of X_6 against X_8 is shown in Figure 4.1. Using the notation and methods of Chapter 2, the estimates of the regression coefficients are given by

$$
\begin{bmatrix} b_0 \\ b_8 \end{bmatrix} = \left\{ \begin{bmatrix} 1 & 1 & \cdots & 1 \\ 35.3 & 29.7 & \cdots & 28.6 \end{bmatrix} \begin{bmatrix} 1 & 35.3 \\ 1 & 29.7 \\ \cdot & \cdot \\ \cdot & \cdot \\ \cdot & \cdot \\ 1 & 28.6 \end{bmatrix} \right\}^{-1}
$$

$$
\times \begin{bmatrix} 1 & 1 & \cdots & 1 \\ 35.3 & 29.7 & \cdots & 28.6 \end{bmatrix} \begin{bmatrix} 20 \\ 20 \\ \cdot \\ \cdot \\ \cdot \\ 22 \end{bmatrix}
$$

Thus, $\hat{X}_6 = 22.1685 - 0.0367X_8$, and the residuals are shown in Table 4.1.

Table 4.1   Residuals: $X_{6i} - \hat{X}_{6i}$

Observation Number i	X_{6i}	$\hat{X}_{6i}$	$X_{6i} - \hat{X}_{6i}$
1	20	20.87	−0.87
2	20	21.08	−1.08
3	23	21.04	1.96
4	20	20.01	−0.01
5	21	19.92	1.08
6	22	19.55	2.45
7	11	19.44	−8.44
8	23	19.36	3.64
9	21	19.58	1.42
10	20	20.06	−0.06
11	20	20.47	−0.47
12	21	21.11	−0.11
13	21	21.14	−0.14
14	19	20.73	−1.73
15	23	20.45	2.55
16	20	20.39	−0.39
17	22	19.99	2.01
18	22	19.60	2.40
19	11	19.60	−8.60
20	23	19.44	3.56
21	20	19.53	0.47
22	21	20.04	0.96
23	20	20.53	−0.53
24	20	20.94	−0.94
25	22	21.12	0.88

We note that there are two residuals −8.44 and −8.60 which have absolute values considerably greater than the other residuals. They arise from months in which the number of operating days was unusually small, eleven in each case. We can, of course, take the attitude that these are "outliers" and that months with so few operating days should not even be considered in the analysis. However, if we wish to obtain a satisfactory prediction equation which will be valid for *all* months, irrespective of the number of operating days, then it is important to take account of these particular results and develop an equation which makes use of the information they contain. As can be seen from the data and from Figure 4.1 and Table 4.2, if these particular months were ignored, the apparent effect of the number of operating days on the response would be small. This would *not* be because the variable did not affect the response but because the variation actually observed in the variable

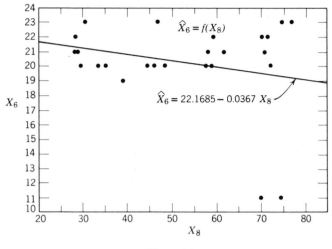

Figure 4.1

was so slight that the variable could not exert any appreciable effect on the response. If a variable appears to have a significant effect on the response in one analysis but not in a second, it may well be that it varied over a wider range in the first set of data than in the second. This, incidentally, is one of the drawbacks of using plant data "as it comes." Quite often the normal operating range of a variable is so slight that no effect on response is revealed, even when the variable does, over larger ranges of operation, have an appreciable effect. Thus designed experiments, which assign levels wider than normal operating ranges, often reveal effects which had not been previously noticed.

4. We now regress $Y - \hat{Y}$ against $X_6 - \hat{X}_6$ by fitting the model

$$(Y_i - \hat{Y}_i) = \beta(X_{6i} - \hat{X}_{6i}) + \epsilon_i.$$

Note that no "β_0" term is required in this first-order model since we are using two sets of residuals whose sums are zero, and thus the line must pass through the origin. (If we did put a β_0 term in, we should find $b_0 = 0$, in any case.) For convenience the two sets of residuals used as data are extracted from Tables 1.2 and 4.1 and are given in Table 4.2. A plot of these residuals is shown in Figure 4.2.

Using a result of Chapter 1,

$$b = \frac{\sum (Y_i - \hat{Y}_i)(X_{6i} - \hat{X}_{6i})}{\sum (X_{6i} - \hat{X}_{6i})^2} = \frac{42.0821}{208.8523} = 0.2015.$$

Table 4.2   Deviations of $\hat{Y}_i = f(X_8)$ and
$\hat{X}_{6i} = f(X_8)$ from Y_i and X_{6i},
Respectively

Observation Number i	$Y_i - \hat{Y}_i$	$X_{6i} - \hat{X}_{6i}$
1	0.17	−0.87
2	−0.12	−1.08
3	1.34	1.96
4	−0.53	−0.01
5	0.55	1.08
6	0.80	2.45
7	−1.32	−8.44
8	1.00	3.64
9	−0.16	1.42
10	0.11	−0.06
11	−1.68	−0.47
12	0.87	−0.11
13	0.50	−0.14
14	−0.93	−1.73
15	1.05	2.55
16	−0.17	−0.39
17	1.20	2.01
18	0.07	2.40
19	−1.21	−8.60
20	1.20	3.56
21	−0.19	0.47
22	−0.51	0.96
23	−1.20	−0.53
24	−0.60	−0.94
25	−0.26	0.88

Thus the equation of the fitted line is

$$(\widehat{Y - \hat{Y}}) = 0.2015(X_6 - \hat{X}_6).$$

Within the parentheses we can substitute for $\hat{Y}$ and $\hat{X}_6$ as functions of X_8, and the large caret on the left-hand side can then be attached to Y to represent the overall fitted value $\hat{Y} = \hat{Y}(X_6, X_8)$ as follows:

$$[\hat{Y} - (13.6215 - 0.0798 X_8)] = 0.2015[X_6 - (22.1685 - 0.0367 X_8)]$$

or

$$\hat{Y} = 9.1545 - 0.0724 X_8 + 0.2015 X_6.$$

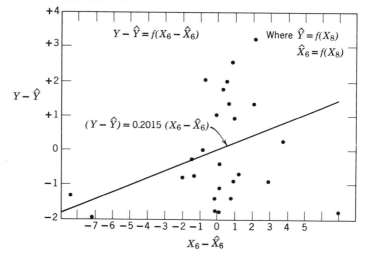

Figure 4.2

The previous result was

$$\hat{Y} = 9.1266 - 0.0724X_8 + 0.2029X_6.$$

In theory these two results are identical; practically, as we can see, slight discrepancies have occurred due to rounding errors. Ignoring rounding errors for the moment we shall show, geometrically, through a simple example, why the two methods should provide us with identical results. (The rest of this section could be omitted at first reading, if desired.)

Consider an example in which we have $n = 3$ observations of the response Y, namely Y_1, Y_2, and Y_3 taken at the three sets of conditions (X_1, Z_1), (X_2, Z_2), (X_3, Z_3). We can plot in three dimensions on axes labeled 1, 2, and 3, with origin at 0, the points $Y \equiv (Y_1, Y_2, Y_3)$, $X \equiv (X_1, X_2, X_3)$, and $Z \equiv (Z_1, Z_2, Z_3)$. The geometrical interpretation of regression is as follows. To regress Y on X we drop a perpendicular YP onto OX. The coordinates of the point P are the fitted values $\hat{Y}_1$, $\hat{Y}_2$, $\hat{Y}_3$. The length OP^2 is the sum of squares due to the regression, OY^2 is the total sum of squares, and YP^2 is the residual sum of squares. By Pythagoras, $OP^2 + YP^2 = OY^2$, which provides the analysis of variance breakup of the sums of squares (see Figure 4.3).

If we complete the parallelogram which has OY as diagonal and OP and PY as sides, we obtain the parallelogram $OP'YP$ as shown. Then the coordinates of P' are the values of the residuals from the regression of variable Y on variable X. In vector terms we could write

$$\overrightarrow{OP} + \overrightarrow{OP'} = \overrightarrow{OY},$$

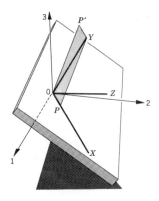

Figure 4.3

or, in "statistical" vector notation,

$$\hat{\mathbf{Y}} + (\mathbf{Y} - \hat{\mathbf{Y}}) = \mathbf{Y}.$$

This result is true in general for n dimensions. (The only reason we take $n = 3$ is so we can provide a diagram.)

Suppose we wish to regress variable Y on variables X and Z simultaneously. The lines OX and OZ define a plane in three dimensions. We drop a perpendicular YT onto this plane. Then the coordinates of the point T are the fitted values $\hat{Y}_1$, $\hat{Y}_2$, $\hat{Y}_3$ for *this* regression. OT^2 is the regression sum of squares, YT^2 is the residual sum of squares, and OY^2 is the total sum of squares. Again, by Pythagoras, $OY^2 = OT^2 + YT^2$ which, again, gives the sum of squares breakup we see in the analysis of variance table. Completion of the parallelogram $OT'YT$ with diagonal OY and sides OT, TY provides OT', the vector of residuals of this regression, and the coordinates of T' give the residuals $\{(Y_1 - \hat{Y}_1), (Y_2 - \hat{Y}_2), (Y_3 - \hat{Y}_3)\}$ of the regression of Y on X and Z simultaneously. Again, in vector notation,

$$\overrightarrow{OT} + \overrightarrow{OT'} = \overrightarrow{OY}$$

or, in "statistical" vector notation,

$$\hat{\mathbf{Y}} + (\mathbf{Y} - \hat{\mathbf{Y}}) = \mathbf{Y}$$

for this regression (see Figure 4.4).

As we saw in the numerical example above, the same final residuals should arise (ignoring rounding) if we do the regressions (1) Y on X, and

(2) Z on X, and then regress the residuals of (1) on the residuals of (2). That this is true can be seen geometrically as follows. Figure 4.5 shows three parallelograms in three dimensional space.

1. $OP'YP$ from the regression of Y on X,
2. $OQ'ZQ$ from the regression of Z on X, and
3. $OT'YT$ from the regression of Y on X and Z simultaneously.

Now the regression of the residuals of (1) onto the residuals of (2) is achieved by dropping the perpendicular from P' onto OQ'. Suppose the point of impact is R. Then a line through O parallel to RP' and of length RP' will be the residual vector of the two-step regression of Y on X and Z. However, the points O, Q', Z, P, Q, X, and T all lie in the plane π defined by OZ and OX. Thus so does the point R. Since $OP'YP$ is a parallelogram, and $P'R$ and YT are perpendicular to plane π, $P'R = YT$ in length. Since $TY = OT'$, it follows that $OT' = RP'$. But OT', RP', and TY are all parallel and perpendicular to plane π. Hence $OT'P'R$ is a parallelogram from which it follows that $\overrightarrow{OT'}$ is the vector of residuals from the two-step regression. Since it originally resulted from the regression of Y on Z and X together the two methods must be equivalent. Thus we can see that the planar regression of Y on X and Z together can be regarded as the totality of successive straight-line regressions of

1. Y on X,
2. Z on X, and
3. residuals of (1) on the residuals of (2).

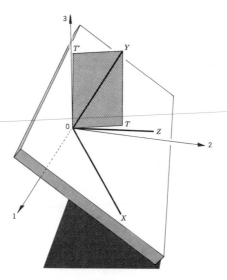

Figure 4.4

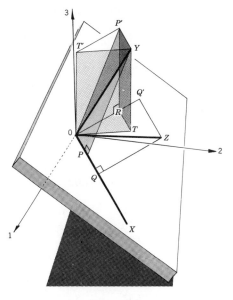

Figure 4.5

The same result is obtained if the roles of Z and X are interchanged. All linear regressions can be broken down into a series of simple regressions in this way. (For an application, see page 180.)

4.2. Examining the Regression Equation

How Useful Is the Equation, $Y = f(X_8, X_6)$?

Utilizing the work given in Chapters 1 and 2, we shall now consider the equation obtained for $\hat{Y}$ as a function of X_8 and X_6. We can calculate the residuals, using the fitted equation and the observed points. These residuals are shown in Table 4.3. The regression analysis of variance is as follows:

ANOVA

Source of Variation	Degrees of Freedom	Sum of Squares	Mean Square	F
Total (uncorrected)	25	2284.1102		
Mean (b_0)	1	2220.2944		
Total (corrected)	24	63.8158		
Regression $\mid b_0$	2	54.1871	27.0936	61.8999
Residual	22	9.6287	0.4377	

On the basis of an α risk of 0.05, the least squares equation

$$\hat{Y} = 9.1266 - 0.0724X_8 + 0.2029X_6$$

is a good predictor; the calculated $F = 61.8999$ for regression is greater than the tabulated $F(2, 22, 0.95) = 3.44$.

Table 4.3

Obs. No.	X_8	X_6	Y	$\hat{Y}$	Residual
1	35.3	20	10.98	10.63	0.35
2	29.7	20	11.13	11.03	0.10
3	30.8	23	12.51	11.56	0.95
4	58.8	20	8.40	8.93	−0.53
5	61.4	21	9.27	8.94	0.33
6	71.3	22	8.73	8.43	0.30
7	74.4	11	6.36	5.97	0.39
8	76.7	23	8.50	8.24	0.26
9	70.7	21	7.82	8.27	−0.45
10	57.5	20	9.14	9.02	0.12
11	46.4	20	8.24	9.82	−1.58
12	28.9	21	12.19	11.29	0.90
13	28.1	21	11.88	11.35	0.53
14	39.1	19	9.57	10.15	−0.58
15	46.8	23	10.94	10.40	0.54
16	48.5	20	9.58	9.67	−0.09
17	59.3	22	10.09	9.30	0.79
18	70.0	22	8.11	8.52	−0.41
19	70.0	11	6.83	6.29	0.54
20	74.5	23	8.88	8.40	0.48
21	72.1	20	7.68	7.96	−0.28
22	58.1	21	8.47	9.18	−0.71
23	44.6	20	8.86	9.96	−1.10
24	33.4	20	10.36	10.77	−0.41
25	28.6	22	11.08	11.52	−0.44
			235.60		$\Sigma(Y_i - \hat{Y}_i) = 0$
			$\bar{Y} = 9.424$		$\Sigma(Y_i - \hat{Y}_i)^2 = 9.6432$

A graph of the observed values of Y and the fitted $\hat{Y}$'s is shown in Figure 4.6. The graph indicates that the fitted model is a good predictor of monthly steam usage. However, has the addition of X_6 to the model been useful?

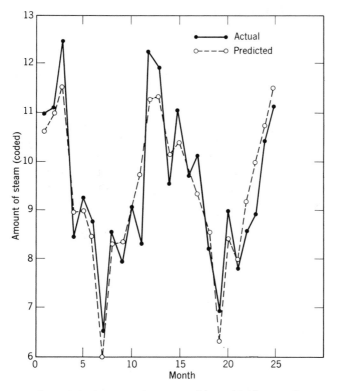

Figure 4.6   Amount of steam used in a plant by month.

What Has Been Accomplished by the Addition of a Second Independent Variable (Namely X_6)?

There are several useful criteria which can be applied to answer this question, and we now discuss them.

THE SQUARE OF THE MULTIPLE CORRELATION COEFFICIENT, R^2.   The square of the multiple correlation coefficient R^2 is defined as (see Eq. 2.6.11)

$$R^2 = \frac{\text{Sum of squares due to regression} \mid b_0}{\text{Total (corrected) sum of squares}}.$$

It is often stated as a percentage, $100R^2$. The larger it is, the better the fitted equation explains the variation in the data. We can compare the value of R^2 at each stage of the regression problem:

STEP 1. $Y = f(X_8)$.

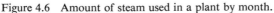

	Regression equation	$100R^2$
	$\hat{Y} = 13.6215 - 0.0798 X_8$	71.44%

STEP 2. $Y = f(X_8, X_6)$.

Regression equation	$100R^2$
$\hat{Y} = 9.1266 - 0.0724X_8 + 0.2029X_6$	84.89%

Thus, we see a substantial increase in R^2. However, this statistic must be used with caution, since one can always make $R^2 = 1$ as described on page 63.

In addition, given that the number of observations is much greater than the number of potential X variables under consideration, the addition of a new variable will always increase R^2 but it will not necessarily increase the precision of the estimate of the response. This is because the reduction in the residual sum of squares may be less than the original residual mean square. Since one degree of freedom is removed from the residual degrees of freedom as well, the resulting mean square may get larger. An example of this can be seen in Appendix B (pp. 387, 395) which we have not yet discussed. We can make the following comparison:

			Residual		
Page	R^2	Variables in Regression Model	Sum of Squares	Degrees of Freedom	Mean Square
387	98.23	1, 2, 3	48.11	9	5.35
395	98.24	1, 2, 3, 4	47.86	8	5.98

We see that although an extra variable has been included in the regression model, the residual mean square has increased since the extra variable produced a residual sum of squares reduction of $48.11 - 47.86 = 0.25 < 5.35$ for the loss of one degree of freedom. The value of R^2 has increased slightly, however.

THE STANDARD ERROR OF ESTIMATE, s.   The residual mean square s^2 is an estimate of $\sigma^2_{Y \cdot X}$, the variance about the regression. Before and after adding a variable to the model, we can check

$$s = \sqrt{\text{residual mean square.}}$$

Examination of this statistic indicates that the smaller it is the better, that is, the more precise will be the predictions. Since s can be *made* zero by including enough parameters in the model—just as R^2 can be made unity—this criterion must also be used cautiously. Provided there are few repeats and many degrees of freedom for error remaining, reduction of s is desirable. In our example at Step 1,

$$s = \sqrt{0.7926} = 0.89.$$

At Step 2,
$$s = \sqrt{0.4377} = 0.66.$$

Thus, the addition of X_6 has decreased s and improved the precision of estimation.

THE STANDARD ERROR OF ESTIMATE s, AS A PERCENTAGE OF THE MEAN RESPONSE.   Another way of looking at the decrease in s is to consider it in relation to the response. In our example, at Step 1, s as a percentage of mean $\bar{Y}$ is

$$0.89/9.424 = 9.44\%.$$

At Step 2, s as a percentage of mean $\bar{Y}$ is

$$0.66/9.424 = 7.00\%.$$

Thus, the addition of X_6 has reduced the standard error of estimate down to about 7% of the mean response. Whether this level of precision is satisfactory or not is a matter for the experimenter to decide, on the basis of his prior knowledge and personal feelings.

THE SEQUENTIAL F-TEST CRITERION (SHOWING THE ADDITIONAL CONTRIBUTION OF X_6 GIVEN THAT X_8 IS ALREADY IN THE EQUATION).   This method of assessing the value of X_6 as a variable added to $Y = f(X_8)$ consists in breaking down the sum of squares due to regression into two parts as follows:

ANOVA

Source of Variation	Degrees of Freedom	Sum of Squares	Mean Square	F
Total (uncorrected)	25	2284.1102		
Mean (b_0)	1	2220.2944		
Total (corrected)	24	63.8158		
Regression $\mid b_0$	2	54.1871	27.0936	61.8999
due to $b_8 \mid b_0$	1	45.5924	45.5924	104.1636
due to $b_6 \mid b_8, b_0$	1	8.5947	8.5947	19.6361
Residual	22	9.6287	0.4377	

Since 19.6361 exceeds $F(1,22, 0.95) = 4.30$, the addition of the variable X_6 has been worthwhile. This F-test is usually called the "sequential F-test" (see section 2.9).

THE PARTIAL F TEST CRITERION (see Section 2.9).   Another way of assessing the value of X_6 is to consider the order of the two variables in the least squares procedure. For example, the following questions could be asked:

1. If we had put X_6 into the equation first what would its contribution have been?
2. Given that X_6 was used first, what contribution does X_8 make when added to regression?

These questions are answered by performing the calculations shown above, but in reverse order. The results are as follows:

ANOVA

Source of Variation	Degrees of Freedom	Sum of Squares	Mean Square	F
Total (uncorrected)	25	2284.1102		
Mean (b_0)	1	2220.2944		
Total (corrected)	24	63.8158		
Regression | b_0	2	54.1871	27.0936	61.8999
due to b_6 | b_0	1	18.3424	18.3424	41.9063
due to b_8 | b_6, b_0	1	35.8447	35.8447	81.8933
Residual	22	9.6287	0.4377	

Note that the contribution of X_6 above is more important than is its contribution after X_8 has been introduced. Note also that this is reflected in the observed value of F for X_8 in the two steps; that is,

Step 1   104.1636,

Step 2   81.8933.

However, X_8 is still the more important variable in both cases, since its contribution in reducing the residual sum of squares is the larger, regardless of the order of introduction of the variables.

Standard Error of b_i

Using the result given in Section 2.6, the variance-covariance matrix of $\mathbf{b}$ is $(\mathbf{X'X})^{-1}\sigma^2$.

Thus, variance of $b_i = V(b_i) = c_{ii}\sigma^2$, where c_{ii} is the diagonal element in $(\mathbf{X'X})^{-1}$ corresponding to the ith variable.

The covariance of b_i, $b_j = c_{ij}\sigma^2$, where c_{ij} is the off-diagonal element in $(\mathbf{X'X})^{-1}$ corresponding to the intersection of the ith row and jth column, or j^{th} row and i^{th} column, since $(\mathbf{X'X})^{-1}$ is symmetric.

Thus the s.e. of b_i is $\sigma\sqrt{c_{ii}}$. For example, using figures from pages 106 and 120 the estimated standard error of b_8 is obtained as follows:

$$\text{est. var } b_8 = s^2 c_{88}$$
$$= (0.4377)(0.146207 \times 10^{-3})$$
$$= 0.639948 \times 10^{-4}.$$

Thus   est. s.e. $b_8 = \sqrt{\text{var } b_8} = \sqrt{0.639948 \times 10^{-4}} = 0.008000.$

Confidence Limits for the True Mean Value of Y, Given a Specific Set of X's

The predicted value $\hat{Y} = b_0 + b_1 X_1 + \cdots + b_p X_p$ is an estimate of
$$E(Y) = \beta_0 + \beta_1 X_1 + \cdots + \beta_p X_p.$$
The variance of $\hat{Y}$, $V[b_0 + b_1 X_1 + \cdots + b_p X_p]$, is
$$V(b_0) + X_1{}^2 V(b_1) + \cdots + X_p{}^2 V(b_p) + 2X_1 \text{ covar } (b_0, b_1) + \cdots$$
$$+ 2X_{p-1}X_p \text{ covar } (b_{p-1}, b_p).$$

This expression can be written very conveniently in matrix notation as follows, where $\mathbf{C} = (\mathbf{X'X})^{-1}$.

$$V(\hat{Y}) = \sigma^2 (\mathbf{X_0'CX_0})$$

$$= \sigma^2 [1 \quad X_1 \quad \cdots \quad X_p] \begin{bmatrix} c_{00} & c_{01} & \cdots & c_{0p} \\ c_{10} & c_{11} & \cdots & c_{1p} \\ \cdot & \cdot & & \cdot \\ \cdot & \cdot & & \cdot \\ \cdot & \cdot & & \cdot \\ c_{p1} & & c_{pp} \end{bmatrix} \begin{bmatrix} 1 \\ X_1 \\ \cdot \\ \cdot \\ \cdot \\ X_p \end{bmatrix}$$

Thus, the $1 - \alpha$ confidence limits on the true mean value of Y at $\mathbf{X_0}$ are given by
$$\hat{Y} \pm t\{(n - p - 1), 1 - \tfrac{1}{2}\alpha\} \cdot s\sqrt{\mathbf{X_0'CX_0}}.$$

For example, the variance of $\hat{Y}$ for the point in the X space ($X_8 = 32$, $X_6 = 22$) is obtained as follows:

$$\text{var}(\hat{Y}) = s^2 (\mathbf{X_0'CX_0})$$
$$= (0.4377)(1, 32, 22)$$
$$\times \begin{bmatrix} 2.778747 & -0.011242 & -0.106098 \\ -0.011242 & 0.146207 \times 10^{-3} & 0.175467 \times 10^{-3} \\ -0.106098 & 0.175467 \times 10^{-3} & 0.478599 \times 10^{-2} \end{bmatrix} \begin{bmatrix} 1 \\ 32 \\ 22 \end{bmatrix}$$
$$= (0.4377)(0.104140) = 0.045582.$$

The 95% confidence limits on the true mean value of Y at $X_8 = 32$, $X_6 = 22$ are given by

$$\hat{Y} \pm t(22, 0.975) \cdot s\sqrt{X_0'CX_0} = 11.2736 \pm (2.074)(0.213499)$$
$$= 11.2736 \pm 0.4418$$
$$= 10.8318, 11.7154$$

These limits are interpreted as follows. Suppose repeated samples of Y's are taken of the same size each time and at the same fixed values of (X_8, X_6) as were used to determine the fitted equation obtained above. Then of all the 95% confidence intervals constructed for the mean value of Y for $X_8 = 32$, $X_6 = 22$, 95% of these intervals will contain the true mean value of Y at $X_8 = 32$, $X_6 = 22$. From a practical point of view we can say that there is a 0.95 probability that the statement, the true mean value of Y at $X_8 = 32$, $X_6 = 22$ lies between 10.8318 and 11.7154, is correct.

Confidence Limits for the Mean of g Observations Given a Specific Set of X's

These limits are calculated from

$$\hat{Y} \pm t(v, 1 - \tfrac{1}{2}\alpha) \cdot s\sqrt{1/g + X_0'CX_0}.$$

For example, the 95% confidence limits for an individual observation for the point $(X_8 = 32, X_6 = 22)$ are

$$\hat{Y} \pm t(22, 0.975) \cdot s\sqrt{1 + X_0'CX_0}$$
$$= 11.2736 \pm (2.074)(0.661589)\sqrt{1 + 0.10413981}$$
$$= 11.2736 \pm (2.074)(0.661589)(1.050781)$$
$$= 11.2736 \pm 1.4418$$
$$= 9.8318, 12.7154$$

Examining the Residuals

The residuals shown in Table 4.3, page 116, could be examined to see if they provide any indication that the model is inadequate. We leave this as an exercise except for the following comments:
1. Residual versus $\hat{Y}$ (Figure 4.7). No unusual behavior is indicated by this plot.
2. Residual versus Y (Figure 4.8). There is some evidence that the larger observed values were underpredicted by the model; that is, six out of seven largest values of Y have positive residuals. This means that the model should be amended to provide better prediction at higher steam

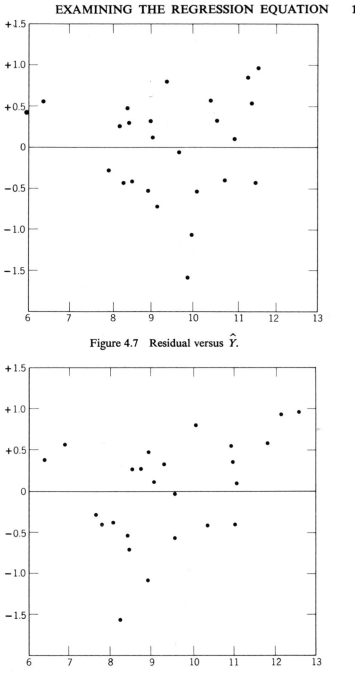

Figure 4.7   Residual versus $\hat{Y}$.

Figure 4.8   Residual versus Y.

levels. Although we do not follow up this point here, additional effort would normally be made to find one or more independent variables which might be added to the model.
3. The runs test indicates no evidence of time dependent nonrandomness.

EXERCISES

A. *Multiple Regression Problem*

Data

X_0	X_1	X_2	Y
1	1	8	6
1	4	2	8
1	9	−8	1
1	11	−10	0
1	3	6	5
1	8	−6	3
1	5	0	2
1	10	−12	−4
1	2	4	10
1	7	−2	−3
1	6	−4	5

Requirements

1. Using least squares procedures, estimate the β's in the model:

$$Y = \beta_0 X_0 + \beta_1 X_1 + \beta_2 X_2 + \epsilon$$

(*Hint:* Use the normal equations for ease of computation.)
2. Write out the Analysis of Variance table.
3. Using $\alpha = 0.05$, test to determine if the overall regression is statistically significant.
4. Calculate the square of the multiple correlation coefficient, namely R^2. What portion of the total variation is explained by the two variables?
5. The inverse of the $\mathbf{X'X}$ matrix for this problem is as follows:

$$\begin{bmatrix} 4.3705 & -0.8495 & -0.4086 \\ -0.8495 & 0.1690 & 0.0822 \\ -0.4086 & 0.0822 & 0.0422 \end{bmatrix}$$

Using the results of the Analysis of Variance table with this matrix, calculate the following:
   a. Variance of b_1.
   b. Variance of b_2.
   c. The variance of the predicted value of Y for the point $X_1 = 3$, $X_2 = 5$.
6. How useful is the regression using X_1 alone? What does X_2 contribute, given that X_1 is already in the regression?
7. How useful is the regression using X_2 alone? What does X_1 contribute, given that X_2 is already in the regression?
8. What are your conclusions?

B. The table below gives twelve sets of observations on three variables X, Y, Z. Find the regression plane of X on Y and Z—that is, the linear combination of Y and Z that best predicts the value of X when only Y and Z are given. By constructing an analysis of variance table for X, or otherwise, test whether it is advantageous to include both Y and Z in the prediction formula.

X	Y	Z
1.52	98	77
1.41	76	139
1.16	58	179
1.45	94	95
1.24	73	142
1.21	57	186
1.63	97	82
1.38	91	100
1.37	79	125
1.36	92	96
1.40	92	99
1.03	54	190

(Cambridge Diploma, 1949)

C. The data below are selected from a much larger body of data referring to candidates for the General Certificate of Education who were being considered for a special award. Here, Y denotes the candidate's total mark, out of 1000, in the G.C.E. examination. Of this mark the subjects selected by the candidate account for a maximum of 800; the remainder, with a maximum of 200, is the mark in the compulsory papers—"General" and "Use of English"—this mark is shown as X_1. X_2 denotes the candidate's mark, out of 100, in the compulsory School Certificate English Language paper taken on a previous occasion.

Compute the multiple regression of Y on X_1 and X_2, and make the necessary tests to enable you to comment intelligently on the extent to which current performance in the compulsory papers may be used to predict aggregate performance in the G.C.E. examination, and on whether previous performance in School Certificate English Language has any predictive value independently of what has already emerged from the current performance in the compulsory papers.

Candidate	Y	X_1	X_2	Candidate	Y	X_1	X_2
1	476	111	68	9	645	117	59
2	457	92	46	10	556	94	97
3	540	90	50	11	634	130	57
4	551	107	59	12	637	118	51
5	575	98	50	13	390	91	44
6	698	150	66	14	562	118	61
7	545	118	54	15	560	109	66
8	574	110	51				

(Cambridge Diploma, 1953)

D. Eight runs were made at various conditions of saturation (X_1) and trans-isomers (X_2). The response, SCI, is listed below as Y for the corresponding levels of X_1 and X_2.

Y	X_1	X_2
66.0	38	47.5
43.0	41	21.3
36.0	34	36.5
23.0	35	18.0
22.0	31	29.5
14.0	34	14.2
12.0	29	21.0
7.6	32	10.0

Sums and uncorrected sums of squares and cross products are given below.

$$\Sigma Y = 223.6 \qquad \Sigma X_1 = 274 \qquad \Sigma X_2 = 198$$

$$\Sigma Y^2 = 8911.76 \qquad \Sigma X_1^2 = 9488 \qquad \Sigma X_2^2 = 5979.08$$

$$\Sigma X_1 Y = 8049.2 \qquad \Sigma X_2 Y = 6954.7 \qquad \Sigma X_1 X_2 = 6875.6$$

1. Fit the model $Y = \beta_0 + \beta_1 X_1 + \beta_2 X_2 + \epsilon$.
2. Is the overall regression significant? (Use $\alpha = 0.05$.)
3. How much of the variation in Y is explained by X_1 and X_2?

E. The effect of sealer plate temperature and sealer plate clearance in a soap wrapping machine affects the percentage of wrapped bars which pass inspection. Some data on these variables were collected and are shown as follows:

Sealer Plate Clearance	Sealer Plate Temperature	% Sealed Properly
X_1	X_2	Y
130	190	35.0
174	176	81.7
134	205	42.5
191	210	98.3
165	230	52.7
194	192	82.0
143	220	34.5
186	235	95.4
139	240	56.7
188	230	84.4
175	200	94.3
156	218	44.3
190	220	83.3
178	210	91.4
132	208	43.5
148	225	51.7

Requirements

1. Assume a linear model $Y = \beta_0 + \beta_1 X_1 + \beta_2 X_2 + \epsilon$ and determine least squares estimates of β_0, β_1, and β_2.
2. Is the overall regression significant? (Use $\alpha = 0.05$.)
3. Is one of the two variables more useful than the other in predicting the percentage sealed properly?
4. What recommendations would you make concerning the operation of the wrapping machine?

F.   Using the 17 observations given below:

1. Fit the model $Y = \beta_0 + \beta_1 X_1 + \beta_2 X_2 + \epsilon$.
2. Test for lack of fit, using pure error.
3. Examine the residuals.
4. Assess the value of including each of the variables X_1 and X_2 in the regression model.

X_1	X_2	Y
17	42	90
19	45	71, 76
20	29	63, 63, 80, 80
21	93	80, 64, 82, 66
25	34	75, 82
27	98	99
28	9	73
30	73	67, 74

CHAPTER 5

MORE COMPLICATED MODELS

5.0. Introduction

In Chapter 1, the first-order linear model with one independent variable was discussed in some detail. The concept of model adequacy and a statistical test for lack of fit were also presented. In Chapter 2, the mathematical analysis of Chapter 1 was put into matrix format so that the extension from a first-order model with one independent variable to a general model, linear in the parameters to be estimated and containing several independent variables, could be made more efficiently. The regression formulation in the general case was provided in matrix form in the later sections of Chapter 2. The work in these sections is applied throughout the rest of the book. In Chapter 4, a first-order model with two independent variables was discussed, both algebraically and geo- metrically. Several criteria for examining a multiple regression equation were given, and formulae for confidence intervals for the β's and for predicted values of Y were shown.

Up to this point the models discussed have been first-order, linear, with one or two independent variables. In this chapter, examples of more complicated models will be given. Some of these models involve trans- formations on the variables and the use of *dummy variables*. We shall also discuss, in this chapter, some aspects of preparing the data for the regression analysis.

We can write the most general type of linear model in variables X_1, $X_2, \ldots, X_k$ in the form

$$Y = \beta_0 Z_0 + \beta_1 Z_1 + \beta_2 Z_2 + \cdots + \beta_p Z_p + \epsilon. \qquad (5.0.1)$$

$Z_0 = 1$ is a dummy variable which is always unity and will in general not be shown. However it is sometimes mathematically convenient to have a Z_0 in the model. For example, if

$$(Z_{1i}, Z_{2i}, \ldots, Z_{pi}) \qquad i = 1, 2, \ldots, n$$

are n settings of the variables $Z_j, j = 1, 2, \ldots, p$ corresponding to observations Y_i, $i = 1, 2, \ldots, n$, then when $j \neq 0$, and $Z_{0i} = 1$,

$$\sum_{i=1}^{n} Z_{ji} = \sum_{i=1}^{n} Z_{ji}Z_{0i}$$

and thus can be represented by the general cross-product expression

$$\sum_{i=1}^{n} Z_{ji}Z_{li}$$

if the normal equations are written out. Note that $\sum_{i=1}^{n} Z_{0i}^2 = n$. Each Z_j, $j = 1, 2, \ldots, p$ is a general function of $X_1, X_2, \ldots, X_k$,

$$Z_j = Z_j(X_1, X_2, \ldots, X_k)$$

and can take any form. In some examples, each Z_j may involve only one X-variable.

Any model that can be written, perhaps after rearrangement or transformation, in the form of Eq. (5.0.1) can be analyzed by the general methods given in Sections 2.6 to 2.12. We now provide some specific examples of models that can be treated by these methods and relate them to the general form of Eq. (5.0.1).

5.1. Polynomial Models of Various Orders in the X_j

First-Order Models

1. If $p = 1$ and $Z_1 = X$ in Eq. (5.0.1), we obtain the simple first-order model with one independent variable:

$$Y = \beta_0 + \beta_1 X + \epsilon. \qquad (5.1.1)$$

2. If $p = k$ and $Z_j = X_j$, we obtain a first-order model with k independent variables:

$$Y = \beta_0 + \beta_1 X_1 + \beta_2 X_2 + \cdots + \beta_k X_k + \epsilon. \qquad (5.1.2)$$

Second-Order Models

1. If $p = 2$, $Z_1 = X$, $Z_2 = X^2$ and $\beta_2 = \beta_{11}$, we obtain a second-order model with one independent variable:

$$Y = \beta_0 + \beta_1 X + \beta_{11} X^2 + \epsilon. \qquad (5.1.3)$$

2. If $p = 5$, $Z_1 = X_1$, $Z_2 = X_2$, $Z_3 = X_1^2$, $Z_4 = X_2^2$, $Z_5 = X_1 X_2$ $\beta_3 = \beta_{11}$, $\beta_4 = \beta_{22}$, and $\beta_5 = \beta_{12}$ we obtain a second-order model with two independent variables:

$$Y = \beta_0 + \beta_1 X_1 + \beta_2 X_2 + \beta_{11} X_1^2 + \beta_{22} X_2^2 + \beta_{12} X_1 X_2 + \epsilon. \quad (5.1.4)$$

A full second-order model in k variables can be obtained in similar fashion when $p = k + k + \frac{1}{2}k(k - 1) = \frac{1}{2}(k^2 + 3k)$. Second-order models are used particularly in response surface studies where it is desired to graduate, or approximate to, the characteristics of some unknown response surface by a polynomial of low order. Note that all possible second-order terms are in the model. This is sensible because omission of terms implies possession of definite knowledge that certain types of surface (those which cannot be represented *without* the omitted terms) cannot possibly occur. Knowledge of this sort is not often available. When it is, it would usually enable a more theoretically based study to be made.

Third-Order Models

1. If $p = 3$, $Z_1 = X$, $Z_2 = X^2$, $Z_3 = X^3$, $\beta_2 = \beta_{11}$, and $\beta_3 = \beta_{111}$, we obtain a third-order model with one independent variable:

$$Y = \beta_0 + \beta_1 X + \beta_{11} X^2 + \beta_{111} X^3 + \epsilon. \quad (5.1.5)$$

2. If $p = 9$ and proper identification of the β_i and Z_i is made (we omit the details now since the examples above should have made the idea clear), the model (5.0.1) can represent a third-order model with two independent variables given by

$$Y = \beta_0 + \beta_1 X_1 + \beta_2 X_2 + \beta_{11} X_1^2 + \beta_{12} X_1 X_2 + \beta_{22} X_2^2$$
$$+ \beta_{111} X_1^3 + \beta_{112} X_1^2 X_2 + \beta_{122} X_1 X_2^2 + \beta_{222} X_2^3 + \epsilon. \quad (5.1.6)$$

The general third-order model for k factors $X_1, X_2, \ldots, X_k$ can be obtained similarly. Third-order models are also used in response surface work though much less frequently than second-order models. Note the method of labeling the β's. This may seem confusing at first but it is done to enable the coefficients to be readily associated with their corresponding powers of the X's. For example, $X_1 X_2^2 = X_1 X_2 X_2$ has a coefficient β_{122}, and so on. A similar notation is used above for second-order models and is standard in response surface work.

Models of *any* desired order can be represented by Eq. (5.0.1) by continuing the process illustrated above.

5.2. Models Involving Transformations Other Than Integer Powers

The polynomial models of Section 5.1 involved powers, and cross products of powers, of the independent variables $X_1, X_2, \ldots, X_k$. Here we provide examples of other types of transformations which are often useful in forming regression models.

Models Obtained by Transforming X_j Only

THE RECIPROCAL TRANSFORMATION.   If in Eq. (5.0.1) we take $p = 2$, $Z_1 = 1/X_1$, $Z_2 = 1/X_2$, we obtain the model

$$Y = \beta_0 + \beta_1\left(\frac{1}{X_1}\right) + \beta_2\left(\frac{1}{X_2}\right) + \epsilon. \qquad (5.2.1)$$

THE LOGARITHMIC TRANSFORMATION.   By taking $p = 2$, $Z_1 = \ln X_1$, $Z_2 = \ln X_2$, Eq. (5.0.1) can represent

$$Y = \beta_0 + \beta_1 \ln X_1 + \beta_2 \ln X_2 + \epsilon. \qquad (5.2.2)$$

THE SQUARE ROOT TRANSFORMATION.   For example,

$$Y = \beta_0 + \beta_1 X_1^{\frac{1}{2}} + \beta_2 X_2^{\frac{1}{2}} + \epsilon. \qquad (5.2.3)$$

Clearly there are many possible transformations, and models can be postulated which contain few or many such terms. Several different transformations could occur in the same model, of course. The choice of what, if any transformation to make is often difficult to decide. The choice would often be made on the basis of previous knowledge of the variables under study. The purpose of making transformations of this type is to be able to use a regression model of simple form in the transformed variables, rather than a more complicated one in the original variables.

An iterative procedure for deciding on transformations of single variables indicated by the data is given by G. E. P. Box and P. W. Tidwell in "Transformation of the independent variables," *Technometrics*, **4**, 1962, 531–550. Transformations could also involve several X_j variables simultaneously, for example, $Z_1 = X_1^{\frac{1}{2}} \ln X_2$. Transformations of this type are sometimes suggested by the form of the fitted equation in untransformed variables. A simple example of this is given in "An analysis of transformations" (which discusses, principally, transformations on the dependent variable, however) by G. E. P. Box and D. R. Cox, *Journal of the Royal Statistical Society, Series B*, **26**, 1964, 211–243, discussion 244–252 (see pages 222–223). Since it involves a model of a type not yet mentioned, we give the example on page 133.

Suitable transformations of the independent variable are also sometimes suggested by plotting the data in various ways. See, for example, "Fitting curves to data" by A. E. Hoerl in the *Chemical Business Handbook*, McGraw-Hill, New York, 1954. Other references are "A quick method for choosing a transformation" by J. L. Dolby, *Technometrics*, **5**, 1963, 317–325 and "On the comparative anatomy of transformations" by J. W. Tukey, *Annals of Mathematical Statistics*, **28**, 1957, 602–632.

Basic information on the reasons for some transformations which can be made on the dependent variable can be found in the paper "The use of transformations" by M. S. Bartlett, *Biometrics*, **3**, 1947, 39–52.

Nonlinear Models That Are Intrinsically Linear

We can divide nonlinear models (i.e., nonlinear in the parameters to be estimated) into two types which can be called *intrinsically linear* and *intrinsically nonlinear* models. If a model is intrinsically linear it can be expressed, by suitable transformation of the variables, in the standard linear model form of Eq. (5.0.1). If a nonlinear model cannot be expressed in this form then it is intrinsically nonlinear (i.e. nonlinear, period). Our interest in this section is in models which are intrinsically linear and so can be handled by the matrix procedures described in Chapter 2. The transformations necessary usually involve both dependent and independent variables. Some examples follow:

THE MULTIPLICATIVE MODEL.

$$Y = \alpha X_1^{\beta} X_2^{\gamma} X_3^{\delta} \epsilon \tag{5.2.4}$$

where α, β, γ, and δ are unknown parameters, ϵ is a multiplicative random error. Taking logarithms to the base e in Eq. (5.2.4) converts the model into the linear form

$$\ln Y = \ln \alpha + \beta \ln X_1 + \gamma \ln X_2 + \delta \ln X_3 + \ln \epsilon. \tag{5.2.5}$$

The transformed model (5.2.5) is now in the form of Eq. (5.0.1) and so can be handled by the standard linear regression procedures described in Chapter 2. It must be emphasized, however, that the requirements for valid tests of significance and confidence interval estimates are now $\ln \epsilon \sim N(0, I\sigma^2)$ (rather than ϵ). Thus, the experimenter must be prepared to check this assumption carefully by examining the residuals from the fitted equation as described in Chapter 3.

An alternative model which is often considered applicable in this case is

$$Y = \alpha X_1^{\beta} X_2^{\gamma} X_3^{\delta} + \epsilon. \tag{5.2.4A}$$

The general linear regression procedures given in Chapter 2 do not hold for this model since it is intrinsically nonlinear. Application of the least squares method will require that iterative procedures be used to estimate α, β, γ, and δ. These procedures are discussed briefly in Chapter 10.

We spoke of an example of transforming the independent variables, given by G. E. P. Box and D. R. Cox. This involved a model of the type of Eq. (5.2.5) and estimates b, c, and d of β, γ, and δ were obtained from data as follows:

$$b = 4.96 \pm 0.20, \qquad c = -5.27 \pm 0.30, \qquad d = -3.15 \pm 0.30.$$

The figures following the plus and minus signs are standard errors. These figures suggested that the assumption $\beta = 5 = -\gamma$, $\delta = -3$ might not be unreasonable. If we set $Z = X_1/X_2$ so that $\ln Z = \ln X_1 - \ln X_2$ and reconvert to Eq. (5.2.4) we obtain a model, suggested by the analysis, of

$$Y \propto Z^{+5} X_3^{-3} \epsilon.$$

Box and Cox remark that this model "fits the data remarkably well."

THE EXPONENTIAL MODEL.

$$Y = e^{\beta_0 + \beta_1 X_1 + \beta_2 X_2} \cdot \epsilon. \tag{5.2.6}$$

Taking natural logarithms of both sides,

$$\ln Y = \beta_0 + \beta_1 X_1 + \beta_2 X_2 + \ln \epsilon. \tag{5.2.7}$$

A RECIPROCAL MODEL.

$$Y = \frac{1}{\beta_0 + \beta_1 X_1 + \beta_2 X_2 + \epsilon}. \tag{5.2.8}$$

Taking reciprocals on both sides,

$$\frac{1}{Y} = \beta_0 + \beta_1 X_1 + \beta_2 X_2 + \epsilon. \tag{5.2.9}$$

In this case, the experimenter will be using the reciprocal of the dependent variable as the response variable.

A MORE COMPLICATED EXPONENTIAL MODEL.

$$Y = \frac{1}{1 + e^{\beta_0 + \beta_1 X_1 + \beta_2 X_2 + \epsilon}}. \tag{5.2.10}$$

Taking reciprocals, subtracting 1, and then taking the natural logarithm of both sides,

$$\ln\left(\frac{1}{Y} - 1\right) = \beta_0 + \beta_1 X_1 + \beta_2 X_2 + \epsilon. \tag{5.2.11}$$

This is an example of an iterated transformation on the dependent variable in order to reduce a complicated nonlinear model to a linear one.

In all cases, where the models are transformed as in these examples, the least squares analysis is applied to the transformed model of Eq. (5.0.1), and so the estimated coefficients are "least squares estimates" only as far as the transformed model is concerned, of course.

Two other examples of models which are intrinsically nonlinear are

$$Y = \beta_0 + \beta_1 e^{-\beta_2 X} + \epsilon$$

and

$$Y = \beta_0 + \beta_1 X + \beta_2(\beta_3)^X + \epsilon.$$

(The latter model is discussed in "A method of fitting the regression curve $E(y) = \alpha + \delta x + \beta \rho^x$" by B. K. Shah and C. G. Khatri, *Technometrics*, 7, 1965, 59–65.)

Summary

The transformations discussed in this section are but a few of the many currently being used to reduce complex models to linear ones. When, as we assume here, the independent variables are not subject to error, there are no problems in transforming them. However, for transformations on the dependent variable, Y, one must be especially careful to check that the least squares assumptions (errors independent, $N(0, \sigma^2)$) are not violated by making the transformation. Often one can avoid transforming the dependent variable by searching for suitable transformations in the X's (as in the paper, mentioned earlier, by G. E. P. Box and P. W. Tidwell "Transformation of the independent variables," *Technometrics*, **4**, 1962, 531–550).

5.3. The Use of "Dummy" Variables in Multiple Regression

The General Concept of a "Dummy" Variable and Its Construction

The variables considered in regression equations usually can take values over some continuous range. Occasionally we must introduce a factor which has two or more distinct levels. For example, data may arise from three machines, or two factories, or six operators. In such a case we cannot set up a continuous scale for the variable "machine" or "factory" or "operator." We must assign to these variables some levels in order to take account of the fact that the various machines or factories or operators may have separate deterministic effects on the response. Variables of this sort are usually called *dummy variables*. They are usually (but not always) unrelated to any physical levels that might exist in the factors themselves.

One example of a dummy variable is found in the attachment of a variable X_0 (whose value is always unity) to the term β_0 in a regression model. The X_0 is unnecessary but provides a notational convenience at times. Other dummy variables are somewhat more than a mere convenience as we shall see.

Suppose we wish to introduce into a model the idea that there are two types of machines (types A and B, say) that produce different levels of response, in addition to the variation which occurs due to other variables. One way of doing this is to add to the model a dummy variable Z and a regression coefficient α (say) so that an additional term αZ appears in the the model. The coefficient α must be estimated at the same time the β's are estimated. Values can be assigned to Z as follows:

$$Z = 0 \text{ if the observation is from machine } A,$$

$$Z = 1 \text{ if the observation is from machine } B.$$

Any two distinct values of Z would actually be suitable, though the above is usually best. However, other assignments are sometimes convenient; for example, suppose of a total of n observations, n_1 comes from type A machines and $n_2 = n - n_1$ from type B machines. If we choose levels

$$Z = \frac{-n_2}{\sqrt{n_1 n_2 (n_1 + n_2)}} \text{ for machine } A,$$

$$Z = \frac{n_1}{\sqrt{n_1 n_2 (n_1 + n_2)}} \text{ for machine } B,$$

it will be found that the corresponding column of the $\mathbf{X}$ matrix is orthogonal to the "β_0 column" and has sum of squares unity, which may be convenient.

Note. If it were desired to take account of three distinct machines, two dummy variables Z_1 and Z_2 would be required. Then we should set

$$(Z_1, Z_2) = (1, 0) \quad \text{for machine } A$$

$$(0, 1) \quad \text{for machine } B$$

$$(0, 0) \quad \text{for machine } C$$

and the model would include extra terms $\alpha_1 Z_1 + \alpha_2 Z_2$, with coefficients α_1, α_2 to be estimated. Again, many different allocations of levels are possible. If desired, columns which are orthogonal to the "β_0 column" and which have sum of squares unity can be achieved by setting

$$(Z_1, Z_2) = \left(\frac{-n_3}{\sqrt{n_1 n_3 (n_1 + n_3)}}, \qquad 0 \right) \text{ for machine } A,$$

$$= \left(0, \qquad \frac{-n_3}{\sqrt{n_2 n_3 (n_2 + n_3)}} \right) \text{ for machine } B,$$

$$= \left(\frac{n_1}{\sqrt{n_1 n_3 (n_1 + n_3)}}, \qquad \frac{n_2}{\sqrt{n_2 n_3 (n_2 + n_3)}} \right) \text{ for machine } C,$$

where n_1, n_2, and n_3 are, respectively, the numbers of observations from machines A, B, and C.

In general, by an extension of this procedure we can deal with r levels by the introduction of $(r - 1)$ dummy variables. Let us now give a few examples of the use of dummy variables in regression analysis.

Examples of the Use of Dummy Variables

EXAMPLE OF BLOCKING.   Three manufacturing plants make identical products but the amount of water used differs markedly from plant to plant. It has been suggested that the amount of water used by a plant is a linear function of three variables, the average daily temperature (°F), the amount of daily production, and the number of people on the daily payroll. Five sets of data on the three variables and the water used were chosen at random from the records of each of the three plants. To take account of the fact that there are three distinct blocks of data, and to prevent any differences in average levels of water used at the three plants from obscuring the effects of the three independent variables, we can introduce two dummy block variables X_1 and X_2 as follows. The postulated model for the prediction of the amount of water used was

$$Y = \beta_0 X_0 + \beta_1 X_1 + \beta_2 X_2 + \beta_3 X_3 + \beta_4 X_4 + \beta_5 X_5 + \epsilon \quad (5.3.1)$$

where $X_0 = 1$ for all observations, and where X_1 and X_2 are used for plant identification as follows:

X_1	X_2	
1	0	= plant no. 1
0	1	= plant no. 2
0	0	= plant no. 3

and where X_3 = average daily temperature (°F),

X_4 = amount of daily production (pounds)

X_5 = number of people on the day's payroll,

ϵ = random error term, and

Y = amount of water used daily.

The data matrix $(\mathbf{X})$ is as follows:

X_0	X_1	X_2	X_3	X_4	X_5	Y
1	1	0	$X_{3,11}$	$X_{4,11}$	$X_{5,11}$	Y_{11}
1	1	0	$X_{3,12}$	$X_{4,12}$	$X_{5,12}$	Y_{12}
1	1	0	$X_{3,13}$	$X_{4,13}$	$X_{5,13}$	Y_{13}
1	1	0	$X_{3,14}$	$X_{4,14}$	$X_{5,14}$	Y_{14}
1	1	0	$X_{3,15}$	$X_{4,15}$	$X_{5,15}$	Y_{15}
1	0	1	$X_{3,21}$	$X_{4,21}$	$X_{5,21}$	Y_{21}
1	0	1	$X_{3,22}$	$X_{4,22}$	$X_{5,22}$	Y_{22}
1	0	1	$X_{3,23}$	$X_{4,23}$	$X_{5,23}$	Y_{23}
1	0	1	$X_{3,24}$	$X_{4,24}$	$X_{5,24}$	Y_{24}
1	0	1	$X_{3,25}$	$X_{4,25}$	$X_{5,25}$	Y_{25}
1	0	0	$X_{3,31}$	$X_{4,31}$	$X_{5,31}$	Y_{31}
1	0	0	$X_{3,32}$	$X_{4,32}$	$X_{5,32}$	Y_{32}
1	0	0	$X_{3,33}$	$X_{4,33}$	$X_{5,33}$	Y_{33}
1	0	0	$X_{3,34}$	$X_{4,34}$	$X_{5,34}$	Y_{34}
1	0	0	$X_{3,35}$	$X_{4,35}$	$X_{5,35}$	Y_{35}

In this table, for example, $X_{4,12}$ represents the observed value of variable X_4 at the first plant when the second response Y_{12} is recorded. The first three columns are examples of the use of dummy variables. Note that for given values X_{30}, X_{40}, and X_{50} of X_3, X_4 and X_5, the estimated response Y_{12} at plant no. 3 is
$$\hat{Y}_{03} = b_0 + b_3 X_{30} + b_4 X_{40} + b_5 X_{50}.$$
The estimated response at plant no. 1 is then
$$\hat{Y}_{03} + b_1$$
and the estimated response at plant no. 2 is then
$$\hat{Y}_{03} + b_2$$
so that b_1 and b_2 estimate the differences in plant levels which do not depend on the variables X_3, X_4, and X_5.

LINEAR TIME TRENDS. There are many situations where predictions are to be made for a future time period, and while this is always a hazardous process, it is a very real and necessary one. In studying information on such a problem one often finds linear trends in the data across time, and one easy way to remove such trends is by use of dummy variables. Two cases will be illustrated. (1) Data in which a single linear time trend occurs, and (2) Data in which two distinct linear time trends occur.

CASE 1. Data in which a single linear time trend occurs. Data are shown on the parity price in cents per pound of live weight of chickens at equal intervals of time in Table 5.1. Two alternative dummy variables used for removing a linear time trend are shown as columns X and X'. Either will

Table 5.1   Parity Price (¢) per Pound of Live
Weight of Chickens

Date	Y Parity Price	X	or	X'
Jan. 1955	29.1	1		−10
May. 1955	29.0	2		−9
Sept. 1955	28.6	3		−8
Jan. 1956	28.1	4		−7
May 1956	28.6	5		−6
Sept. 1956	28.7	6		−5
Jan. 1957	28.2	7		−4
May 1957	28.6	8		−3
Sept. 1957	28.6	9		−2
Jan. 1958	28.1	10		−1
May 1958	28.7	11		0
Sept. 1958	28.6	12		1
Jan. 1959	26.9	13		2
May 1959	27.0	14		3
Sept. 1959	26.8	15		4
Jan. 1960	25.7	16		5
May 1960	25.9	17		6
Sept. 1960	25.6	18		7
Jan. 1961	25.1	19		8
May 1961	25.2	20		9
Sept. 1961	25.1	21		10

do, although the centered column X' is better since it is orthogonal to the column of 1's in the **X**-matrix. The appropriate models in the two cases are

$$Y = \beta_0 + \beta_1 X + \text{(other terms with independent variables)} + \epsilon$$

(5.3.2)

and (since $X_i' = X_i - \bar{X}, i = 1, 2, \ldots, n$)

$$Y = (\beta_0 + \beta_1 \bar{X}) + \beta_1 X' + \text{(other terms)} + \epsilon$$
$$= \beta_0' + \beta_1 X' + \text{(other terms)} + \epsilon.$$ (5.3.3)

Note.   Here, since $n = 21$ is odd, the quantities $X_i' = X_i - \bar{X}$ are all integers. When n is even we can use instead $X_i' = 2(X_i - \bar{X})$ to avoid fractions. For example,

$$X_i = \quad 1 \quad\quad 2 \quad 3 \quad 4 \quad\quad (\bar{X} = 2\tfrac{1}{2})$$
$$X_i - \bar{X} = -1\tfrac{1}{2} \quad -\tfrac{1}{2} \quad \tfrac{1}{2} \quad 1\tfrac{1}{2}$$
$$2(X_i - \bar{X}) = -3 \quad\quad -1 \quad 1 \quad 3$$

CASE 2.   Data in which two distinct linear time trends occur. Economic data often appear to have two or more linear trends in them. Suppose, for

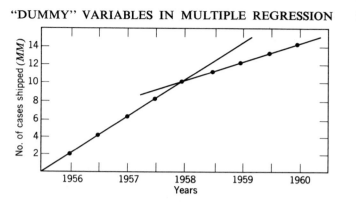

Figure 5.1   Shipments of Product A (1956–1960).

example, that a graph of the shipments of a particular product appears as shown in Figure 5.1. The data plotted here are unrealistic and without error, and are chosen only to demonstrate the use of dummy variables. We wish to determine the slopes of the two distinct lines and the point of intersection of the two trend lines, when it is unknown. We shall ignore the effect of other variables. If present these would cause additional terms to be added to the model.

Example 1.   The point of intersection is known, but the two slopes are unknown.

The values of Y plotted are successively 2, 4, 6, 8, 10, 11, 12, 13, 14, and it is known that the point of intersection occurs at $Y = 10$. To take account of the two slopes, two dummy variables are needed, X_1 and X_2, say. Let X_1 take equally spaced values for all points on the first line and be constant for all points on the second line. Let X_2 do the exact opposite. Then, for example, we can take account of the two linear time trends by using the data matrix given below.

Obs. No.	X_0	X_1	X_2	Y
1	1	−4	0	2
2	1	−3	0	4
3	1	−2	0	6
4	1	−1	0	8
5	1	0	0	10
6	1	0	1	11
7	1	0	2	12
8	1	0	3	13
9	1	0	4	14

Note that the entries for X_1 and X_2 at the known point of intersection where $Y = 10$ are zeros. Consider the model

$$Y = \beta_0 + \beta_1 X_1 + \beta_2 X_2 + \epsilon. \tag{5.3.4}$$

Then b_1 represents the slope of the first trend line, and b_2 represents the slope of the second trend line, which starts at observation five and continues through observation nine. The estimate b_0 is the value of $\hat{Y}$ at $X_1 = X_2 = 0$, that is, $\hat{Y}$ at the point of intersection of the lines.

The normal equations for the model of Eq. (5.3.4) are

$$9b_0 - 10b_1 + 10b_2 = 80$$
$$-10b_0 + 30b_1 \qquad = -40$$
$$10b_0 \qquad + 30b_2 = 130$$

Solving these equations, we have
1. The value of $\hat{Y}$ at the point of intersection is given by $b_0 = 10$.
2. The slope of the first line is given by $b_1 = 2$.
3. The slope of the second line is given by $b_2 = 1$.
(Since no error is involved in this example, the model is exact.)

Example 2. An alternative method when the point of intersection is known, but the two slopes are unknown.

Using the same data as before, the data matrix can be constructed as follows:

Obs. No.	X_0	X_1	X_2	Y
1	1	1	0	2
2	1	2	0	4
3	1	3	0	6
4	1	4	0	8
5	1	5	0	10
6	1	5	1	11
7	1	5	2	12
8	1	5	3	13
9	1	5	4	14

This setup will provide estimates of the slopes as before but the constant term b_0 will be the value of Y when $X_1 = X_2 = 0$, that is, the intercept of the first trend line.

Example 3. The point of intersection and the slopes are unknown. Another dummy variable X_3 is needed to take care of the unknown point of intersection as given below. (Note that we are using new, constructed data without error for this example.)

Obs. No.	X_0	X_1	X_2	X_3	Y
1	1	1	0	0	2
2	1	2	0	0	4
3	1	3	0	0	6
4	1	4	0	0	8
5	1	5	0	1	9.5
6	1	5	1	1	10.5
7	1	5	2	1	11.5
8	1	5	3	1	12.5
9	1	5	4	1	13.5

Consider the model

$$Y = \beta_0 X_0 + \beta_1 X_1 + \beta_2 X_2 + \beta_3 X_3 + \epsilon. \tag{5.3.5}$$

The value β_3 represents a step change which comes into effect at the fifth observation point and is the vertical distance the second line lies *above* the first at this point. (If the second line lies below the first, β_3 is negative.)
The normal equations for the model of Eq. (5.3.5) are

$$9b_0 + 35b_1 + 10b_2 + 5b_3 = 77.5$$
$$35b_0 + 155b_1 + 50b_2 + 25b_3 = 347.5$$
$$10b_0 + 50b_1 + 30b_2 + 10b_3 = 125$$
$$5b_0 + 25b_1 + 10b_2 + 5b_3 = 57.5$$

Solving these equations, we have

$$b_0 = 0 \quad \text{(intercept of line no. 1),}$$
$$b_1 = 2 \quad \text{(slope of line no. 1),}$$
$$b_2 = 1 \quad \text{(slope of line no. 2),}$$
$$b_3 = -\tfrac{1}{2} \quad \text{(the vertical distance between line 2 and line 1 at the fifth observation point.)}$$

The situation is shown graphically in Figure 5.2. The negative sign of b_3 and the fact that $b_1 > b_2$ indicates that the point of intersection of the two lines is to the left of the fifth observation point. In fact, it occurs when $X_1 = 4\tfrac{1}{2}$.

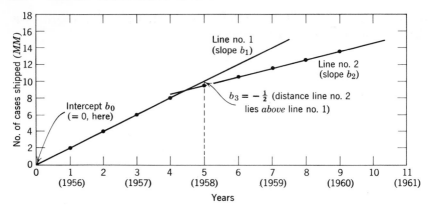

Figure 5.2   Use of dummy variables; two lines, point of intersection unknown.

5.4.   Preparing the Input Data Matrix for General Regression Problems

Suppose we wish to fit by least squares a general linear model of the form

$$Y = \beta_0 + \beta_1 Z_1 + \beta_2 Z_2 + \cdots + \beta_p Z_p + \epsilon \qquad (5.4.1)$$

where the $Z_i = Z_i(X_1, X_2, \ldots, X_k)$ are some (specified) functions of the independent variables $X_1, X_2, \ldots, X_k$. Write

$$\boldsymbol{\beta}' = (\beta_0, \beta_1, \beta_2, \ldots, \beta_p)$$

for the vector of parameters, and

$$\mathbf{Y}' = (Y_1, Y_2, \ldots, Y_i, \ldots, Y_n)$$

for the vector of n observations. Then the estimate of $\boldsymbol{\beta}$, namely

$$\mathbf{b} = (b_0, b_1, b_2, \ldots, b_p)',$$

is given by $\mathbf{b} = (\mathbf{X}'\mathbf{X})^{-1}\mathbf{X}'\mathbf{Y}$ applying the formula given in Chapter 2 where

$$\mathbf{X} = \begin{bmatrix} 1 & Z_{11} & Z_{21} & \cdots & Z_{p1} \\ 1 & Z_{12} & Z_{22} & \cdots & Z_{p2} \\ & \cdots & & \cdots & \\ 1 & Z_{1i} & Z_{2i} & \cdots & Z_{pi} \\ & \cdots & & \cdots & \\ 1 & Z_{1n} & Z_{2n} & \cdots & Z_{pn} \end{bmatrix} \qquad (5.4.2)$$

and where Z_{ji} is the observed value of Z_j which corresponds to the observation Y_i. To take a very simple example, suppose we are using the model

$$Y = \beta_0 + \beta_1 X + \beta_{11} X^2 + \epsilon \qquad (5.4.3)$$

so that $Z_1 = X$, $Z_2 = X^2$, $\beta_2 = \beta_{11}$ in the general form above. If the data available are

$$X = 1 \quad 2 \quad 3 \quad 4 \quad 5$$
$$Y = Y_1 \quad Y_2 \quad Y_3 \quad Y_4 \quad Y_5$$

then

$$
\mathbf{X} =
\begin{array}{c}
1 \quad Z_1 = X \quad Z_2 = X^2 \\
\begin{bmatrix}
1 & 1 & 1 \\
1 & 2 & 4 \\
1 & 3 & 9 \\
1 & 4 & 16 \\
1 & 5 & 25
\end{bmatrix}
\end{array}
\qquad
\mathbf{Y} =
\begin{array}{c}
Y \\
\begin{bmatrix}
Y_1 \\
Y_2 \\
Y_3 \\
Y_4 \\
Y_5
\end{bmatrix}
\end{array}
$$

(Alternatively, if $Z_2 = \sqrt{X}$, the entries for the Z_2 column would be 1, $\sqrt{2}$, $\sqrt{3}$, 2, $\sqrt{5}$, and so on for more general cases.) When data are being prepared for computer calculations we often write the $\mathbf{X}$ matrix and $\mathbf{Y}$ vector side by side without divisions and call it the *data matrix* or the *input data matrix*. For example, the data matrix for the simple example above is

X_0	X	X^2	Y
1	1	1	Y_1
1	2	4	Y_2
1	3	9	Y_3
1	4	16	Y_4
1	5	25	Y_5

(5.4.4)

	X_0	X	X^2	Y
Column total	—	15	55	$\sum_{i=1}^{n} Y_i$
Column mean	—	3	11	$\bar{Y}$
Sum of squares of column	—	55	979	

When only one or two independent variables are involved in a regression model, the straightforward calculation of $\mathbf{b} = (\mathbf{X'X})^{-1}\mathbf{X'Y}$, as illustrated by the two variable example of Chapter 4, usually causes no difficulty provided that sufficient figures are carried in the computation. In problems with several independent variables and with large bodies of data, results can be entirely invalidated due to roundoff errors. A typical way in which roundoff errors occur is the following: Suppose $(a/b) - (c/d) = e$ is required. If (a/b) is large, and (c/d) is large but e is small, rounding the numbers (a/b) and (c/d) too severely may cause all significant figures in the result e to be lost. *Example:* $a = 100, b = 3, c = 166.663, d = 5$. Suppose in computing by hand we rounded always to three figures. Then $a/b = 33.333$, $c/d = 33.333$, so $e = 0$. (More accurately $e = (ad - bc)/bd = (500 - 499.989)/15 = 0.011/15$). Thus if, at a later stage, e were multiplied by, say, 1,000,000, the result would be zero (instead of the correct $11,000/15 = 733.\dot{3}$—an enormous difference).

Although a digital computer carries many more figures than a human computer, errors of this type frequently arise and often go undetected entirely or for a considerable time. Supposedly sound conclusions are sometimes based entirely on the vagaries of roundoff error.

In "A warning of round-off errors in regression," *American Statistician,* **17**, December 1963, 13–15, R. J. Freund reported an example in which five different regression calculations using four different regression programs produced considerable differences in the values of the estimated coefficients due to roundoff errors. To try to overcome this, some computers can provide (by special request) *double precision arithmetic*. This means that, on request, the computer will work with numbers twice the length it normally handles. To use this as a standard technique is wasteful of computer time and is often an unnecessary precaution. It is far better to recognize that roundoff errors can arise and to try to take preliminary steps which will minimize them and perhaps succeed in avoiding them altogether.

Two prime causes of roundoff errors are the following.

1. The numbers involved in a regression calculation are of widely differing orders—for example, numbers like 52,793, −943, and 6 involved in the same computation.

2. The matrix which must be inverted is very close to being singular. We can see from Eqs. (2.1.10) and (2.1.11) that the determinant of a matrix is involved in each element of the inverse. When the determinant of a matrix is small compared with other numbers in the calculation, rounding trouble is likely to arise, and this is true not only for 2×2 and 3×3 matrices but in general. When det $(\mathbf{X'X})$ is very small, compared with other numbers in the calculation, the matrix $\mathbf{X'X}$ is said to be *ill* (or *badly*) *conditioned.* When det $(\mathbf{X'X}) = 0$ the matrix $\mathbf{X'X}$ is said to be *singular;* if this happens in a computer calculation, overflow will occur and/or the computer will stop.

We shall now discuss some steps which can be taken to improve the form of the computations and which are useful for other reasons as well. These are *centering* of the data and use of the correlation matrix instead of the $\mathbf{X'X}$ matrix. Orthogonalization of the columns of the $\mathbf{X}$ matrix will be discussed in Section 5.6 after a brief discussion of orthogonal polynomials in Section 5.5. Centering and use of the correlation matrix are standard in most linear regression programs. Orthogonalization is a useful procedure which can be employed for checking the $X'X$ matrix for singularity.

"Centering" the Data

Suppose we have an input data matrix of the form given below, together with the column means shown.

	Z_0	Z_1	Z_2	$\cdots$	Z_p	Y	
	1	Z_{11}	Z_{21}	$\cdots$	Z_{p1}	Y_1	
	1	Z_{12}	Z_{22}	$\cdots$	Z_{p2}	Y_2	(5.4.5)
	$\cdots$						
	1	Z_{1n}	Z_{2n}	$\cdots$	Z_{pn}	Y_n	
Column sum	—	$\sum Z_{1i}$	$\sum Z_{2i}$	$\cdots$	$\sum Z_{pi}$	$\sum Y_i$	(summed over $i = 1, 2, \ldots, n$)
Column mean	—	$\bar{Z}_1$	$\bar{Z}_2$	$\cdots$	$\bar{Z}_p$	$\bar{Y}$	

Our model is

$$Y = \beta_0 + \beta_1 Z_1 + \beta_2 Z_2 + \cdots + \beta_p Z_p + \epsilon. \qquad (5.4.6)$$

We can rewrite this in the form

$$Y = \{\beta_0 + \beta_1 \bar{Z}_1 + \beta_2 \bar{Z}_2 + \cdots + \beta_p \bar{Z}_p\}$$
$$+ \beta_1(Z_1 - \bar{Z}_1) + \beta_2(Z_2 - \bar{Z}_2) + \cdots + \beta_p(Z_p - \bar{Z}_p) + \epsilon$$

where $\bar{Y}, \bar{Z}_1, \bar{Z}_2, \ldots, \bar{Z}_p$ are actual numerical values obtained from the data. If we write $z_j = Z_j - \bar{Z}_j, j = 1, 2, \ldots, p, \beta_0' = \beta_0 + \beta_1 \bar{Z}_1 + \beta_2 \bar{Z}_2 + \cdots + \beta_p \bar{Z}_p$, the model can be expressed as

$$Y = \beta_0' + \beta_1 z_1 + \beta_2 z_2 + \cdots + \beta_p z_p + \epsilon. \qquad (5.4.7)$$

We can make the same transformations on the data as made on the variables above so that $z_{ji} = Z_{ji} - \bar{Z}_j, j = 1, 2, \ldots, p$ for $i = 1, 2, \ldots, n$. It follows that $\bar{z}_j = 0, j = 1, 2, \ldots, n$. The first normal equation obtained by differentiating the residual sum of squares with respect to β_0' is easily seen to reduce to

$$b_0' + b_1 \bar{z}_1 + b_2 \bar{z}_2 + \cdots + b_p \bar{z}_p = \bar{Y}$$

which means

$$b_0' = \bar{Y}$$

no matter what values $b_1, b_2, \ldots, b_p$ may have.

Since this will always be true we can omit β_0' from the model and use the form

$$Y - \bar{Y} = \beta_1(Z_1 - \bar{Z}_1) + \beta_2(Z_2 - \bar{Z}_2) + \cdots + \beta_p(Z_p - \bar{Z}_p) + \epsilon' \quad (5.4.8)$$

as an alternative form to Eq. (5.4.6). The apparent gain in now having to estimate one less parameter is compensated for by the fact that the n (new) dependent variable observations, namely $(Y_1 - \bar{Y}), (Y_2 - \bar{Y}), \ldots,$ $(Y_n - \bar{Y})$, are now connected by the restriction

$$\sum_{i=1}^{n}(Y_i - \bar{Y}) = 0$$

and so one degree of freedom is lost overall.

The data matrix for Eq. (5.4.8) will take the following form:

z_1	z_2	$\cdots$	z_p	y
$Z_{11} - \bar{Z}_1$	$Z_{21} - \bar{Z}_2$		$Z_{p1} - \bar{Z}_p$	$Y_1 - \bar{Y}$
$Z_{12} - \bar{Z}_1$	$Z_{22} - \bar{Z}_2$		$Z_{p2} - \bar{Z}_p$	$Y_2 - \bar{Y}$
$\cdots$				
$Z_{1n} - \bar{Z}_1$	$Z_{2n} - \bar{Z}_n$		$Z_{pn} - \bar{Z}_p$	$Y_n - \bar{Y}$

To illustrate this in a very simple case we apply it to the example used above with input data matrix (5.4.4). We set $y_i = Y_i - \bar{Y}$, $z_{1i} = X_i - 3$, $z_{2i} = X_i^2 - 11$, which provides a new data matrix as follows:

	z_1	z_2	y
	-2	-10	$y_1 = Y_1 - \bar{Y}$
	-1	-7	$y_2 = Y_2 - \bar{Y}$
	0	-2	$y_3 = Y_3 - \bar{Y}$
	1	5	$y_4 = Y_4 - \bar{Y}$
	2	14	$y_5 = Y_5 - \bar{Y}$
Column total	0	0	0
Sum of squares of column	10	374	

Note that the effect of centering the X and X^2 columns has reduced the absolute sizes of the numbers in the calculation and has emphasized the spread and the distribution of the column numbers about their mean rather than their absolute values. Centering is also a necessary preliminary for obtaining the correlation matrix of the variables which is a very important feature of the selection procedures to be discussed in Chapter 6.

The Correlation Matrix

Suppose we wish to fit the model

$$Y = \beta_0 + \beta_1 Z_1 + \beta_2 Z_2 + \epsilon.$$

By centering the data as shown above we can convert this into the model of form

$$Y - \bar{Y} = \beta_1(Z_1 - \bar{Z}_1) + \beta_2(Z_2 - \bar{Z}_2) + \epsilon$$

as in Eq. (5.4.8) with $p = 2$.

When the model is written in this form, the "$\mathbf{X'X}$ matrix" of the problem takes the form

$$\mathbf{S} = \begin{bmatrix} S_{11} & S_{12} \\ S_{21} & S_{22} \end{bmatrix}$$

where $S_{jl} = \sum_{i=1}^{n}(Z_{ji} - \bar{Z}_j)(Z_{li} - \bar{Z}_l)$, $j, l = 1, 2$, and where Z_{ji}, $i = 1, 2, \ldots, n$ are the n observations taken on Z_j. These numbers S_{jl} will often be of different sizes. In larger $\mathbf{S}$ matrices, say, 5×5 and higher this may often lead to the occurrence of roundoff errors, when the matrix $\mathbf{S}$ is inverted, even when the work is performed in an electronic computer. Let us transform the centered data by

$$x_{ji} = \frac{(Z_{ji} - \bar{Z}_j)}{S_{jj}^{\frac{1}{2}}}, \quad j = 1, 2, \qquad y_i = \frac{(Y_i - \bar{Y})}{S_{yy}^{\frac{1}{2}}}$$

where $S_{yy} = \sum_{i=1}^{n}(Y_i - \bar{Y})^2$. We can make similar transformations on the variables Z_1, Z_2, and Y by dropping the i-suffix throughout. This will bring the "centered" model above into the new form

$$y S_{yy}^{\frac{1}{2}} = \beta_1 S_{11}^{\frac{1}{2}} x_1 + \beta_2 S_{22}^{\frac{1}{2}} x_2 + \epsilon$$

or

$$y = \alpha_1 x_1 + \alpha_2 x_2 + \epsilon'$$

where $\alpha_1 = \beta_1(S_{11}/S_{yy})^{\frac{1}{2}}$ and $\alpha_2 = \beta_2(S_{22}/S_{yy})^{\frac{1}{2}}$ are new coefficients to be estimated from the transformed data (y_i, x_{1i}, x_{2i}) $i = 1, 2, \ldots, n$, and represent *scaled* forms of the original coefficients β_1 and β_2. When the model is written in this form the "$\mathbf{X'X}$ matrix" for the problem takes the new form:

$$\begin{bmatrix} 1 & r_{12} \\ r_{21} & 1 \end{bmatrix}$$

called the *correlation matrix of the Z's*, where

$$r_{12} = \frac{S_{12}}{(S_{11}S_{22})^{\frac{1}{2}}} = r_{21}$$

is the correlation between Z_1 and Z_2 as described in Section 1.6. Since, we can also write

$$r_{jy} = \frac{S_{jy}}{(S_{jj}S_{yy})^{\frac{1}{2}}}$$

where $S_{jy} = \sum_{i=1}^{n}(Z_{ji} - \bar{Z}_j)(Y_i - \bar{Y})$, as the correlation between Z_j ($j = 1$ or 2) and Y (again see Section 1.6), the normal equations for the new model are simply

$$\begin{bmatrix} 1 & r_{12} \\ r_{21} & 1 \end{bmatrix} \begin{bmatrix} a_1 \\ a_2 \end{bmatrix} = \begin{bmatrix} r_{1y} \\ r_{2y} \end{bmatrix}$$

where a_1 and a_2 are the least squares estimates of α_1 and α_2. The solution of these equations is easily seen (using the fact that $r_{21} = r_{12}$) to be

$$a_1 = \frac{(r_{1y} - r_{12}r_{2y})}{D}$$

$$a_2 = \frac{(r_{2y} - r_{12}r_{1y})}{D}$$

where D = the determinant of the correlation matrix
$$= 1 - r_{12}^2.$$

The following points should be noticed here. Transforming the regression problem into a form in which it involves correlations is good in general because it makes all the numbers in the calculation lie between -1 and 1. When numbers are all of this order the adverse effects of roundoff error are minimized. While the dangers are slight when only two variables are involved, this is very important when many independent variables are being manipulated in a computer. (As a general rule, the use of correlations as shown here is not needed if the problem is small enough to do by hand. It is an essential part of a good computer routine, however.) In determining the values of both the estimates a_1 and a_2 above we must divide by $D = 1 - r_{12}^2$, the determinant of the correlation matrix. Thus if r_{12}^2 is very close to unity, D is very close to zero, and the formal solution approaches indeterminancy with a_1 and a_2 approximately equal and opposite and both very large. Now r_{12} is the correlation between Z_1 and Z_2. Thus if Z_1 and Z_2 either depend totally or almost totally one on the other, or simply vary together, or almost together, r_{12} will be unity, or almost unity, and D will be zero, or almost zero. Actually, of course, if D is zero we cannot derive the solutions for a_1 and a_2 as before, properly speaking, since the two normal equations actually represent only one equation. The overall implication of this is that it is essential to compute the value of the determinant $D = 1 - r_{12}^2$ so that the dependence in the normal equations can be detected. In more general problems the correlation matrix has the form

$$\begin{bmatrix} 1 & r_{12} & r_{13} & \cdots & r_{1p} \\ r_{21} & 1 & r_{23} & \cdots & r_{2p} \\ r_{31} & r_{32} & 1 & \cdots & r_{3p} \\ & \cdots & & & \\ r_{p1} & r_{p2} & r_{p3} & \cdots & 1 \end{bmatrix}$$

but similar remarks apply, and evaluation of the determinant of this matrix is an essential part of a good computer regression routine. Of course, when only one independent variable is involved no problem occurs since the matrix is a single element—unity.

We recall that the transformations made above to produce the 2×2 correlation matrix from the "$\mathbf{X'X}$ matrix" involve a reparameterization $\alpha_j = \beta_j (S_{jj}/S_{yy})^{1/2}$, $j = 1, 2$. We also recall (page 145) that $b_0' = \bar{Y}$ estimates the quantity $\beta_0 + \beta_1 \bar{Z}_1 + \beta_2 \bar{Z}_2$.   Estimates b_0, b_1, b_2 of the original coefficients $\beta_0, \beta_1, \beta_2$ are obtained as follows:

$$b_1 = a_1 \left(\frac{S_{yy}}{S_{11}} \right)^{1/2}$$

$$b_2 = a_2 \left(\frac{S_{yy}}{S_{22}} \right)^{1/2}$$

$$b_0 = \bar{Y} - b_1 \bar{Z}_1 - b_2 \bar{Z}_2.$$

Some computer routines provide printouts of both sets of coefficients.

Correlations have an important role to play in some of the variable selection procedures discussed in Chapter 6. In some of these, we wish to add independent variables one by one to the postulated model. The first independent variable placed in the postulated model is chosen as the one which is most correlated with Y, that is, the variable Z_j whose r_{jy} is the largest of all the r_{ly}, $l = 1, 2, \ldots, p$. Suppose this is Z_1 for simplicity. The model

$$Y = \beta_0 + \beta_1 Z_1 + \epsilon$$

is fitted. New variables $Z_2^*, Z_3^*, \ldots, Z_p^*$ are then constructed by finding the residuals of Z_2 after regressing it on Z_1, that is, the residuals from fitting the model $Z_2 = \alpha_0 + \alpha_1 Z_1 + \epsilon'$, the residuals of Z_3 after regressing it on $Z_1, \ldots$, the residuals of Z_p after regressing it on Z_1, respectively. A new dependent variable Y^* has values which consist of the residuals of Y regressed on Z_1 (i.e., using the model $Y = \beta_0 + \beta_1 Z_1 + \epsilon$ above). Similar work was given in Section 4.1 when $p = 2$.

The values of the new dependent variable Y^* and the new independent variables $Z_2^*, Z_3^*, \ldots, Z_p^*$ represent those portions of the corresponding original data vectors which have no dependence on the values of the variable Z_1. We can now find a new set of correlations which involve the starred variables. These are called *partial correlations* and can be written, for example, as $r_{2y \cdot 1}$, meaning the correlation of variables Z_2^* and Y^* and read as "the partial correlation of variables Z_2 and Y after both have been adjusted for variable Z_1." In the second stage of the selection procedure we should add to the model the variable Z_j whose partial correlation coefficient with Y, adjusted for Z_1 (namely $r_{jy \cdot 1}$), was the greatest; that is, we should choose the variable Z_j most correlated with Y after the effect of Z_1 had been removed both from Y and from Z_j. If the second Z variable selected in this way is Z_2 say, the third stage of the selection procedure involves partial correlations of the form $r_{jy \cdot 12}$; that is, the correlations between (1) the residuals of Z_j regressed on Z_1 and Z_2 and (2) the residuals of Y regressed on Z_1 and Z_2. We shall not go farther into the details of partial correlation calculations in this book.

5.5. Orthogonal Polynomials

Orthogonal polynomials are used to fit a polynomial model of any order in one variable. The idea is as follows. Suppose we have n observations (X_i, Y_i) $i = 1, 2, \ldots, n$ where X is an independent variable and Y is a dependent variable, and we wish to fit the model

$$Y = \beta_0 + \beta_1 X + \beta_2 X^2 + \cdots + \beta_p X^p + \epsilon. \tag{5.5.1}$$

In general the columns of the resulting $\mathbf{X}$ matrix will not be orthogonal. If later we wish to add another term $\beta_{p+1} X^{p+1}$, changes will usually occur in the estimates of all the other coefficients. However, if we can construct polynomials of the form

$$\psi_0(X_i) = 1 \qquad \text{zero-order polynomial}$$

$$\psi_1(X_i) = P_1 X_i + Q_1 \qquad \text{first order}$$

$$\psi_2(X_i) = P_2 X_i^2 + Q_2 X_i + R_2 \qquad \text{second order}$$

$$\cdots$$

$$\psi_r(X_i) = P_r X_i^r + Q_r X_i^{r-1} + \cdots + T_r \qquad r\text{-th order}$$

$$\cdots$$

with the property that they are *orthogonal polynomials*, that is,

$$\sum_{i=1}^{n} \psi_j(X_i)\psi_l(X_i) = 0 \qquad (j \neq l), \tag{5.5.2}$$

for all $j, l < n - 1$, we can rewrite the model as

$$Y = \alpha_0 \psi_0(X) + \alpha_1 \psi_1(X) + \cdots + \alpha_p \psi_p(X) + \epsilon. \quad (5.5.3)$$

In this case we shall have

$$\mathbf{X} = \begin{bmatrix} 1 & \psi_1(X_1) & \psi_2(X_1) & \cdots & \psi_p(X_1) \\ 1 & \psi_1(X_2) & \psi_2(X_2) & \cdots & \psi_p(X_2) \\ & \cdots & & & \\ 1 & \psi_1(X_n) & \psi_2(X_n) & \cdots & \psi_p(X_n) \end{bmatrix} \quad (5.5.4)$$

so that

$$\mathbf{X'X} = \begin{bmatrix} A_{00} & & & & & \\ & A_{11} & & & \mathbf{0} & \\ & & A_{22} & & & \\ & & & \cdot & & \\ \mathbf{0} & & & & \cdot & \\ & & & & & \cdot \\ & & & & & A_{pp} \end{bmatrix} \quad (5.5.5)$$

where $A_{jj} = \sum_{i=1}^{n} \{\psi_j(X_i)\}^2$ since all off-diagonal terms vanish, by Eq. (5.5.2). Since the inverse matrix $(\mathbf{X'X})^{-1}$ is also diagonal and is obtained by inverting each element separately (see Eq. (2.1.13)) the least squares procedure provides, as an estimate of α_j, the quantity

$$a_j = \frac{\sum\limits_{i=1}^{n} Y_i \psi_j(X_i)}{\sum\limits_{i=1}^{n} [\psi_j(X_i)]^2}, \quad j = 0, 1, 2, \ldots, p$$

$$= \frac{A_{jY}}{A_{jj}} \quad (5.5.6)$$

with an obvious notation. Since $V(\mathbf{b}) = (\mathbf{X'X})^{-1}\sigma^2$ for the general regression model, it follows that the variance of a_j is

$$V(a_j) = \frac{\sigma^2}{A_{jj}} \quad (5.5.7)$$

where σ^2 is usually estimated from the analysis of variance table. To obtain the entries in the table we evaluate the sum of squares due to a_j from

$$\text{SS}(a_j) = \frac{A_{jY}^2}{A_{jj}} \quad (5.5.8)$$

and so obtain an analysis of variance table as follows:

Source	SS	df	MS
a_0(mean)	SS(a_0)	1	—
a_1	SS(a_1)	1	SS(a_1)
a_2	SS(a_2)	1	SS(a_2)
. . .	. . .	. . .	. . .
a_p	SS(a_p)	1	SS(a_p)
Residual	By subtraction	$n - p - 1$	s^2
Total	$\sum_{i=1}^{n} Y_i^2$	n	

If the model is correct, s^2 estimates σ^2. Usually we pool, with the residual, sums of squares whose mean squares are not significantly larger than s^2 to obtain an estimate of σ^2 based on more degrees of freedom. Note that if it were desired to add another term $\alpha_{p+1}\psi_{p+1}(X)$ to Eq. (5.5.3) no recomputation would be necessary for the coefficients already obtained, due to the orthogonality of the polynomials. Thus higher and higher order polynomials can be fitted with ease and the process terminated when a suitably fitting equation is found.

The $\psi_j(X_i)$ can be constructed and the procedure above can be carried out for any values of the X_i. However, when the X_i are *not* equally spaced, polynomials must be specially constructed. (See, for example, "Orthogonal polynomial fitting" by J. Wishart and T. Metakides, *Biometrika*, **40**, 1953, 361–369, and "A simple method for constructing orthogonal polynomials when the independent variable is unequally spaced," by D. S. Robson, *Biometrics*, **15**, 1959, 187–191.) If the X_i are equally spaced, tables can be employed. The actual numerical values of $\psi_j(X_i)$ and A_{jj} as well as the general functional forms of $\psi_j(X)$, for $j = 1, 2, \ldots, 6$ and for $n \leq 52$ are given in *Biometrika Tables for Statisticians Volume 1*, by E. S. Pearson and H. O. Hartley, Cambridge University Press, 1958. This means that provided $p \leq 6$ we have the elements of the **X** and **X'X** matrices. Similar values for $j \leq 5$ and $n \leq 75$ appear in *Statistical Tables for Biological, Agricultural, and Medical Research*, by R. A. Fisher and F. Yates, Hafner Publishing Co., New York, 1964 (6th ed.). When $p > 6$, more extensive tables such as *Values and Integrals of the Orthogonal Polynomials up to $n = 26$*, by D. B. DeLury, University of Toronto Press, 1960, are needed.

Although orthogonal polynomials are often recommended only when desk calculators are used, J. W. Bright and G. S. Dawkins have concluded, in "Some aspects of curve fitting using orthogonal polynomials," *Industrial and Engineering Chemistry Fundamentals*, **4**, 1965, February, 93–97 that even when a computer is available, and especially when the polynomial is of high order, orthogonal polynomials are worthwhile. Using them provides greater computing accuracy and reduced computing times. We illustrate the application of orthogonal polynomials by an example.

Example. The Gillette Company reported net income per share for the years 1957–1964 as follows:

Year (Z_i)	1957	1958	1959	1960	1961	1962	1963	1964
Income per share ($ Y_i)	0.93	0.99	1.11	1.33	1.52	1.60	1.47	1.33

Fit a polynomial of suitable order which will provide a satisfactory approximating function for these data.

Solution. We ignore the year values Z_i for the moment except to note that they are equally spaced. From a table of orthogonal polynomials we find values $\psi_j(X_i)$ corresponding to $n = 8$ observations as shown in Table 5.2. To provide slightly easier calculations we work with $(Y - 0.93)$ rather than Y. Note that $A_{00} = n = 8$
We consider the model

$$Y - 0.93 = \sum_{j=0}^{6} \alpha_j \psi_j(X).$$

Using Eq. (5.5.6) we obtain

$$a_0 = \frac{[0(1) + 0.06(1) + \cdots + 0.40(1)]}{8} = \frac{2.84}{8}$$

$$= 0.355$$

$$a_1 = \frac{[0(-7) + 0.06(-5) + \cdots + 0.40(7)]}{168} = \frac{6.86}{168}$$

$$= 0.040833$$

$$a_2 = \frac{-4.10}{168} = -0.024405$$

$$a_3 = \frac{-3.60}{264} = -0.013636$$

$$a_4 = \frac{1.36}{616} = 0.002208$$

$$a_5 = \frac{2.94}{2,184} = 0.001346$$

$$a_6 = \frac{[0(1) + 0.06(-5) + \cdots + 0.40(1)]}{264} = \frac{0.10}{264}$$

$$= 0.000379$$

Table 5.2   The Calculation Table for the Example

$Y - 0.93$	ψ_0	ψ_1	ψ_2	ψ_3	ψ_4	ψ_5	ψ_6
0	1	−7	7	−7	7	−7	1
0.06	1	−5	1	5	−13	23	−5
0.18	1	−3	−3	7	−3	−17	9
0.40	1	−1	−5	3	9	−15	−5
0.59	1	1	−5	−3	9	15	−5
0.67	1	3	−3	−7	−3	17	9
0.54	1	5	1	−5	−13	−23	−5
0.40	1	7	7	7	7	7	1
A_{jj}	8	168	168	264	616	2184	264

The fitted model is thus

$$\hat{Y} - 0.93 = \sum_{j=0}^{6} a_j \psi_j(X)$$

where the a_j are as shown and the $\psi_j(X)$ are found in tables. (We shall do this in a moment, after reducing the equation.) To evaluate the entries in the analysis of variance table we need the following values

$j = 0$	1	2	3	4	5	6
$A_{jY} = 2.84$	6.86	−4.10	−3.60	1.36	2.94	0.10
$A_{jj} = 8$	168	168	264	616	2184	264

By applying Eq. (5.5.8) we obtain the analysis of variance table:

Source	SS	df	MS
a_0 (mean)	1.008	1	
a_1	0.280	1	(same
a_2	0.100	1	as SS)
a_3	0.049	1	
a_4	0.003	1	
a_5	0.004	1	
a_6	0.000	1	
Residual	0.001	1	
Total	1.445	8	

Note. When the order of the fitted polynomial is $p = N - 1$, the maximum order, the model fits the data exactly and no residual occurs. Here $p = N - 2$ and so the residual is $SS(a_7)$ in fact.

If an external estimate of σ^2 was available we could compare all the mean squares with it. As it is, we observe that terms of third and lower order account for most of the variation in the data. We can thus adopt

$$\hat{Y} - 0.93 = 0.355 + 0.041\psi_1(X) - 0.024\psi_2(X) - 0.014\psi_3(X)$$

as our fitted model and rearrange the analysis of variance table as follows.

Source	SS	df	MS
a_0 (mean)	1.008	1	1.008
a_1	0.280	1	0.280
a_2	0.100	1	0.100
a_3	0.049	1	0.049
Residual	0.008	4	$s^2 = 0.002$
Total	1.445	8	

In order to obtain the fitted equation in terms of the original variables we have first to substitute for the ψ_j and relate these to the Z's. From the table of orthogonal polynomials (e.g., *Biometrika Tables*, page 212, with $n = N = 8$),

$$\psi_0(X) = 1, \quad \psi_1(X) = 2X, \quad \psi_2(X) = X^2 - \tfrac{21}{4}$$

$$\psi_3(X) = \tfrac{2}{3}[X^3 - \tfrac{37}{4}X.]$$

Z and X correspond as follows:

$\psi_1(X_i) = 2X_i$:	-7	-5	-3	-1	1	3	5	7
X_i:	$-\tfrac{7}{2}$	$-\tfrac{5}{2}$	$-\tfrac{3}{2}$	$-\tfrac{1}{2}$	$\tfrac{1}{2}$	$\tfrac{3}{2}$	$\tfrac{5}{2}$	$\tfrac{7}{2}$
Z_i:	1957	1958	1959	1960	1961	1962	1963	1964

Clearly the required coding is

$$X = Z - 1960\tfrac{1}{2}.$$

The fitted polynomial is thus

$$\hat{Y} = 1.285 + 0.082(Z - 1960\tfrac{1}{2})$$
$$- 0.024[(Z - 1960\tfrac{1}{2})^2 - \tfrac{21}{4}]$$
$$- \frac{0.028}{3}[(Z - 1960\tfrac{1}{2})^3 - \tfrac{37}{4}(Z - 1960\tfrac{1}{2})].$$

This can be rearranged as a cubic in Z but the above form is easier to substitute into to obtain fitted values and residuals, in this case. It is recommended that the reader plot the data and fitted values and also examine the residuals as described in Chapter 3.

5.6.   Transforming X Matrices to Obtain Orthogonal Columns

The X matrix in a regression problem must be such that none of the columns can be expressed as a linear combination of the other columns. This effectively implies also that there must be at least as many rows not dependent on other rows, as there are parameters to estimate, or else a dependence will appear in the columns also. As an example, suppose observations Y are recorded at only three levels of X, namely $X = a, b,$ and c, but that the model $Y = \beta_0 + \beta_1 X + \beta_2 X^2 + \beta_3 X^3 + \epsilon$ is postulated. The X matrix takes the form

$$\begin{bmatrix} 1 & a & a^2 & a^3 \\ 1 & b & b^2 & b^3 \\ 1 & c & c^2 & c^3 \end{bmatrix}$$

and the columns are dependent since (column 4) $- (a + b + c)$(column 3) $+ (ab + bc + ca)$(column 2) $- abc$(column 1) $= 0$. To spot such a dependence in a general regression problem is often very difficult. When it exists the $X'X$ matrix will always be singular and thus cannot be inverted. When the columns of the X-matrix are almost dependent the $X'X$ matrix will be almost singular and difficulties in inversion, including large round-off errors are likely.

One procedure that can be programmed and used as a routine check on X matrices (either in all cases or in suspected cases) consists of successively transforming the columns so that each new column is orthogonal to all previously transformed columns. If a column dependence exists we shall eventually obtain a new column which consists entirely of zeros. If the columns are nearly dependent, a new column will contain all very small numbers perhaps with some zeros. The procedure is completely general. The column transformation takes the following form:

$$\begin{aligned} \mathbf{Z}_{iT} &= (\mathbf{I} - \mathbf{Z}(\mathbf{Z}'\mathbf{Z})^{-1}\mathbf{Z}')\mathbf{Z}_i \\ &= \mathbf{Z}_i - \mathbf{Z}(\mathbf{Z}'\mathbf{Z})^{-1}\mathbf{Z}'\mathbf{Z}_i \end{aligned} \tag{5.6.1}$$

where $\mathbf{Z} =$ the matrix of column vectors already transformed,

$\mathbf{Z}_i =$ the next column vector of X to be transformed, and

$\mathbf{Z}_{iT} =$ the transformed vector orthogonal to vectors already in $\mathbf{Z}$.

Note that $\mathbf{Z}_{iT}$ is actually the residual vector of $\mathbf{Z}_i$ after $\mathbf{Z}_i$ has been regressed on the columns of $\mathbf{Z}$.

To illustrate this process we shall use a special case which will lead us to obtain the orthogonal polynomials for $N = 5$. Suppose values of Y are recorded at $X = 1, 2, 3, 4,$ and 5 and the model

$$Y = \beta_0 + \beta_1 X + \beta_2 X^2 + \beta_3 X^3 + \epsilon \tag{5.6.2}$$

is postulated. The original $\mathbf{X}$ matrix is

$$\mathbf{X} = \begin{array}{cccc} 1 & X & X^2 & X^3 \end{array}$$

$$\mathbf{X} = \begin{bmatrix} 1 & 1 & 1 & 1 \\ 1 & 2 & 4 & 8 \\ 1 & 3 & 9 & 27 \\ 1 & 4 & 16 & 64 \\ 1 & 5 & 25 & 125 \end{bmatrix}$$

We set

$$\mathbf{Z}_{1T} = \begin{bmatrix} 1 \\ 1 \\ 1 \\ 1 \\ 1 \end{bmatrix} = \mathbf{Z}$$

at this stage. (One column vector must be chosen to begin the process.) Choose

$$\mathbf{Z}_2 = \begin{bmatrix} 1 \\ 2 \\ 3 \\ 4 \\ 5 \end{bmatrix}$$

Then by Eq. (5.6.1)

$$\mathbf{Z}_{2T} = \begin{bmatrix} 1 \\ 2 \\ 3 \\ 4 \\ 5 \end{bmatrix} - \begin{bmatrix} 1 \\ 1 \\ 1 \\ 1 \\ 1 \end{bmatrix} (5)^{-1}(15)$$

$$= \begin{bmatrix} 1 \\ 2 \\ 3 \\ 4 \\ 5 \end{bmatrix} - \begin{bmatrix} 3 \\ 3 \\ 3 \\ 3 \\ 3 \end{bmatrix} = \begin{bmatrix} -2 \\ -1 \\ 0 \\ 1 \\ 2 \end{bmatrix}$$

At this stage

$$\mathbf{Z} = \begin{bmatrix} 1 & -2 \\ 1 & -1 \\ 1 & 0 \\ 1 & 1 \\ 1 & 2 \end{bmatrix} = [\mathbf{Z}_{1T}, \mathbf{Z}_{2T}]$$

and the third column of $\mathbf{X}$ is used as $\mathbf{Z}_3$. We find that

$$(\mathbf{Z'Z})^{-1} = \begin{bmatrix} 5 & 0 \\ 0 & 10 \end{bmatrix}^{-1} = \begin{bmatrix} \frac{1}{5} & 0 \\ 0 & \frac{1}{10} \end{bmatrix}$$

$$\mathbf{Z'Z}_i = \begin{bmatrix} 55 \\ 60 \end{bmatrix}, \text{ so } (\mathbf{Z'Z})^{-1}\mathbf{Z'Z}_i = \begin{bmatrix} 11 \\ 6 \end{bmatrix}.$$

Hence

$$\mathbf{Z}_{3T} = \begin{bmatrix} 1 \\ 4 \\ 9 \\ 16 \\ 25 \end{bmatrix} - \begin{bmatrix} 11 - 12 \\ 11 - 6 \\ 11 + 0 \\ 11 + 6 \\ 11 + 12 \end{bmatrix} = \begin{bmatrix} 2 \\ -1 \\ -2 \\ -1 \\ 2 \end{bmatrix}.$$

We leave the evaluation of $\mathbf{Z}_{4T}$ as an exercise. The final, orthogonal matrix is as below.

$$\begin{bmatrix} 1 & -2 & 2 & -1.2 \\ 1 & -1 & -1 & 2.4 \\ 1 & 0 & -2 & 0 \\ 1 & 1 & -1 & -2.4 \\ 1 & 2 & 2 & 1.2 \end{bmatrix}$$

Note that the first three columns are ψ_0, ψ_1, and ψ_2—the orthogonal polynomials of zero, first, and second order for $n = 5$. The fourth column is 1.2 times ψ_3, the orthogonal polynomial of third order for $n = 5$.

As remarked above, the process is completely general. It is true that column dependence can also be detected by noting that the determinant of the $\mathbf{X'X}$ matrix (or the correlation matrix) is zero. The transformation procedure has the additional advantage of showing in what columns the dependence occurs, however.

EXERCISES

A. The following data are observed at equally spaced intervals of time.

Observation Number	Response Y
1	1
2	4
3	6
4	7
5	9.5
6	11
7	11.5
8	13
9	13.5

Requirements

Assume that one time trend holds for the first four observations and that another time trend holds for the last five observations.

1. Using dummy variables as in Section 5.3, Example 3, determine the slopes of the two lines. Interpret the β coefficients.
2. What is the best estimate of β_3? What does this coefficient represent? Can you say anything about the point of intersection of the two lines?
3. Determine if the fitted model is statistically significant.

B. Using the generalized transformation procedure in Section 5.6 for obtaining a matrix with orthogonal columns, show that the process of fitting the model $Y = \beta_0 + \beta_1 X_1 + \beta_2 X_2 + \beta_3 X_3 + \epsilon$ to the following data by least squares leads to a singular $\mathbf{X'X}$ matrix.

X_1	X_2	X_3	Y
1	-2	4	81
2	-7	11	88
4	3	5	94
7	1	13	95
8	-1	17	123

C. Eighteen observations were obtained on four independent variables and one dependent variable in a process. It is suggested that the model

$$Y = \beta_0 X_0 + \beta_1 X_1 + \beta_2 X_2 + \beta_3 X_3 + \beta_4 X_4 + \beta_{11} X_1^2 + \beta_{12} X_1 X_2 + \beta_{22} X_2^2$$
$$+ \beta_{13} X_1 X_3 + \beta_{14} X_1 X_4 + \epsilon$$

is a reasonable one. The data are as shown at the top of page 160.

Requirements

1. Examine the data and the model. Is it possible to fit the proposed model to the data? Why or why not?
2. Determine an estimate of the variance of the random error, σ^2.

X_1	X_2	X_3	X_4	Y
20	50	75	15	27
27	55	60	20	23
22	62	68	16	18
27	55	60	20	26
24	75	72	8	23
30	62	73	18	27
32	79	71	11	30
24	75	72	8	23
22	62	68	16	22
27	55	60	20	24
40	90	78	32	16
32	79	71	11	28
50	84	72	12	31
40	90	78	32	22
20	50	75	15	24
50	84	72	12	31
30	62	73	18	29
27	55	60	20	22

D. Four levels, coded as -3, -1, 1, and 3 were chosen for each of two variables X_1 and X_2, to provide a total of sixteen experimental conditions when all possible combinations (X_1, X_2) were taken. It was decided to use the resulting sixteen observations to fit a regression equation including a constant term, all possible first-order, second-order, third-order and fourth-order terms in X_1 and X_2. The data were fed into a computer routine which usually obtains a vector estimate

$$b = (X'X)^{-1}X'Y.$$

The computer refused to obtain the estimates. Why?

The experimenter, who had meanwhile examined the data, decided at this stage to ignore the levels of variable X_2 and fit a fourth-order model in X_1 only to the *same* observations. The computer again refused to obtain the estimates. Why?

E. Over a 15-year period, 1950–1964, the Gillette Company reported per share net income of 0.64, 0.60, 0.56, 0.73, 0.92, 1.04, 1.13, 0.93, 0.99, 1.11, 1.33, 1.52, 1.60, 1.47, and 1.33 (part of these data were used in Section 5.5, as an example). Fit a polynomial of suitable order to these data using orthogonal polynomials and plot both the observed and fitted values on a graph. Evaluate the residuals = observed values − fitted values and examine them using the methods given in Chapter 3.

F. A newly born baby was weighed weekly, the figure adopted in each case being the average of the weights on three successive days. Twenty such weights are shown at the top of page 161, in ounces. Fit to the data, using orthogonal polynomials, a polynomial model of a degree justified by the accuracy of the figures, that is, test as you go along for the significance of the linear, quadratic, etc., terms.

No. of week	1	2	3	4	5	6	7	8	9	10
Weight	141	144	148	150	158	161	166	170	175	181
No. of week	11	12	13	14	15	16	17	18	19	20
Weight	189	194	196	206	218	229	234	242	247	257

(*Cambridge Diploma*, 1950)

G. Bars of soap are scored for their appearance in a manufacturing operation. These scores are on a 1-10 scale, and the higher the score the better. The difference between operator performance and the speed of the manufacturing line are believed to measurably affect the quality of the appearance. The following data were collected on this problem:

Operator	Line Speed	Appearance (Sum for 30 Bars)
1	150	255
1	175	246
1	200	249
2	150	260
2	175	223
2	200	231
3	150	265
3	175	247
3	200	256

Requirements

1. Using dummy variables, fit a multiple regression model to this data.
2. Using $\alpha = 0.05$, determine whether operator differences are important in bar appearance. Using the regression model, demonstrate that the average appearance score for operator no. 1 is 250, operator no. 2 is 238, and operator no. 3 is 256.
3. Does line speed affect appearance? (Use $\alpha = 0.05$.)
4. What model would you use to predict bar appearance?

H. A finished product is known to lose weight after it is produced. The following data demonstrate this drop in weight.

Time After Production	Weight Difference (in 1/16 oz)
t	*Y*
0	0.21
0.5	−1.46
1.0	−3.04
1.5	−3.21
2.0	−5.04
2.5	−5.37
3.0	−6.03
3.5	−7.21
4.0	−7.46
4.5	−7.96

Requirements

1. Using orthogonal polynomials, develop a second-order fitted equation which represents the loss in weight as a function of time after production.
2. Analyze the residuals from this model and draw conclusions about its adequacy.

I. The cloud point of a liquid is a measure of the degree of crystallization in a stock that can be measured by the refractive index. It has been suggested that the percentage of I-8 in the base stock is an excellent predictor of cloud point using the second order model:

$$Y = \beta_0 + \beta_1 X + \beta_{11} X^2 + \epsilon.$$

The following data were collected on stocks with known percentage of I-8.

% I-8	Cloud Point
X	Y
0	22.1
1	24.5
2	26.0
3	26.8
4	28.2
5	28.9
6	30.0
7	30.4
8	31.4
0	21.9
2	26.1
4	28.5
6	30.3
8	31.5
10	33.1
0	22.8
3	27.3
6	29.8
9	31.8

Requirements

1. Determine the best fitting second-order model.
2. Using $\alpha = 0.05$, check the overall regression.
3. Test for lack of fit.
4. Would the first-order model, $Y = \beta_0 + \beta_1 X + \epsilon$, have been sufficient? Use the residuals from this simpler model to support your conclusions.
5. Comment on the use of the fitted second-order model as a predictive equation.

CHAPTER 6

SELECTING THE "BEST" REGRESSION EQUATION

6.0. Introduction

We shall defer discussion of the general model-building process to Chapter 8, and in this chapter deal only with the use of specific statistical procedures for selecting variables in regression. Suppose we wish to establish a linear regression equation for a particular response Y in terms of "independent" or predictor variables $X_1, X_2, \ldots, X_k$. We assume this is the complete set of variables from which the equation is to be chosen and includes any functions, such as squares and cross products, thought to be desirable and necessary. Two opposed criteria of selecting a resultant equation are usually involved. They are as follows:

1. To make the equation useful for predictive purposes we should want our model to include as many X's as possible so that reliable fitted values can be determined.
2. Because of the costs involved in obtaining information on a large number of X's and subsequently monitoring them, we should like the equation to include as few X's as possible.

The compromise between these extremes is what is usually called *selecting the best regression equation*. There is no unique statistical procedure for doing this, and personal judgment will be a necessary part of any of the statistical methods discussed. In this chapter we shall describe several procedures which have been proposed; all of these appear to be in current use. To add to the confusion they do not all necessarily lead to the same solution when applied to the same problem, although for many problems they *will* achieve the same answer. The following procedures will be discussed: (1) all possible regressions, (2) backward elimination, (3) forward selection, (4) stepwise regression, (5) two variations on the four previous methods, and (6) stagewise regression. After each discussion, we state our opinion of the procedure and, in Section 6.7, we compare the various equations fitted to a particular set of data.

6.1. All Possible Regressions

This procedure is a rather cumbersome one and is quite impossible without access to a high-speed computer. Thus it has come into use only since fast computers have become generally available. The procedure first requires the fitting of every possible regression equation which involves X_0 plus any number of the variables $X_1, \ldots, X_k$ (where we have added a dummy variable $X_0 = 1$ to the set of X's as usual). Since each X_i can either be, or not be, in the equation (two possibilities) and this is true for every X_i, $i = 1, 2, \ldots, k$ (kX's), there are altogether 2^k equations. (The X_0 term is always in the equation.) If $k = 10$, a not unusually excessive number, $2^k = 1024$ equations must be examined for the second step in the procedure! Here the regressions are divided into sets of runs which involve p variables, $p = 1, 2, \ldots, k$ and each set is ordered according to some criterion; usually the criterion is the value of R^2 achieved by the least squares fit. The leaders in this ordering within each set are then selected for further examination and a decision is made, after viewing the R^2 values, on which equation is best to use. An example will help make this clear. We shall use the data in a four-variable ($k = 4$) problem given by A. Hald on page 647 of his book *Statistical Theory with Engineering Applications*, published by John Wiley & Sons. This particular problem has been chosen because it illustrates some typical difficulties which occur in regression analysis. The data are given on the computer sheets in Appendix B. The independent variables here are X_1, X_2, X_3, and X_4. The response variable is X_5. A β_0 term is *always* included. Thus there are $2^4 = 16$ possible regression equations, which involve X_0 and the X_i, $i = 1, 2, 3, 4$. These also appear in Appendix B. We now apply the procedure given above.

1. Divide the runs into five sets:

   Set A consists of the run with only the mean value. (Model $E(Y) = \beta_0$)
   Set B consists of the four 1-variable runs. (Model $E(Y) = \beta_0 + \beta_i X_i$)
   Set C consists of all the 2-variable runs. (Model $E(Y) = \beta_0 + \beta_i X_i + \beta_j X_j$)

   Set D consists of all the 3-variable runs. (And so on . . .)
   Set E consists of the run with 4 variables.

2. Order the runs within each set by the value of the square of the multiple correlation coefficient, R^2.

3. Examine the leaders and see if there is any consistent pattern of variables in the leading equations in each set. For this example, we have

Set	Variables in Equation	R^2
Set B	$\hat{Y} = f(X_4)$	67.5%
Set C	$\hat{Y} = f(X_1, X_2)$	97.9%
	$\hat{Y} = f(X_1, X_4)$	97.2%
Set D	$\hat{Y} = f(X_1, X_2, X_4)$	98.234%
Set E	$\hat{Y} = f(X_1, X_2, X_3, X_4)$	98.237%

(Notice that in set C there are two leaders with practically the same size R^2 value.) If we view these results we see that after two variables have been introduced, further gain in R^2 is minor. Examination of the correlation matrix for the data (Appendix B) reveals that (X_1 and X_3) and (X_2 and X_4) are highly correlated, since

$$r_{13} = -0.82413372 \quad \text{and} \quad r_{24} = -0.97295516.$$

Thus, the addition of further variables when X_1 and X_2 or when X_1 and X_4 are already in the regression equation will remove very little of the unexplained variation in the response. This is clearly shown by the slight increase in R^2 from set C to set D. The gain in R^2 from set D to set E is extremely small. This is simply explained by the observation that the X's are mixture ingredients and the sum of the X values for any specific point is nearly a constant (actually between 95 and 99).

What equation should be selected for further attention? One of the equations in set C is clearly indicated but which one? If $f(X_1, X_2)$ is chosen, there is some inconsistency because the best single variable equation involves X_4. For this reason many workers would prefer to use $f(X_1, X_4)$. The examination of all possible regressions does not provide a clear-cut answer to the problem. Other information, such as knowledge of the characteristics of the product studied and the physical role of the X-variables, must as always be added to enable a decision to be made.

If all regressions are done on a large problem, an assessment of the magnitude of the residual mean square as the number of variables in regression increases sometimes indicates the best cutoff point for the number of variables in regression. For example, in the Hald data, the average residual mean square for all sets of "r" variables can be calculated and plotted against r. Using the computer sheets in Appendix B, the residual mean squares are shown as follows:

r	Residual Mean Squares	Average $s^2(r)$
1	115.06, 82.39, 176.31, 80.35	113.53
2	5.79, 122.71, 7.48, 41.54, 86.89, 17.57	47.00
3	5.35, 5.33, 5.65, 8.20	6.13
4	5.98	5.98

The plot of the average $s^2(r)$ against r is shown in Figure 6.1. It appears here that three variables should be included. However, this procedure yields only a guideline to selecting the number of variables to enter regression. It does not pick out the specific set of "r" variables, and the procedure does not guarantee that there is not a better set of "k" variables ($k < r$). For example, in the Hald data, one of the runs with $r = 2$ variables had a residual mean square of 5.79, indicating that there existed a better run with two variables than was indicated by the average residual mean square for $r = 2$, namely, 47.00.

GENERAL REMARKS. It was mentioned earlier that the data used in this example contained a theoretical restriction $X_1 + X_2 + X_3 + X_4 = $ constant. This means that X_4 is, theoretically, dependent on X_1, X_2, and X_3. Thus if all four X's are included in the model, the $\mathbf{X'X}$ matrix would, theoretically, be singular and have zero determinant, both before and after transformation. As we see from the appropriate calculation sheet the transformed determinant actually has the very small value 0.0010677. When the determinant does have such a small value it sometimes happens that the calculations involve primarily rounding errors and are meaningless. While this has not happened in the present case, the occurrence of a small determinant must always be regarded as a danger signal.

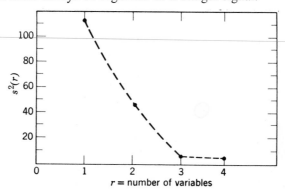

Figure 6.1 Plot of average residual mean square against r.

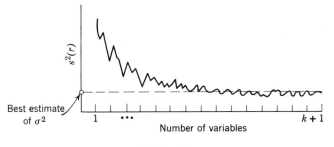

Figure 6.2

When the number of potential variables in the model is large, say greater than $k = 10$, and the number of data points much larger than k, say $5k$ to $10k$, the plot of $s^2(r)$ is worth considering when one does all regressions. The fitting of regression equations that involve more independent variables than are necessary to obtain a satisfactory fit to the data is called *overfitting*. As more and more independent variables are added to an already overfitted equation the residual mean square will tend to stabilize and approach the true value of σ^2 as the number of variables increases, provided that all important variables have been included, and the number of observations greatly exceeds the number of variables in the fitted equation—five to ten times as many as indicated above. This situation is illustrated in Figure 6.2.

OPINION. In general the analysis of all regressions is quite unwarranted. While it means that the statistician has "looked at all possibilities" it also means he has examined a large number of regression equations which intelligent thought would often reject out of hand. The amount of computer time used is wasteful and the sheer physical effort of examining all the computer printouts is enormous when more than a few variables are being examined. Some sort of selection procedure which shortens this task is preferable.

6.2. The Backward Elimination Procedure

The backward elimination method is an improvement on the "all regressions" method in that it attempts to permit the examination not of *all* regressions but of only the "best" regression containing a certain number of variables. The basic steps in the procedure are these:
1. A regression equation containing all variables is computed.
2. The partial F-test value is calculated for every variable treated as though *it were the last variable to enter the regression equation.*
3. The lowest partial F-test value, F_L say, is compared with a preselected significance level F_0, say.

(a) If $F_L < F_0$, remove the variable X_L, which gave rise to F_L, from consideration and recompute the regression equation in the remaining variables; re-enter stage 2.

(b) If $F_L > F_0$, adopt the regression equation as calculated.

Using the same data (Hald, 1952) as we used in the previous section, we can illustrate this procedure.

First, do the complete regression on all independent variables. In the example referred to in Section 6.1, find the least squares equation, $\hat{Y} = f(X_1, X_2, X_3, X_4)$. The regression procedures force an ordering of the variables into regression. In Appendix B, p. 395, note that the complete model was obtained by fitting X_4 first, then X_3, then X_2, and finally X_1. In order to eliminate variables at this point, one must determine the contribution of each of the variables X_1, X_2, X_3, and X_4 to the regression sum of squares as if each were in the last position. The partial F-values shown in the last column of this printout provide measures of these contributions.

Using the partial F-test, choose the smallest value and compare it to some critical value of F based on a predetermined α-risk. In this case, the critical F value for, say, $\alpha = 0.10$ is $F(1, 8, 0.90) = 3.46$. The smallest partial F is for variable, X_3; namely calculated $F = 0.0182345$. Since the calculated F is smaller than the critical value 3.46 reject X_3.

Next, find the least squares equation, $\hat{Y} = f(X_1, X_2, X_4)$. This is shown on page 389. The overall F value for the equation is $F = 166.83$, which is statistically significant and in fact exceeds $F(3, 9, 0.999) = 13.90$. Examining this equation for potential elimination, one sees that X_4 should be removed. The procedure for this elimination is similar to the preceding elimination with one change; namely, the critical F value is $F(1, 9, 0.90) = 3.36$.

We now find the least squares equation $\hat{Y} = f(X_1, X_2)$, shown on page 375. This provides a statistically significant overall equation with an F value of 229.50 which exceeds $F(2, 10, 0.999) = 14.91$. Both variables X_1 and X_2 are significant regardless of position, as indicated by the significant partial F's. Thus, the backward elimination selection procedure is terminated and yields the equation:

$$\hat{Y} = 52.58 + 1.47X_1 + 0.66X_2.$$

OPINION. This is a very satisfactory procedure in general especially for statisticians who like to see all the variables in the equation once in order "not to miss anything." It is much more economical of computer time and manpower than the "all regressions" method. However, if the input data yields an $X'X$ matrix which is ill-conditioned—that is, nearly singular—then this procedure may yield nonsense due to rounding errors. With new computing equipment this is not usually a serious problem.

We believe this method to be slightly inferior to another selection procedure to be given. On the whole, though, it is an excellent procedure. Note that some programs based on this procedure use a t-test on the square root of the partial F-value instead of an F-test as given above.

6.3. The Forward Selection Procedure

The backward elimination method begins with the largest regression, using all variables, and subsequently reduces the number of variables in the equation until a decision is reached on the equation to use. The forward selection procedure is an attempt to achieve a similar conclusion working from the other direction, that is, to *insert* variables in turn until the regression equation is satisfactory. The order of insertion is determined by using the partial correlation coefficient as a measure of the importance of variables not yet in the equation. The basic procedure is as follows. First we select the X most correlated with Y (suppose it is X_1) and find the first-order, linear regression equation $\hat{Y} = f(X_1)$. We next find the partial correlation coefficient of X_j ($j \neq 1$) and Y (after allowance for X_1). Mathematically this is equivalent to finding the correlation between (a) the residuals from the regression $\hat{Y} = f(X_1)$ and (b) the residuals from another regression $\hat{X}_j = f_j(X_1)$ (which we have not actually performed). The X_j with the highest partial correlation coefficient with Y is now selected, (suppose this is X_2) and a second regression equation $\hat{Y} = f(X_1, X_2)$ is fitted. This process continues. After $X_1, X_2, \ldots, X_q$ are in the regression the partial correlation coefficients are the correlations between (a) the residuals from the regression $\hat{Y} = f(X_1, X_2, \ldots, X_q)$ and (b) the residuals from a regression $\hat{X}_j = f_j(X_1, X_2, \ldots, X_q)$ ($j > q$). As each variable is entered into the regression, the following values are examined:

1. R^2, the multiple correlation coefficient.
2. The partial F-test value for the variable most recently entered, which shows whether the variable has taken up a significant amount of variation over that removed by variables previously in the regression.

As soon as the partial F value related to the most recently entered variable becomes nonsignificant the process is terminated.

We shall use the Hald data once again to illustrate the workings of the forward selection procedure. The analysis would proceed as follows. (Refer to the printout where indicated.)

1. Calculate the correlation of all the independent variables with the response. Select, as the first variable to enter the regression the one most highly correlated with the response. Examination of the correlation matrix in Appendix B shows that X_4 is most highly correlated with the

response Y or X_5; $r_{45} = -0.82130513$ or -0.821. (We shall round off these figures to three decimal places.) Thus X_4 is the first variable to enter the regression equation.

2. Regress Y on X_4 and obtain the least squares equation shown on page 373. The overall F-test shows that the regression equation is significant.
3. Calculate the partial correlation coefficients of all variables not in. regression with the response. Choose as the next variable to enter into the regression the one with the highest partial correlation coefficient. This is variable X_1. Shown at the bottom of page 373 is the square of the partial correlation coefficient of X_1 with X_5 ($= Y$), given that X_4 is in regression; the value is

$$r_{15.4}^2 = 0.915.$$

4. With X_1 as well as X_4 in the regression, the least squares equation $\hat{Y} = f(X_4, X_1)$ is that shown on page 379. This equation has a percentage R^2 of 97.2% and is significant, since we have an overall $F = 176.63$ which exceeds $F(2, 10, 0.999) = 14.91$; that the new variable X_1 provides a significant decrease in the residual sum of squares is shown by its partial F value, 108.22, which exceeds $F(1, 10, 0.999) = 21.04$.
5. The squared partial correlation coefficients of all variables not in regression with the response X_5 are shown at the bottom of page 379. The variable to be entered next is X_2.
6. The new equation $\hat{Y} = f(X_4, X_1, X_2)$ is shown on page 389. The percentage square of the multiple correlation coefficient, R^2, has increased from 97.2% to 98.2%; the addition of X_2 to the regression is statistically significant if we take $\alpha = 0.10$ as shown by its partial F value 5.03 which exceeds $F(1, 9, 0.90) = 3.36$.
7. Since each variable has entered accompanied by a significant reduction in the error mean square, we continue and enter the last variable, X_3. The analysis is shown in Appendix B, page 395. The partial F value for X_3 is 0.018 which is not significant. Thus X_3 would be rejected and the procedure terminated.

The complete analysis of variance table at this final stage is as follows:

Source of Variation	df	SS	MS
Total	12	2715.76	
Due to Regression	4	2667.90	
due to X_4	1	1831.90	1831.90*
due to $X_1 \mid X_4$	1	809.10	809.10*
due to $X_2 \mid X_4, X_1$	1	26.79	26.79*
due to $X_3 \mid X_4, X_1, X_2$	1	0.11	0.11
Due to error	8	47.86	5.98

This procedure then would lead to the choice of a regression equation as

$$\hat{Y} = 71.65 - 0.24X_4 + 1.45X_1 + 0.42X_2.$$

OPINION.   The forward selection procedure is basically a good idea. It is more economical of computer facilities than the methods previously discussed, and it avoids working with more X's than are necessary while improving the equation at every stage. One of its drawbacks is that it makes no effort to explore the effect that the introduction of a new variable may have on the role played by a variable which entered at an earlier stage. This deficiency is overcome by the stepwise procedure which is thus an improvement on the forward-selection procedure.

6.4.   The Stepwise Regression Procedure

In spite of its entirely different name, this procedure is, in fact, an improved version of the forward-selection procedure discussed in the previous section. The improvements involve the re-examination at every stage of the regression of the variables incorporated into the model in previous stages. A variable which may have been the best single variable to enter at an early stage may, at a later stage, be superfluous because of the relationships between it and other variables now in the regression. To check on this, the partial F criterion for each variable in the regression at any stage of calculation is evaluated and compared with a preselected percentage point of the appropriate F distribution. This provides a judgment on the contribution made by each variable as though it had been the most recent variable entered, irrespective of its actual point of entry into the model. Any variable which provides a nonsignificant contribution is removed from the model. This process is continued until no more variables will be admitted to the equation and no more are rejected. The complete stepwise solution for the Hald data is shown in Appendix B, pages 397-402.

STEP 1.   The stepwise procedure starts with the simple correlation matrix and enters into regression the X variable most highly correlated with the response. Here X_4 is entered as in the forward-selection procedure.

STEP 2.   Using the partial correlation coefficients as before, it now selects, as the next variable to enter regression, that X variable whose partial correlation with the response is highest. In this problem it is X_1, again as in the forward selection procedure (p. 399).

STEP 3.   Given the regression equation $\hat{Y} = f(X_4, X_1)$, the method now examines the contribution X_4 would have made if X_1 had been entered first and X_4 entered second. (The forward selection procedure *does not* do this.) Since the value of the partial F (shown on p. 399) is 159.295, which is statistically significant at $\alpha = 0.05$, X_4 is retained. The stepwise method now selects as the next variable to enter, the one most highly partially correlated with the response (given that variables X_4 and X_1 are already in regression). This is seen to be variable X_2. (The squared partial correlation coefficient of X_2 with the response is 0.358 shown at the bottom of page 399.)

STEP 4.   A regression equation of form $\hat{Y} = f(X_4, X_1, X_2)$ is now determined by least squares. The variable X_2 enters with a significant sequential F value of 5.026 which exceeds $F(1, 9, 0.90) = 3.36$. At this point partial F-tests for the variables X_1 and X_4 are made to determine if they should remain in the regression equation. As a consequence, X_4 is rejected since its partial F value of 1.863 given on page 400 is less than $F(1, 9, 0.90) = 3.36$.

STEP 5.   The only remaining variable is X_3. Since this variable is immediately rejected, the stepwise regression procedure terminates and chooses as its best regression equation $\hat{Y} = f(X_1, X_2)$ as shown on page 401, namely,

$$\hat{Y} = 52.58 + 1.47X_1 + 0.66X_2.$$

OPINION.   We believe this to be the best of the variable selection procedures discussed and recommend its use. However, stepwise regression can easily be abused by the "amateur" statistician. As with all the procedures discussed, sensible judgment is still required in the initial selection of variables and in the critical examination of the model through examination of residuals. It is easy to rely too heavily on the automatic selection performed in the computer. A discussion of the method is given by M. A. Efroymson in an article, "Multiple Regression Analysis" in the book *Mathematical Methods for Digital Computers* (see also Section 6.8).

6.5.   Two Variations on the Four Previous Methods

While the procedures discussed do not necessarily select the absolute best model, they usually select an acceptable one. However, some alternative, combined procedures have been suggested in attempts to improve the model selection. We now discuss two of these.

1. One proposal has been as follows: Run the stepwise regression procedure with given levels for acceptance and rejection. When the selection procedure stops, determine the number of variables in the final selected model. Using this number of variables, say r, do all possible sets of r variables from the k original variables and choose the best set.

OPINION. This procedure would reveal the situation pointed out in the discussion of the Hald data for the two-variable case: namely, that there are two candidates for the model instead of one. When this happens, one could say that there is insufficient information in the data to make a clear-cut single choice. Other *a priori* considerations and the experimenter's judgment would be needed to select a final predictive model. This procedure also fails when a larger set of variables would be better, but was never examined by the stepwise algorithm. In our experience, the added advantages of this procedure are minor, and much extra computation is involved.

2. A second proposal is to use the stepwise regression, with less restrictive acceptance and rejection levels, which forces the program to accept several additional variables beyond what would be accepted with less conservative levels. This allows the investigation of additional variables that would not enter under the usual stepwise procedure and may lead to a different model.

OPINION. This procedure has proved helpful occasionally; that is, it has provided an alternative model with about the same predictive characteristics. In our experience, however, this occurs when the problem is almost impossible to solve because of *very high* intercorrelations among the independent (X) variables and thus requires more than just statistical screening procedures.

6.6. The Stagewise Regression Procedure

This procedure does not provide a true least squares solution for the variables included in the final equation. The basic idea is as follows. After a regression equation in the X variable most correlated with Y has been fitted, the residuals $Y_i - \hat{Y}_i$ are found. These residuals are now considered as response values and regressed against the X (of those which remain) most correlated with this new response. The process is continued to any desired stage. Since at each stage

Response = Fitted Response + (Response − Fitted Response)

the regression equations can be back-substituted stage by stage until the final stagewise equation is attained. It is *not* the least squares solution for the variables involved. We shall again use the Hald example for our illustration. The calculations are quite easy to perform on a hand calculator without the aid of an electronic computer. (Our printouts in Appendix B are unsuitable beyond the first stage of this method, of course.)

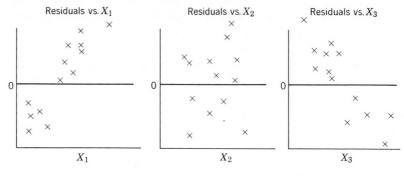

Figure 6.3

STEP 1.   Plot the response Y against each potential X in turn. This corresponds to inspecting the complete correlation matrix and choosing, as before, that X most highly correlated with the response.

STEP 2.   Regress Y on X_4, which gives the equation $\hat{Y} = 117.57 - 0.74X_4$, and evaluate the residuals $Z_i = Y_i - \hat{Y}_i$ for each X_4 value. These are shown on page 374.

STEP 3.   Use these residuals Z_i as responses and select from the remaining X variables the one most highly correlated with these residuals. An initial plot is often useful (see Figure 6.3).

STEP 4.   Although both X_1 and X_3 are possible candidates, calculation of the correlation coefficients reveals X_1 to be the X most correlated with the residuals. Thus we now regress the residuals Z_i from Step 2 against X_1. The data for this regression are:

X_1	Z
7	5.22
1	-4.88
11	1.50
11	4.73
7	2.69
11	7.87
3	-10.44
1	-12.59
2	-8.22
21	17.52
1	-8.67
11	4.59
10	0.69

The best fitting line is as follows:

$$\hat{Z} = -10.10 + 1.35X_1.$$

At the first stage we could express

$$Y_i = \hat{Y}_i + (Y_i - \hat{Y}_i)$$
$$= \hat{Y}_i + Z_i.$$

Now, since

$$Z_i = \hat{Z}_i + (Z_i - \hat{Z}_i),$$

we can write

$$Y_i = \hat{Y}_i + \hat{Z}_i + (Z_i - \hat{Z}_i).$$

Thus at this second stage our fitted equation is represented by $\hat{Y}_i + \hat{Z}_i$, that is,

$$117.57 - 0.74X_4 - 10.10 + 1.35X_1$$

or

$$107.47 - 0.74X_4 + 1.35X_1.$$

The residuals are now $Z_i - \hat{Z}_i = Y_i - \hat{Y}_i - \hat{Z}_i$ as expected. Other variables can be added in similar fashion—at each stage regressing the current residuals against a new variable until a regression is nonsignificant, when the process is terminated, without including the final variable. Terminating our example at the second stage, we see that the final equation is *not* the least squares equation involving X_1 and X_4.

By comparison, the least squares equation for this model is, from page 379 of Appendix B,

$$\hat{Y} = 103.10 - 0.61X_4 + 1.44X_1.$$

Mathematically, the differences in the b coefficients are as follows:

Least Squares	Stagewise
$b_4 = \dfrac{\left[\sum\limits_{i=1}^{n} x_{4i}y_i - \dfrac{\sum\limits_{i=1}^{n} x_{4i}x_{1i} \sum\limits_{i=1}^{n} x_{1i}y_i}{\sum\limits_{i=1}^{n} x_{1i}^2}\right]}{\sum\limits_{i=1}^{n} x_{4i}^2 - \dfrac{\left(\sum\limits_{i=1}^{n} x_{4i}x_{1i}\right)^2}{\sum\limits_{i=1}^{n} x_{1i}^2}}$	$b_4 = \dfrac{\sum\limits_{i=1}^{n} x_{4i}y_i}{\sum\limits_{i=1}^{n} x_{4i}^2}$
$b_1 = \dfrac{\left[\sum\limits_{i=1}^{n} x_{1i}y_i - \dfrac{\sum\limits_{i=1}^{n} x_{4i}x_{1i} \sum\limits_{i=1}^{n} x_{4i}y_i}{\sum\limits_{i=1}^{n} x_{4i}^2}\right]}{\sum\limits_{i=1}^{n} x_{1i}^2 - \dfrac{\left(\sum\limits_{i=1}^{n} x_{4i}x_{1i}\right)^2}{\sum\limits_{i=1}^{n} x_{4i}^2}}$	$b_1 = \dfrac{\sum\limits_{i=1}^{n} x_{1i}y_i - \dfrac{\sum\limits_{i=1}^{n} x_{4i}x_{1i} \sum\limits_{i=1}^{n} x_{4i}y_i}{\sum\limits_{i=1}^{n} x_{4i}^2}}{\sum\limits_{i=1}^{n} x_{1i}^2}$

where $x_{ji} = X_{ji} - \bar{X}_j$,
$y_i = Y_i - \bar{Y}$.

Comparing b_1 obtained by least squares (call it $b_{1,\text{LS}}$) and b_1 obtained by the stagewise process (call it $b_{1,\text{SW}}$),

$$b_{1,\text{SW}} = b_{1,\text{LS}} \left[\frac{\sum x_1^2 - \dfrac{(\sum x_4 x_1)^2}{\sum x_4^2}}{\sum x_1^2} \right]$$

$$= b_{1,\text{LS}} \left[1 - \frac{(\sum x_4 x_1)^2}{\sum x_1^2 \sum x_4^2} \right].$$

We can rewrite this in the form

$$b_{1,\text{SW}} = b_{1,\text{LS}}[1 - r_{14}^2]$$

where $r_{14}^2 =$ the square of the simple correlation coefficient for X_1, X_4. Thus, the stagewise regression estimate is smaller, in absolute value than the least squares estimate by a factor proportional to the square of the correlation coefficient. This, in turn, implies that the variable being added is more important than the regression on the residuals would indicate.

While this procedure will always yield less precision, that is, larger residual mean square, than the least squares procedure it has the following advantage. It enables one to select a first variable because of reasons other than its correlation with Y. For example, given a set of X's highly correlated with each other, the selection procedures will choose first that X most highly correlated with Y, say X_1. At the next stage, all other X-variables not in regression are adjusted for X_1. Thus, if X_2 is highly correlated with X_1, it might be rejected as a possible variable. However, X_2 might be just the variable that the experimenter wants to use; for example, X_2 might be a variable directly under the control of the experimenter while X_1 might be a variable that is not under his control. In a marketing situation, for example, X_2 might be one's own advertising expenditures while X_1 might be competitive advertising expenditures. Thus, the experimenter could force a regression $\hat{Y} = f(X_2)$ and then continue the problem by using the residuals from this fitting as the dependent variable in further regressions.

In addition, there are situations where a trend in the data should be removed before trying to write a prediction equation. Economists often correct the data for trend or seasonality and then proceed to analyze the resultant deviations by least squares procedures.

OPINION. One feature of the stagewise regression is that variables can be entered in such a way that the expected direction of any effects can be preserved (otherwise they need not be entered). A straightforward least squares equation does not always achieve this due to correlations between the variables and the particular regions in the X-space in which our data have occurred. There is no harm in this provided we do not attempt to isolate specific terms in the model. The true least squares fitting usually provides better overall prediction than the stagewise fitting in spite of this apparent disadvantage. For this reason, stagewise fitting is not recommended for typical industrial problems. For its suitability in economic situations, see papers mentioned in the Bibliography.

6.7. A Summary of the Least Squares Equations from the Methods Described

Using the Hald data, the following equations have been selected by the various methods:

1. $\hat{Y} = 52.58 + 1.47X_1 + 0.66X_2$

    This equation was selected by the all-regressions method, the backward-elimination procedure, and the stepwise procedure.
2. The all regressions method also provided

$$\hat{Y} = 103.10 + 1.44X_1 - 0.61X_4$$

    as a close second choice, which was in one sense a more consistent choice, as discussed. The equation
3. $\hat{Y} = 71.65 + 1.45X_1 + 0.42X_2 - 0.24X_4$

    was given by the forward-selection procedure.

Although in many cases the same equation will arise from all procedures, this is not guaranteed as we can see in the cases above, and as illustrated by H. C. Hamaker in "On multiple regression analysis," *Statistica Neerlandica*, **16**, 1962, 31–56.

In a theoretical sense the all-regressions procedure is best in that it enables us to "look at everything." However, it is interesting to note that both the backward elimination and stepwise procedures picked the same equation. Much would depend on the selection of rejection levels for the various F-tests, however, and also on the statistician's feelings about the level of increase desired in R^2 when all regressions are viewed. Inconsistent choices might lead to entirely different equations being reached by the various methods and this is not surprising.

Our own preference for a practical regression method is the stepwise procedure. Our second choice is the backward elimination. In cases of doubt all regressions could be viewed if the physical effort were reasonable. The other methods given are, in our opinion, of lesser value, but each has its use in special cases.

6.8. Computational Method for Stepwise Regression

There are many computational methods available for performing multiple regression calculations. For example, an excellent, well-written exposition of the Abbreviated Doolittle method is found in the R. L. Anderson and T. A. Bancroft textbook, *Statistical Theory in Research*, 1952, McGraw-Hill Book Co., New York. Several other procedures can be found in P. S. Dwyer's book, *Linear Computations*, 1960, John Wiley and Son, New York. Since the stepwise procedure of Efroymson has proven useful in practice, the computations for this method will be illustrated using the Hald data recorded in Appendix B.

Computational Procedure

I. Construct the correlation matrix from the raw data.

DATA

X_1	X_2	X_3	X_4	Y
7	26	6	60	78.5
1	29	15	52	74.3
11	56	8	20	104.3
11	31	8	47	87.6
7	52	6	33	95.9
11	55	9	22	109.2
3	71	17	6	102.7
1	31	22	44	72.5
2	54	18	22	93.1
21	47	4	26	115.9
1	40	23	34	83.8
11	66	9	12	113.3
10	68	8	12	109.4

A. Calculate the corrected sums of squares and cross-products matrix. The matrix of uncorrected SS's and uncorrected sums of cross-products is

	X_1	X_2	X_3	X_4	Y
Totals	97	626	153	390	1240.5

$$\mathbf{X'X} = \begin{pmatrix} 1139 & 4922 & 769 & 2620 & 10032.0 \\ & 33050 & 7201 & 15739 & 62027.8 \\ & & 2293 & 4628 & 13981.5 \\ & & & 15062 & 34733.3 \\ & & & & 121088.09 \end{pmatrix}$$

After correction for the means, we obtain

$$
X_c'X_c = \begin{pmatrix}
415.230769 & 251.076924* & -372.615384 & -290 & 775.961540 \\
 & 2905.692310 & -166.538461 & -3041 & 2292.953850 \\
 & & 492.307693 & 38 & -618.230760 \\
 & & & 3362 & -2481.700000 \\
 & & & & 2715.763100
\end{pmatrix}
$$

*where for example $\sum x_1 x_2 = \sum X_1 X_2 - \dfrac{(\sum X_1)(\sum X_2)}{n}$; that is,

$$
251.076924 = 4922 - \frac{(97)(626)}{13} .
$$

Later we use the notation s_i^2 for the ith diagonal term of $X_c'X_c$.

B. Using $X_c'X_c$, calculate the correlation coefficients making up the correlation matrix. For example, using abbreviated summation notation as above

$$
r_{12} = \frac{\sum x_1 x_2}{\sqrt{(\sum x_1^2)(\sum x_2^2)}} = \frac{251.076924}{\sqrt{(415.230769)(2905.692310)}} = 0.228579
$$

C. Assemble the correlation coefficients into a matrix. For the Hald data, there are $k = 4$ independent variables, and in notational form, the correlation matrix is written as:

$$
R = \begin{bmatrix}
r_{11} & r_{12} & r_{13} & r_{14} \\
r_{21} & r_{22} & r_{23} & r_{24} \\
r_{31} & r_{32} & r_{33} & r_{34} \\
r_{41} & r_{42} & r_{43} & r_{44}
\end{bmatrix}
$$

D. Augment the correlation matrix, R, in the following manner:

$$
A = \begin{bmatrix}
R(k \times k) & T'(k \times 1) & I(k \times k) \\
T(1 \times k) & S(1 \times 1) & 0(1 \times k) \\
-I(k \times k) & 0(k \times 1) & 0(k \times k)
\end{bmatrix}
$$

where $R(k \times k)$ = correlation matrix for the k independent variables,

$T(1 \times k)$ = correlation vector of the k independent variables with the response, Y,

$T'(k \times 1)$ = transpose of T,
$S(1 \times 1)$ = correlation of the response with itself ($\equiv 1$),
$I(k \times k)$ = identity matrix,
$-I(k \times k)$ = negative identity matrix.

The **A** matrix for the Hald problem is shown in Tableau I. The elements of this matrix are to be identified from now on by A_{ij} where i refers to the row of the matrix and j refers to the column of the matrix.

II. Stepwise procedure. The basic idea of the method is to perform a regression with several variables as a series of straight line regressions (as in Section 4.1) and to readjust the Tableau stage by stage to effect this. At each stage, appropriate entries in the Tableau provide regression coefficients and entries for the analysis of variance table in coded units. Properly combined, as indicated in the following pages, the entries also provide F-values for two series of tests, one for the entry of a variable into the regression equation, and one for the deletion of a variable from the regression equation. Although the appropriate F-distribution percentage points can be employed at each stage for the tests, it is simpler to choose fixed critical values that do not depend on the (changing) degrees of freedom. The critical F-value for entry is never chosen numerically less than the critical F-value for deletion, and is usually equal to it.

STEP 1

A. Selection of the first variable to enter regression: Using Tableau I, calculate a set of statistics, V_i where $V_i = r_{iY} r_{Yi}/r_{ii} = A_{i,\overline{k+1}} A_{\overline{k+1},i}/A_{ii}$, $i = 1, 2, \ldots, k$. Choose the maximum V_i. For the Hald data in Tableau I,

$$V_1 = r_{1Y} r_{Y1}/r_{11} = (0.730717)(0.730717)/1 = 0.533947$$
$$V_2 = r_{2Y} r_{Y2}/r_{22} = (0.816253)(0.816253)/1 = 0.666269$$
$$V_3 = r_{3Y} r_{Y3}/r_{33} = (-0.534671)(-0.534671)/1 = 0.285873$$
$$V_4 = r_{4Y} r_{Y4}/r_{44} = (-0.821305)(-0.821305)/1 = 0.674542$$

Since V_4 is the maximum, Variable X_4 is the first variable to be considered.

Next it must be determined if X_4 should be entered at all, by applying the standard F-test; at later stages the sequential F-test is used, allowance being made for variables already in regression. For this example we shall arbitrarily set the critical F-value for both entry and deletion of a variable at 3.29 so that it will not be necessary to consult an F-table at every stage. As pointed out above, such an arbitrary specification is not necessary and the appropriate critical point for some desired test level could be found at every stage, using the appropriate number of degrees of freedom.

If X_j is the first variable entered into regression, the resulting analysis of variance table, in terms of correlations, takes the following form:

Source of Variation	df	SS	MS
Total (corrected)	$n - 1$	$r_{YY}^2 = 1$	
Regression	1	r_{jY}^2	r_{jY}^2
Residual	$n - 2$	$r_{YY}^2 - r_{jY}^2$	$(r_{YY}^2 - r_{jY}^2)/(n - 2)$

Tableau I

A_{i1}	A_{i2}	A_{i3}	A_{i4}	A_{i5}	A_{i6}	A_{i7}	A_{i8}	A_{i9}
A_{11} 1.000000	A_{12} 0.228579	A_{13} -0.824134	A_{14} -0.245445	A_{15} 0.730717	A_{16} 1	A_{17} 0	A_{18} 0	A_{19} 0
A_{21} 0.228579	A_{22} 1.000000	A_{23} -0.139242	A_{24} -0.972955	A_{25} 0.816253	A_{26} 0	A_{27} 1	A_{28} 0	A_{29} 0
A_{31} -0.824134	A_{32} -0.139242	A_{33} 1.000000	A_{34} 0.029537	A_{35} -0.534671	A_{36} 0	A_{37} 0	A_{38} 1	A_{39} 0
A_{41} -0.245445	A_{42} -0.972955	A_{43} 0.029537	A_{44} 1.000000	A_{45} -0.821305	A_{46} 0	A_{47} 0	A_{48} 0	A_{49} 1
A_{51} 0.730717	A_{52} 0.816253	A_{53} -0.534671	A_{54} -0.821305	A_{55} 1.000000	A_{56} 0	A_{57} 0	A_{58} 0	A_{59} 0
A_{61} -1	A_{62} 0	A_{63} 0	A_{64} 0	A_{65} 0	A_{66} 0	A_{67} 0	A_{68} 0	A_{69} 0
A_{71} 0	A_{72} -1	A_{73} 0	A_{74} 0	A_{75} 0	A_{76} 0	A_{77} 0	A_{78} 0	A_{79} 0
A_{81} 0	A_{82} 0	A_{83} -1	A_{84} 0	A_{85} 0	A_{86} 0	A_{87} 0	A_{88} 0	A_{89} 0
A_{91} 0	A_{92} 0	A_{93} 0	A_{94} -1	A_{95} 0	A_{96} 0	A_{97} 0	A_{98} 0	A_{99} 0

In our case, since $r_{Y4}^2 = V_4 = 0.674542$, we have

Source	df	SS	MS	F
Total (corrected)	12	1.000000		
Regression (X_4)	1	0.674542	0.674542	22.7986
Residual	11	0.325458	0.029587	

Since 22.7986 exceeds the selected critical value 3.29, variable X_4 is entered into regression.

It can be shown that, at any stage of the procedure, the test ratio for entry of the *next* variable takes the general form

$$\{\phi \cdot V_{max}\}/\{a_{55} - V_{max}\} \tag{6.8.1}$$

where a_{55} is the Tableau element which replaces S at any stage and ϕ is the number of residual degrees of freedom left *after* entry—that is, one less than the number *before* entry. In Tableau I, $a_{55} = r_{YY}^2 = 1$, $V_{max} = V_4 = r_{Y4}^2 = 0.674542$, and $\phi = 12 - 1 = 11$.

B. The correlation matrix must be adjusted for the entrance of X_4 into regression. This is shown in Tableau II. Since Variable 4 is the variable to be entered, Row 4 in Tableau I is divided by A_{44} and entered into Tableau II, all elements of which are B's

$$B_{41} = \frac{A_{41}}{A_{44}}, \qquad B_{42} = \frac{A_{42}}{A_{44}}, \ldots, \text{etc.}$$

The rest of the elements in Tableau II are obtained by applying the following algorithm:

$$B_{ij} = A_{ij} - \frac{A_{il}A_{lj}}{A_{ll}}$$

where l identifies the variable just entered.
Here, $l = 4$, thus,

$$B_{11} = A_{11} - \frac{A_{14}A_{41}}{A_{44}} = 1.000000 - \frac{(-0.245445)(-0.245445)}{1}$$

$$= 0.939757$$

$$B_{12} = A_{12} - \frac{A_{14}A_{42}}{A_{44}} = 0.228579 - \frac{(-0.245445)(-0.972955)}{1}$$

$$= -0.010228, \ldots, \text{and so on.}$$

$$\cdot$$
$$\cdot$$
$$\cdot$$

$$B_{99} = A_{99} - \frac{A_{94}A_{49}}{A_{44}} = 0 - \frac{(-1)(1)}{1} = 1$$

Tableau II

B_{11} 0.939757	B_{12} -0.010228	B_{13} -0.816884	B_{14} 0	B_{15} 0.529132	B_{16} 1	B_{17} 0	B_{18} 0	B_{19} 0.245445
B_{21} -0.010228	B_{22} 0.053359	B_{23} -0.110504	B_{24} 0	B_{25} 0.017160	B_{26} 0	B_{27} 1	B_{28} 0	B_{29} 0.972955
B_{31} -0.816884	B_{32} -0.110504	B_{33} 0.999128	B_{34} 0	B_{35} -0.510412	B_{36} 0	B_{37} 0	B_{38} 1	B_{39} -0.029537
B_{41} -0.245445	B_{42} -0.972955	B_{43} 0.029537	B_{44} 1	B_{45} -0.821305	B_{46} 0	B_{47} 0	B_{48} 0	B_{49} 1
B_{51} 0.529132	B_{52} 0.017160	B_{53} -0.510412	B_{54} 0	B_{55} 0.325458	B_{56} 0	B_{57} 0	B_{58} 0	B_{59} 0.821305
B_{61} -1	B_{62} 0	B_{63} 0	B_{64} 0	B_{65} 0	B_{66} 0	B_{67} 0	B_{68} 0	B_{69} 0
B_{71} 0	B_{72} -1	B_{73} 0	B_{74} 0	B_{75} 0	B_{76} 0	B_{77} 0	B_{78} 0	B_{79} 0
B_{81} 0	B_{82} 0	B_{83} -1	B_{84} 0	B_{85} 0	B_{86} 0	B_{87} 0	B_{88} 0	B_{89} 0
B_{91} -0.245445	B_{92} -0.972955	B_{93} 0.029537	B_{94} 0	B_{95} -0.821305	B_{96} 0	B_{97} 0	B_{98} 0	B_{99} 1

C. Summary of information after Step 1. Using Tableau II, we can summarize the information in the following way:

Analysis of Variance Table

Source of Variation	df	Sum of Squares		Mean Square		F
		Correlation Form	Original Units	Correlation Form	Original Units	
Total (corrected)	12	1	2715.7631			
Regression (X_4)	1	$\begin{array}{c}1 - B_{55}\\0.674542\end{array}$	1831.89627	0.674542	1831.89627	22.7985
Residual	11	$\begin{array}{c}B_{55}\\0.325458\end{array}$	883.86683	0.029587	$s^2 = 80.35153$	

1. The corrected sum of squares is used as a constant for conversion from correlation form to original units.
2. Variable entered: X_4.
3. Sequential F-test for entrance: 22.7986 (calculated before entry).
4. Percentage of variation explained:

$$(0.674542) \times 100 = 67.4542\%.$$

5. Standard deviation of residuals: $s = \sqrt{80.35153} = 8.963902$.
6. Standardized b coefficient for X_4: $B_{45} = -0.821305$.
7. Decoded b coefficient for X_4 (variable admitted):

$$b_4 = B_{45}\sqrt{\frac{\sum x_5^{\,2}}{\sum x_4^{\,2}}} = B_{45} \times \frac{s_5}{s_4}$$

$$= (-0.821305)\sqrt{\frac{2715.7631}{3362}} = -0.738162$$

where the notation s_i here denotes the square root of the ith diagonal term of the matrix $\mathbf{X}_c'\mathbf{X}_c$ on page 179. (Note carefully that s—without subscript—denotes, throughout, the square root of s^2 in the appropriate analysis of variance table.)

8. Standard error of the decoded b coefficient for X_4:

$$s\sqrt{\frac{B_{44}}{\sum x_4^{\,2}}} = \sqrt{\frac{(80.35153) \times 1}{3362}} = 0.154596$$

STEP 2

A. Test for the elimination of variables already in regression. The test for elimination of variables already in regression need not be done at this stage since the sequential F and partial F-tests are identical—that is, only one variable, X_4, is in regression.

B. Selection of the next variable to enter regression. Using Tableau II, determine V_{max} for the variables not in regression.

$$V_1 = \frac{B_{15}B_{51}}{B_{11}} = \frac{(0.529132)(0.529132)}{0.939757} = 0.297930$$

$$V_2 = \frac{B_{25}B_{52}}{B_{22}} = \frac{(0.017160)(0.017160)}{0.053359} = 0.005519$$

$$V_3 = \frac{B_{35}B_{53}}{B_{33}} = \frac{(-0.510412)(-0.510412)}{0.999128} = 0.260748$$

Now $V_{max} = V_1 = 0.297930$.

The F-value for entry of X_1 is

$$\frac{(0.297930)(10)}{(0.325458) - (0.297930)} = 108.22.$$

Since 108.22 exceeds 3.29 (the arbitrary F value) accept X_1 into regression.

C. Create Tableau III by selecting X_1 as the second variable to enter regression.

1. Divide the row of Tableau II corresponding to the variable entering (here row 1), by the leading diagonal element in that row, and enter the resulting row into Tableau III. Thus,

$$C_{11} = \frac{B_{11}}{B_{11}}, \qquad C_{12} = \frac{B_{12}}{B_{11}}$$

$$C_{13} = \frac{B_{13}}{B_{11}}, \ldots, \text{etc.}$$

2. All other elements of Tableau III are obtained by employing the same algorithm used in Step 1. Here,

$$C_{ij} = B_{ij} - \frac{B_{il}B_{lj}}{B_{ll}} \qquad \begin{array}{l} \text{(holds for every element} \\ \text{except those in the row} \\ \text{corresponding to the} \\ \text{variable being admitted)} \end{array}$$

where $l =$ variable being added. For example,

$$C_{21} = B_{21} - \frac{B_{21}B_{11}}{B_{11}} = 0.$$

Tableau III

C_{11}	C_{12}	C_{13}	C_{14}	C_{15}	C_{16}	C_{17}	C_{18}	C_{19}
1	−0.010884	−0.869250	0	0.563052	1.064105	0	0	0.261179

C_{21}	C_{22}	C_{23}	C_{24}	C_{25}	C_{26}	C_{27}	C_{28}	C_{29}
0	0.053248	−0.119395	0	0.022919	0.010833	1	0	0.975626

C_{31}	C_{32}	C_{33}	C_{34}	C_{35}	C_{36}	C_{37}	C_{38}	C_{39}
0	−0.119395	0.289051	0	−0.050463	0.869250	0	1	0.183816

C_{41}	C_{42}	C_{43}	C_{44}	C_{45}	C_{46}	C_{47}	C_{48}	C_{49}
0	−0.975626	−0.183816	1	−0.683107	0.261179	0	0	1.064105

C_{51}	C_{52}	C_{53}	C_{54}	C_{55}	C_{56}	C_{57}	C_{58}	C_{59}
0	0.022919	−0.050463	0	0.027529	−0.563052	0	0	0.683107

C_{61}	C_{62}	C_{63}	C_{64}	C_{65}	C_{66}	C_{67}	C_{68}	C_{69}
0	−0.010884	−0.869250	0	0.563052	1.064105	0	0	0.261179

C_{71}	C_{72}	C_{73}	C_{74}	C_{75}	C_{76}	C_{77}	C_{78}	C_{79}
0	−1	0	0	0	0	0	0	0

C_{81}	C_{82}	C_{83}	C_{84}	C_{85}	C_{86}	C_{87}	C_{88}	C_{89}
0	0	−1	0	0	0	0	0	0

C_{91}	C_{92}	C_{93}	C_{94}	C_{95}	C_{96}	C_{97}	C_{98}	C_{99}
0	−0.975626	−0.183816	0	−0.683107	0.261179	0	0	1.064105

D. Summary of information through Step 2.

Analysis of Variance

| | | Sum of Squares | | Mean Square | | |
| | | | | | | |
Source of Variation	df	Correlation Form	Original Units	Correlation Form	Original Units	F
Total (corrected)	12	1.000000	2715.7635			
Regression (X_4, X_1)	2	$1 - C_{55}$ 0.972471	2641.0012	0.4862355	1320.5006	176.626535
Residual	10	C_{55} 0.027529	74.7623	0.0027529	7.47623	

Variables entered: X_4, X_1.
The partial F-value for X_1, given X_4, is (where ϕ is the number of residual degrees of freedom)

$$\frac{\phi(C_{15})^2}{(C_{55})(C_{66})} = \frac{10(0.563052)^2}{(0.02759)(1.064105)} = 108.22$$

which is, of course, the value found by formula (6.8.1) for the entry test, since X_1 was the variable just entered.
The partial F-value for X_4, given X_1, is

$$\frac{\phi(C_{45})^2}{(C_{55})(C_{99})} = \frac{10(-0.683107)^2}{(0.027529)(1.064105)} = 159.295.$$

Percentage of variation explained: $(0.972471) \times 100 = 97.25\%$.
Standard deviation of residuals $s = \sqrt{7.47623} = 2.73426$.
Standardized b coefficient for X_4: $C_{45} = -0.683107$.
Standardized b coefficient for X_1: $C_{15} = 0.563052$.
Decoded b coefficient for X_4:

$$b_4 = (-0.683107) \times \frac{s_5}{s_4} = (-0.683107)\sqrt{\frac{2715.7631}{3362}} = -0.613954.$$

Decoded b coefficient for X_1:

$$b_1 = (0.563052) \times \frac{s_5}{s_1} = (0.563052)\sqrt{\frac{2715.7631}{415.230769}} = 1.439958.$$

Standard error of $b_4 = 2.734264 \sqrt{\dfrac{1.064105}{3362}}$

$$= 0.0486445.$$

Standard error of $b_1 = 2.734264 \sqrt{\dfrac{1.064105}{415.230769}}$

$$= 0.138417.$$

STEP 3

A. Test for elimination of variables. The test for the elimination of variables is identical to the partial F-test calculated in the previous step. We need not worry about the last variable entered, since the critical F value for the entering of a variable is always greater than or equal to the critical F value for deletion. Thus, only the other variables in regression need be checked.

The partial F value for X_4 given X_1 was 159.295. We compare this with the critical F for deletion, which is arbitrarily set at $F = 3.29$. Since the calculated F exceeds the critical F, do not reject X_4.

B. Test for accepting new variable. Calculate V for all variables not in regression:

$$V_2 = \frac{C_{25}C_{52}}{C_{22}} = \frac{(0.022919)(0.022919)}{0.053248} = 0.00986479$$

$$V_3 = \frac{C_{35}C_{53}}{C_{33}} = \frac{(-0.050463)(-0.050463)}{0.289051} = 0.00880991.$$

Thus, $V_{\max} = V_2 = 0.00986479.$

The F-value for entry of variable X_2 is

$$\frac{(0.009865)(9)}{(0.027529 - 0.009865)}$$

$$= 5.026$$

Since 5.026 exceeds the critical value 3.29, enter X_2 into regression.

C. Create Tableau IV by selecting X_2 as the third variable to enter regression.

1. Divide all elements of Row 2 in Tableau III by the leading diagonal element, 0.053248, and enter this as Row 2 in Tableau IV.

2. All other elements of Tableau IV are obtained by employing the same algorithm used in Step 1 and Step 2. Here

$$D_{ij} = C_{ij} - \frac{C_{il}C_{lj}}{C_{ll}}$$

where $l = 2$.

Tableau IV

D_{i1}	D_{i2}	D_{i3}	D_{i4}	D_{i5}	D_{i6}	D_{i7}	D_{i8}	D_{i9}
D_{11} 1	D_{12} 0	D_{13} -0.893654	D_{14} 0	D_{15} 0.567737	D_{16} 1.066330	D_{17} 0.204391	D_{18} 0	D_{19} 0.460589
D_{21} 0	D_{22} 1	D_{23} -2.242271	D_{24} 0	D_{25} 0.430415	D_{26} 0.204391	D_{27} 18.780350	D_{28} 0	D_{29} 18.322604
D_{31} 0	D_{32} 0	D_{33} 0.021336	D_{34} 0	D_{35} 0.000926	D_{36} 0.893654	D_{37} 2.242271	D_{38} 1	D_{39} 2.371435
D_{41} 0	D_{42} 0	D_{43} -2.371435	D_{44} 1	D_{45} -0.263182	D_{46} 0.460589	D_{47} 18.322604	D_{48} 0	D_{49} 18.940119
D_{51} 0	D_{52} 0	D_{53} 0.000926	D_{54} 0	D_{55} 0.017664	D_{56} -0.567737	D_{57} -0.430415	D_{58} 0	D_{59} 0.263182
D_{61} 0	D_{62} 0	D_{63} -0.893654	D_{64} 0	D_{65} 0.567737	D_{66} 1.066330	D_{67} 0.204391	D_{68} 0	D_{69} 0.460589
D_{71} 0	D_{72} 0	D_{73} -2.242271	D_{74} 0	D_{75} 0.430415	D_{76} 0.204391	D_{77} 18.780350	D_{78} 0	D_{79} 18.322604
D_{81} 0	D_{82} 0	D_{83} -1	D_{84} 0	D_{85} 0	D_{86} 0	D_{87} 0	D_{88} 0	D_{89} 0
D_{91} 0	D_{92} 0	D_{93} -2.371435	D_{94} 0	D_{95} -0.263182	D_{96} 0.460589	D_{97} 18.322604	D_{98} 0	D_{99} 18.940119

For example,

$$D_{13} = C_{13} - \frac{C_{12}C_{23}}{C_{22}} = -0.869250 - \frac{(-0.010884)(-0.119395)}{0.053248}$$

$$= -0.893654.$$

Analysis of Variance

		Sum of Squares		Mean Square		
Source of Variation	df	Correlation Form	Original Units	Correlation Form	Original Units	F
Total (corrected)	12	1.000000	2715.7635			
Regression (X_4, X_1, X_2)	3	$1 - D_{55}$ 0.982336	2667.7923	0.327445	889.2641	166.833953
Residual	9	D_{55} 0.017664	47.9712	0.0019627	5.330133	

Variables entered: X_4, X_1, X_2.
The partial F-value for X_2, given X_1 and X_4 (where ϕ is the number of residual degrees of freedom), is

$$\frac{\phi(D_{25})^2}{(D_{55})(D_{77})} = \frac{9(0.430415)^2}{(0.017664)(18.780350)} = 5.026$$

which is, of course, the value found by formula (6.8.1) for the entry test, since X_2 was the variable just entered.

The partial F-value, for X_4 given X_1 and X_2, is

$$\frac{\phi(D_{45})^2}{D_{55}D_{99}} = \frac{(9)(-0.263182)^2}{(0.017664)(18.940119)} = 1.863.$$

The partial F-value for X_1, given X_4 and X_2, is

$$\frac{\phi(D_{15})^2}{(D_{55})(D_{66})} = \frac{(9)(0.567737)^2}{(0.017664)(1.066330)} = 154.013.$$

(Note that the divisors D_{77}, D_{99}, and D_{66} in the three cases above are in the second, fourth, and first diagonal places of the 4×4 right-hand bottom corner of Tableau IV and correspond to partial F-values involving variables X_2, X_4, and X_1, respectively.)

Percentage of variation explained: $(0.982336) \times 100 = 98.23\%$.

Standard deviation of residuals: $\sqrt{5.330133} = 2.308708$.

Standardized b coefficient for X_4: $D_{45} = -0.263182$.

Standardized b coefficient for X_1: $D_{15} = 0.567737$.

Standardized b coefficient for X_2: $D_{25} = 0.430415$.

Decoded b coefficient for X_4:

$$b_4 = (-0.263182) \times \frac{s_5}{s_4} = (-0.263182)\sqrt{\frac{2715.7631}{3362}} = -0.236539.$$

Decoded b coefficient for X_1:

$$b_1 = (0.567737) \times \frac{s_5}{s_1} = (0.567737)\sqrt{\frac{2715.7631}{415.230769}} = 1.451939.$$

Decoded b coefficient for X_2:

$$b_2 = (0.430415) \times \frac{s_5}{s_2} = (0.430415)\sqrt{\frac{2715.7631}{2905.692310}} = 0.416110.$$

$$\text{Standard error of } b_4 = 2.308708\sqrt{\frac{18.940119}{3362}} = 0.173285.$$

$$\text{Standard error of } b_1 = 2.308708\sqrt{\frac{1.066330}{415.230769}} = 0.116996.$$

$$\text{Standard error of } b_2 = 2.308708\sqrt{\frac{18.78030}{2905.692310}} = 0.185608.$$

STEP 4

A. Test for elimination of variables. Partial F-tests in previous step (1) for X_2, given X_4 and X_1 is 5.026 which exceeds the selected value of 3.29, (2) for X_4, given X_1 and X_2 is 1.863 which does not exceed 3.29, and (3) for X_1, given X_4 and X_2, is 154.013 which exceeds 3.29. Thus we eliminate X_4 from the regression, and create Tableau V.

1. Divide all elements of Row 4 in Tableau IV by $D_{99} = 18.940119$ and enter this as Row 4 in Tableau V.

2. All other elements of Tableau V are obtained by applying the algorithm,

$$E_{ij} = D_{ij} - \frac{D_{ik}D_{qj}}{D_{kk}}$$

where, here, $q = 4$ and the pivotal element has $k = 9$, for example,

$$E_{23} = D_{23} - \frac{D_{29}D_{43}}{D_{99}} = -2.242271 - \frac{(18.322604)(-2.371435)}{18.940119}$$

$$= 0.051847.$$

Tableau V

E_{11} 1	E_{12} 0	E_{13} −0.835985	E_{14} −0.024318	E_{15} 0.574137	E_{16} 1.055129	E_{17} −0.241181	E_{18} 0	E_{19} 0
E_{21} 0	E_{22} 1	E_{23} 0.051847	E_{24} −0.967396	E_{25} 0.685017	E_{26} −0.241181	E_{27} 1.055129	E_{28} 0	E_{29} 0
E_{31} 0	E_{32} 0	E_{33} 0.318256	E_{34} −0.125207	E_{35} 0.033878	E_{36} 0.835985	E_{37} −0.051847	E_{38} 1	E_{39} 0
E_{41} 0	E_{42} 0	E_{43} −0.125207	E_{44} 0.052798	E_{45} −0.013896	E_{46} 0.024318	E_{47} 0.967396	E_{48} 0	E_{49} 1
E_{51} 0	E_{52} 0	E_{53} 0.033878	E_{54} −0.013896	E_{55} 0.021322	E_{56} −0.574137	E_{57} −0.685017	E_{58} 0	E_{59} 0
E_{61} 0	E_{62} 0	E_{63} −0.835985	E_{64} −0.024318	E_{65} 0.574137	E_{66} 1.055129	E_{67} −0.241181	E_{68} 0	E_{69} 0
E_{71} 0	E_{72} 0	E_{73} 0.051847	E_{74} −0.967396	E_{75} 0.685017	E_{76} −0.241181	E_{77} 1.055129	E_{78} 0	E_{79} 0
E_{81} 0	E_{82} 0	E_{83} −1	E_{84} 0	E_{85} 0	E_{86} 0	E_{87} 0	E_{88} 0	E_{89} 0
E_{91} 0	E_{92} 0	E_{93} 0	E_{94} −1	E_{95} 0	E_{96} 0	E_{97} 0	E_{98} 0	E_{99} 0

Analysis of Variance

Source of Variation	df	Sum of Squares Correlation Form	Sum of Squares Original Units	Mean Square Correlation Form	Mean Square Original Units	F
Total (corrected)	12	1.000000	2715.7635			
Regression	2	$1 - E_{55}$ 0.978678	2657.8580	0.489339	1328.9290	229.521107
Residual	10	E_{55} 0.021322	57.9055	0.002132	5.7906	

Variables in regression : X_1, X_2.
Variable leaving regression: X_4.
 The partial F-value for X_1 given X_2 is

$$\frac{\phi(E_{15})^2}{E_{55}E_{66}} = \frac{(10)(0.574137)^2}{(0.021322)(1.055129)} = 146.520.$$

 The partial F-value for X_2 given X_1 is

$$\frac{\phi(E_{25})^2}{E_{55}E_{77}} = \frac{(10)(0.685017)^2}{(0.021322)(1.055129)} = 208.578.$$

Percentage of variation explained: $(0.978678) \times 100 = 97.87\%$.
Standard deviation of residuals: $\sqrt{5.7906} = 2.406367$.
Standardized b coefficient for X_1: $E_{15} = 0.574137$.
Standardized b coefficient for X_2: $E_{25} = 0.685017$.
Decoded b coefficient for X_1:

$$b_1 = (0.574137) \times \frac{s_5}{s_1} = (0.574137)\sqrt{\frac{2715.7631}{415.230769}} = 1.468306.$$

Decoded b coefficient for X_2:

$$b_2 = 0.685017 \times \frac{s_5}{s_2} = (0.685017)\sqrt{\frac{2715.7631}{2905.692310}} = 0.6622507.$$

$$\text{Standard error of } b_1 = 2.406367\sqrt{\frac{1.055129}{415.230769}} = 0.121303.$$

$$\text{Standard error of } b_2 = 2.406367\sqrt{\frac{1.055129}{2905.692310}} = 0.045855.$$

B. Test for accepting new variable. Determine $V_{\max}$ for all variables not in regression:

$$V_3 = \frac{E_{35}E_{53}}{E_{33}} = \frac{(0.033878)(0.033878)}{0.318256} = 0.00360628$$

$$V_4 = \frac{E_{45}E_{54}}{E_{44}} = \frac{(-0.013896)(-0.013896)}{0.052798} = 0.00365731.$$

Thus $V_{\max} = V_4 = 0.00365731$.

There is no need to calculate the F-value for entry, since X_4 has just been eliminated from the regression in the previous step.

CONCLUSION. The final regression equation chosen by the stepwise procedure is (see page 193, and (2) on page 195)

$$\hat{Y} = 52.5773400 + 1.4683057X_1 + 0.6622507X_2.$$

This is obtained by substituting into the formula

$$Y = (Y - b_1X_1 - b_2X_2) + b_1X_1 + b_2X_2.$$

Rounding to two decimal places gives the same equation as on page 172, namely,

$$Y = 52.58 + 1.47X_1 + 0.66X_2.$$

Comments

1. At each step of the regression analysis, the inverse of the correlation matrix is found in the lower right $k \times k$ matrix.

For example, at Tableau III, variables X_1 and X_4 are in regression. The 4×4 matrix is

$$\begin{bmatrix} 1.064105 & 0 & 0 & 0.261179 \\ 0 & 0 & 0 & 0 \\ 0 & 0 & 0 & 0 \\ 0.261179 & 0 & 0 & 1.064105 \end{bmatrix}$$

Extracting the submatrix for the 2 variables in regression we have

$$\mathbf{R}^{-1} = \begin{bmatrix} 1.064105 & 0.261179 \\ 0.261179 & 1.064105 \end{bmatrix}$$

where

$$\mathbf{R} = \begin{bmatrix} 1.000000 & -0.245445 \\ -0.245445 & 1.000000 \end{bmatrix}$$

is the correlation matrix for X_1 and X_4 and $RR^{-1} = I$. Similarly, Tableau IV yields the inverse of the correlation matrix for the 3 variables, X_1, X_2, and X_4.

2. The constant term for the regression equation at each step is calculated from

$$\bar{Y} - \sum b_i \bar{X}_i$$

where i ranges over the variables in regression at each step.

3. The square of the partial correlation coefficient of the variables not in regression with the response Y is calculated as follows, where a_{ij} denotes a Tableau entry,

$$\frac{a_{iY}^2}{a_{ii} a_{YY}} = r_{iY \cdot jk \ldots}^2$$

where $j, k, \ldots$ are already in regression. For example, after Step 2, from Tableau III,

$$r_{2Y \cdot 14}^2 = \frac{(0.022919)^2}{(0.053248)(0.027529)} = 0.358342$$

$$r_{3Y \cdot 14}^2 = \frac{(-0.050463)^2}{(0.289051)(0.027529)} = 0.320023.$$

4. An excellent computer program for this procedure is found in the book *Mathematical Methods for Digital Computers* edited by Anthony Ralston and Herbert S. Wilf (John Wiley & Sons). (Borrowed programs should be checked for automatic computer "cutoffs"; sometimes the resultant model is not what would have been obtained in the standard stepwise procedure just illustrated.)

EXERCISES

A. A production plant cost-control engineer is responsible for cost reduction. One of the costly items in his plant is the amount of water used by the production facilities each month. He decided to investigate water usage by collecting seventeen observations on his plant's water usage and other variables. He had heard about multiple regression, but since he was quite skeptical he added a column of random numbers to his original observations. The complete set of data is shown after Exercise C, with means, standard deviations, and correlation coefficients and seven of the $2^5 = 32$ possible regression runs are provided for your information ($\alpha = 0.05$ for the confidence statements on the β coefficients).

Requirements
1. Demonstrate that the computer printout for the first-order regression $Y \equiv X_6 = f(X_2)$ is correct by doing all the calculations.
2. Using the information on the printout for the first-order regression $Y \equiv X_6 = f(X_2, X_4)$, derive a 95% confidence interval estimate on β_4. Compare with the computer printout estimate.
3. Derive the inverse of the correlation matrix for the first-order regression $Y \equiv X_6 = f(X_1, X_2, X_4)$. Show all calculations.

B. Using the information and computer printouts from problem A: (1) Choose a potential model for the prediction of water usage. (2) Justify your choice and discuss its inadequacies. (3) What can you say concerning the random vector whose elements are denoted by X_5?

C. If the stepwise procedure is used on the problem in Exercise A with a critical value for acceptance and rejection of variables of $F = 3.74$, the procedure will stop with the first order model $Y = f(X_2, X_4)$. If the critical value for acceptance had been $F = 2.00$, then the stepwise procedure would select the first order model $Y = f(X_1, X_2, X_3, X_4)$. The backward elimination approach would select $Y = f(X_1, X_2, X_3, X_4)$.

Requirement

Summarize your analysis and conclusions, using the information in Exercises A, B, and C.

Data Code

X_1 = average monthly temperature (°F),
X_2 = amount of production (M pounds),
X_3 = number of plant operating days in the month,
X_4 = number of persons on the monthly plant payroll,
X_5 = a two digit random number, and
X_6 = Y is the monthly water usage (gallons)

Original Data

	X_1	X_2	X_3	X_4	X_5	$X_6 = Y$
1	58.8	7107	21	129	52	3067
2	65.2	6373	22	141	68	2828
3	70.9	6796	22	153	29	2891
4	77.4	9208	20	166	23	2994
5	79.3	14792	25	193	40	3082
6	81.0	14564	23	189	14	3898
7	71.9	11964	20	175	96	3502
8	63.9	13526	23	186	94	3060
9	54.5	12656	20	190	54	3211
10	39.5	14119	20	187	37	3286
11	44.5	16691	22	195	42	3542
12	43.6	14571	19	206	22	3125
13	56.0	13619	22	198	28	3022
14	64.7	14575	22	192	7	2922
15	73.0	14556	21	191	42	3950
16	78.9	18573	21	200	33	4488
17	79.4	15618	22	200	92	3295

Mean Values of the Variables

64.852940	12900.470	21.470588	181.823520	45.470587	3303.7058

Standard Deviations of the Variables

13.5100930	3526.78600	1.46277340	21.9949850	27.4775310	446.698370

Correlation Matrix

1	1.00000000	−0.02410741	0.43762975	−0.08205777	0.10762982	0.28575758
2	−0.02410741	1.00000000	0.10573055	0.91847987	−0.11145872	0.63074956
3	0.43762975	0.10573055	1.00000000	0.03188120	0.03768543	−0.08882581
4	−0.08205777	0.91847987	0.03188120	1.00000000	−0.15900788	0.41324613
5	0.10762982	−0.11145872	0.03768543	−0.15900788	1.00000000	−0.06562381
6	0.28575758	0.63074956	−0.08882581	0.41324613	−0.06562381	1.00000000

Regressions

1. Analysis of the model, $Y = X_6 = f(X_2)$:

Variable Entering	2
Per cent variation explained $R - Sq$	39.7845000
Standard deviation of residuals	357.9998900
Mean of the response	3303.7058000
Std. dev. as per cent of response mean	10.836%
Degrees of freedom	15
Determinant value	1.0000000

ANOVA

Source	df	SS	MS	Overall F
Total	16	3192631.00		
Regression	1	1270172.00	1270172.00	9.91
Residual	15	1922459.00	128163.92	

B Coefficients and Confidence Limits

Var. No.	Mean	Decoded B Coefficient	Limits Upper/Lower	Standard Error	Partial F-test
2	12900.47	0.0798899	0.1339688 0.0258111	0.0253772	9.91

Constant term in prediction equation 2273.0881000.

Residual Analysis

Obs. No.	Observed Y	Predicted Y	Residual	Normal Deviate
1	3067.000	2840.8659	226.1341	0.6316597
2	2828.000	2782.2266	45.7734	0.1278587
3	2891.000	2816.0201	74.9799	0.2094411
4	2994.000	3008.7146	−14.7146	−0.0411022
5	3082.000	3454.8200	−372.8200	−1.0413969
6	3898.000	3436.6051	461.3949	1.2888129
7	3502.000	3228.8913	273.1087	0.7628737
8	3060.000	3353.6794	−293.6794	−0.8203338
9	3211.000	3284.1751	−73.1751	−0.2043998
10	3286.000	3401.0541	−115.0541	−0.3213803
11	3542.000	3606.5310	−64.5310	−0.1802542
12	3125.000	3437.1644	−312.1644	−0.8719679
13	3022.000	3361.1091	−339.1091	−0.9472324
14	2922.000	3437.4839	−515.4839	−1.4398996
15	3950.000	3435.9660	514.0340	1.4358496
16	4488.000	3756.8839	731.1161	2.0422243
17	3295.000	3520.8091	−225.8091	−0.6307519

2. Analysis of the model, $Y = X_6 = f(X_4)$:

Variable entering	4
Per cent variation explained $R - Sq$	17.0772400
Standard deviation of residuals	420.1124900
Mean of the response	3303.7058000
Std. dev. as per cent of response mean	12.716%
Degrees of freedom	15
Determinant value	1.0000000

ANOVA

Source	df	SS	MS	Overall F
Total	16	3192631.00		
Regression	1	545213.00	545213.00	3.09
Residual	15	2647418.00	176494.50	

B Coefficients and Confidence Limits

Var. No.	Mean	Decoded B Coefficient	Limits Upper/Lower	Standard Error	Partial F-test
4	181.8235200	8.3926569	18.5683810 −1.7830690	4.7750943	3.09

Constant term in prediction equation 1777.7234000.

Residual Analysis

Obs. No.	Observed Y	Predicted Y	Residual	Normal Deviate
1	3067.000	2860.3761	206.6239	0.4918299
2	2828.000	2961.0880	−133.0880	−0.3167913
3	2891.000	3061.7999	−170.7999	−0.4065575
4	2994.000	3170.9044	−176.9044	−0.4210882
5	3082.000	3397.5061	−315.5061	−0.7510039
6	3898.000	3363.9355	534.0645	1.2712416
7	3502.000	3246.4383	255.5617	0.6083173
8	3060.000	3338.7575	−278.7575	−0.6635306
9	3211.000	3372.3282	−161.3282	−0.3840119
10	3286.000	3347.1502	−61.1502	−0.1455567
11	3542.000	3414.2914	127.7086	0.3039867
12	3125.000	3506.6107	−381.6107	−0.9083536
13	3022.000	3439.4694	−417.4694	−0.9937086
14	2922.000	3389.1135	−467.1135	−1.1118772
15	3950.000	3380.7208	569.2792	1.3550637
16	4488.000	3456.2547	1031.7453	2.4558786
17	3295.000	3456.2547	−161.2547	−0.3838370

3. Analysis of the model, $Y = X_6 = f(X_1, X_2)$:

Variables entering	1, 2
Per cent variation explained $R - Sq$	48.8476600
Standard deviation of residuals	341.5412200
Mean of the response	3303.7058000
Std. dev. as per cent of response mean	10.338%
Degrees of freedom	14
Determinant value	0.9994190

ANOVA

Source	df	SS	MS	Overall F
Total	16	3192631.00		
Regression	2	1559525.00	779762.70	6.68
Residual	14	1633106.00	116650.40	

B Coefficients and Confidence Limits

Var. No.	Mean	Decoded B Coefficient	Limits Upper/Lower	Standard Error	Partial F-test
1	64.8529400	9.9568521	23.5174330 −3.6037290	6.3219494	2.48
2	12900.4700000	0.0808094	0.1327561 0.0288628	0.0242176	11.13

Constant term in prediction equation 1615.4950000.

Residual Analysis

Obs. No.	Observed Y	Predicted Y	Residual	Normal Deviate
1	3067.000	2775.2705	291.7295	0.8541561
2	2828.000	2779.6803	48.3197	0.1414755
3	2891.000	2870.6167	20.3833	0.0596804
4	2994.000	3130.2486	−136.2486	−0.3989229
5	3082.000	3600.4065	−518.4065	1.5178446
6	3898.000	3598.9086	299.0914	−0.8757110
7	3502.000	3298.1968	203.8032	0.5967163
8	3060.000	3344.7662	−284.7662	−0.8337682
9	3211.000	3180.8676	30.1324	0.0882248
10	3286.000	3149.7390	136.2610	0.3989592
11	3542.000	3407.3652	134.6348	0.3941978
12	3125.000	3227.0880	−102.0880	−0.2989039
13	3022.000	3273.6224	−251.6224	−0.7367263
14	2922.000	3437.5008	−515.5008	−1.5093369
15	3950.000	3518.6074	431.3926	1.2630762
16	4488.000	3901.9643	586.0357	1.7158564
17	3295.000	3668.1508	−373.1508	−1.0925498

4. Analysis of the model, $Y = X_6 = f(X_2, X_4)$:

Variables entering	2, 4
Per cent variation explained $R - Sq$	57.4219500
Standard deviation of residuals	311.6041600
Mean of the response	3303.7058000
Std. dev. as per cent of response mean	9.432%
Degrees of freedom	14
Determinant value	0.1563949

ANOVA

Source	df	SS	MS	Overall F
Total	16	3192631.00		
Regression	2	1833271.00	916635.40	9.44
Residual	14	1359360.00	97097.15	

B Coefficients and Confidence Limits

Var. No.	Mean	Decoded B Coefficient	Limits Upper/Lower	Standard Error	Partial F-test
2	12900.4700000	0.2034310	0.3232374 0.0836246	0.0558538	13.27
4	181.8235200	−21.5673720	−2.3570090 −40.7777350	8.9558802	5.80

Constant term in prediction equation 4600.8056000.

Residual Analysis

Obs.	Observed Y	Predicted Y	Residual	Normal Deviate
1	3067.000	3264.3988	−197.3988	−0.6334922
2	2828.000	2856.2720	−28.2720	−0.0907305
3	2891.000	2683.5148	207.4852	0.6658615
4	2994.000	2893.8146	100.1854	0.3215150
5	3082.000	3447.4543	−365.4543	−1.1728158
6	3898.000	3487.3415	410.6585	1.3178851
7	3502.000	3260.3641	241.6359	0.7754579
8	3060.000	3340.8823	−280.8823	−0.9014074
9	3211.000	3077.6278	133.3722	0.4280180
10	3286.000	3439.9495	−153.9495	−0.4940547
11	3542.000	3790.6350	−248.6350	−0.7979194
12	3125.000	3122.1202	2.8798	0.0092419
13	3022.000	3100.9929	−78.9929	−0.2535040
14	2922.000	3424.8771	−502.8771	−1.6138330
15	3950.000	3442.5793	507.4207	1.6284143
16	4488.000	4065.6553	422.3447	1.3553885
17	3295.000	3464.5167	−169.5167	−0.5440130

5. Analysis of the model, $Y = X_6 = f(X_1, X_2, X_3)$:

Variables entering	1, 2, 3
Per cent variation explained $R - Sq$	59.2865800
Standard deviation of residuals	316.2070100
Mean of the response	3303.7058000
Std. dev. as per cent of response mean	9.571%
Degrees of freedom	13
Determinant value	0.7944891

ANOVA

Source	df	SS	MS	Overall F
Total	16	3192631.00		
Regression	3	1892802.00	630933.830	6.31
Residual	13	1299829.00	99986.869	

B Coefficients and Confidence Limits

Var. No.	Mean	Decoded B Coefficient	Limits Upper/Lower	Standard Error	Partial F-test
1	64.8529400	15.2339490	29.3340350 1.1338630	6.5278179	5.45
2	12900.4700000	0.0861496	0.1349897 0.0373095	0.0226112	14.52
3	21.4705880	−110.6610800	20.2625400 −241.5847000	60.6127880	3.33

Constant term in prediction equation 3580.3279000.

Residual Analysis

Obs. No.	Observed Y	Predicted Y	Residual	Normal Deviate
1	3067.000	2764.4665	302.5335	0.9567577
2	2828.000	2688.0689	139.9311	0.4425300
3	2891.000	2811.3436	79.6564	0.2519122
4	2994.000	3339.4792	−345.4792	−1.0925728
5	3082.000	3296.1776	−214.1776	−0.6773335
6	3898.000	3523.7554	374.2446	1.1835430
7	3502.000	3493.1207	8.8793	0.0280807
8	3060.000	3173.8316	−113.8316	−0.3599908
9	3211.000	3287.6656	−76.6656	−0.2424538
10	3286.000	3185.1931	100.8069	0.3188003
11	3542.000	3261.6175	280.3825	0.8867055
12	3125.000	3397.2530	−272.2530	−0.8609961
13	3022.000	3172.1564	−150.1564	−0.4748674
14	2922.000	3387.0508	−465.0508	−1.4707162
15	3950.000	3622.5168	327.4832	1.0356607
16	4488.000	4058.4599	429.5401	1.3584142
17	3295.000	3700.8438	−405.8438	−1.2834750

6. Analysis of the model, $Y = X_6 = f(X_1, X_2, X_4)$:

Variables entering	1, 2, 4
Per cent variation explained $R - Sq$	63.1905300
Standard deviation of residuals	300.6647200
Mean of the response	3303.7058000
Std. dev. as per cent of response mean	9.101%
Degrees of freedom	13
Determinant value	0.1527141

ANOVA

Source	df	SS	MS	Overall F
Total	16	3192631.00		
Regression	3	2017440.00	672480.100	7.44
Residual	13	1175191.00	90399.276	

B Coefficients and Confidence Limits

Obs. No.	Mean	Decoded B Coefficient	Limits Upper/Lower	Standard Error	Partial F-test
1	64.8529400	8.0364126	20.1979790 −4.1251550	5.6303554	2.04
2	12900.4700000	0.1933407	0.3107467 0.0759347	0.0543546	12.65
4	181.8235200	−19.6762750	−0.7925920 −38.5599580	8.7424461	5.07

Constant term in prediction equation 3865.9449000.

Residual Analysis

Obs. No.	Observed Y	Predicted Y	Residual	Normal Deviate
1	3067.000	3174.3186	−107.3186	−0.3569378
2	2828.000	2847.7243	−19.7243	−0.0656023
3	2891.000	2739.1996	151.8004	0.5048826
4	2994.000	3001.9824	−7.9824	−0.0265492
5	3082.000	3565.6065	−483.6065	−1.6084577
6	3898.000	3613.8919	284.1081	0.9449333
7	3502.000	3313.5425	188.4575	0.6268028
8	3060.000	3334.8104	−274.8104	−0.9140095
9	3211.000	3012.3566	198.6434	0.6606808
10	3286.000	3233.6966	52.3034	0.1739592
11	3542.000	3613.7407	−71.7407	−0.2386070
12	3125.000	2980.1867	144.8133	0.4816438
13	3022.000	3053.1881	−31.1881	−0.1037305
14	2922.000	3425.9961	−503.9961	−1.6762728
15	3950.000	3508.7012	441.2988	1.4677438
16	4488.000	4155.6790	332.3210	1.1052876
17	3295.000	3588.3755	−293.3755	−0.9757563

7. Analysis of the model, $Y = X_6 = f(X_1, X_2, X_3, X_4)$:

Variables entering	1, 2, 3, 4
Per cent variation explained $R - Sq$	76.7027100
Standard deviation of residuals	248.9639800
Mean of the response	3303.7058000
Std. dev. as per cent of response mean	7.536%
Degrees of freedom	12
Determinant value	0.1198925

ANOVA

Source	df	SS	MS	Overall F
Total	16	3192631.00		
Regression	4	2448834.00	612208.570	9.88
Residual	12	743797.00	61983.083	

B Coefficients and Confidence Limits

Var. No.	Mean	Decoded B Coefficient	Limits Upper/Lower	Standard Error	Partial F-test
1	64.8529400	13.8688640	25.1120960 2.6256320	5.1598130	7.22
2	12900.4700000	0.2117030	0.3109414 0.1124646	0.0455431	21.61
3	21.4705880	−126.6903600	−22.0497400 −231.3309800	48.0223160	6.96
4	181.8235200	−21.8179740	−5.9450100 −37.6909380	7.2845180	8.97

Constant term in prediction equation 6360.3385000.

Residual Analysis

Obs. No.	Observed Y	Predicted Y	Residual	Normal Deviate
1	3067.000	3205.3847	−138.3847	−0.5558423
2	2828.000	2750.2493	77.7507	0.3122970
3	2891.000	2657.0365	233.9635	0.9397484
4	2994.000	3227.5588	−233.5588	−0.9381229
5	3082.000	3213.5221	−131.5221	−0.5282776
6	3898.000	3529.4835	368.5165	1.4802000
7	3502.000	3538.3717	−36.3717	−0.1460922
8	3060.000	3138.0322	−78.0332	−0.3134277
9	3211.000	3116.2823	94.7177	0.3804474
10	3286.000	3283.4248	2.5752	0.0103437
11	3542.000	3469.3447	72.6553	0.2918306
12	3125.000	3148.1257	−23.1257	−0.0928877
13	3022.000	2913.0311	108.9689	0.4376894
14	2922.000	3366.9861	−444.9861	−1.7873513
15	3950.000	3626.5837	323.4163	1.2990485
16	4488.000	4362.4591	125.5409	0.5042533
17	3295.000	3617.1208	−322.1208	−1.2938449

D. A set of data given on page 454 of K. A. Brownlee's *Statistical Theory and Methodology in Science and Engineering* (Second Edition), 1965, John Wiley and Sons, New York, is reproduced below. The figures are from 21 days of operation of a plant oxidizing NH_3 to HNO_3 and the variables are:

X_1 = rate of operation,
X_2 = temperature of the cooling water in the coils of the absorbing tower for the nitric oxides,
X_3 = concentration of HNO_3 in the absorbing liquid (coded by minus 50, times 10),
Y = the per cent of the ingoing NH_3 that is lost by escaping in the unabsorbed nitric oxides ($\times 10$). This is an inverse measure of the yield of HNO_3 for the plant.

Data from Operation of a Plant for the Oxidation of
Ammonia to Nitric Acid

Run No.	Air Flow X_1	Cooling Water Inlet Temperature X_2	Acid Concentration X_3	Stack Loss Y
1	80	27	89	42
2	80	27	88	37
3	75	25	90	37
4	62	24	87	28
5	62	22	87	18
6	62	23	87	18
7	62	24	93	19
8	62	24	93	20
9	58	23	87	15
10	58	18	80	14
11	58	18	89	14
12	58	17	88	13
13	58	18	82	11
14	58	19	93	12
15	50	18	89	8
16	50	18	86	7
17	50	19	72	8
18	50	19	79	8
19	50	20	80	9
20	56	20	82	15
21	70	20	91	15

Investigation of plant operations indicates that the following sets of runs can be considered as replicates: (1, 2), (4, 5, 6), (7, 8), (11, 12), and (18, 19). While the runs in each set are not *exact* replicates, the points *are* sufficiently close to each other in the X-space for them to be used as such. Computer printouts for all possible regressions are shown on the following pages. (*Note*: Roundoff error and printout cut-off instructions cause non-uniformity in fifth decimal places.)

Requirements

1. Assuming that a linear multiple regression model will suitably represent the data, what model would you fit first?
2. Calculate the pure error.
3. Write out a skeleton analysis of variance table for this experiment (i.e., the sources of variation and degrees of freedom, only).
4. Indicate what statistical test could be performed to check on the adequacy of the model.
5. Does the test for the adequacy of the model have sufficient sensitivity for rejecting the null hypothesis?
6. Examine all the regression equations given. Choose the fitted model which would appear to be most satisfactory as a prediction equation.
7. Examine the residuals to check that no obvious variable is missing.
8. Under what conditions would one be willing to use the chosen equation? Make recommendations for further experimentation.

Oxidation of Ammonia to Nitric Acid

Original and/or Transformed Data

	X_1	X_2	X_3	$X_4 = Y$
1	80.00000000	27.00000000	89.00000000	42.00000000⎫
2	80.00000000	27.00000000	88.00000000	37.00000000⎭
3	75.00000000	25.00000000	90.00000000	37.00000000
4	62.00000000	24.00000000	87.00000000	28.00000000⎫
5	62.00000000	22.00000000	87.00000000	18.00000000⎬
6	62.00000000	23.00000000	87.00000000	18.00000000⎭
7	62.00000000	24.00000000	93.00000000	19.00000000⎫
8	62.00000000	24.00000000	93.00000000	20.00000000⎭
9	58.00000000	23.00000000	87.00000000	15.00000000
10	58.00000000	18.00000000	80.00000000	14.00000000
11	58.00000000	18.00000000	89.00000000	14.00000000⎫
12	58.00000000	17.00000000	88.00000000	13.00000000⎭
13	58.00000000	18.00000000	82.00000000	11.00000000
14	58.00000000	19.00000000	93.00000000	12.00000000
15	50.00000000	18.00000000	89.00000000	8.00000000
16	50.00000000	18.00000000	86.00000000	7.00000000
17	50.00000000	19.00000000	72.00000000	8.00000000
18	50.00000000	19.00000000	79.00000000	8.00000000⎫
19	50.00000000	20.00000000	80.00000000	9.00000000⎭
20	56.00000000	20.00000000	82.00000000	15.00000000
21	70.00000000	20.00000000	91.00000000	15.00000000

Means of Transformed Variables

1	60.42857000	21.09523700	86.28571300	17.52380900

Standard Deviations of Transformed Variables

1	9.16826650	3.16077100	5.35857090	10.17162000

Correlation Matrix

1	1.00000000	0.78185250	0.50014295	0.91966375
2	0.78185250	1.00000000	0.39093959	0.87550465
3	0.50014295	0.39093959	1.00000000	0.39982969
4	0.91966375	0.87550465	0.39982969	1.00000000

1. *Control Information*

No. of observations	21
Response variable is no.	04
Risk level for B conf. interval	05%
List of excluded variables	02 03
Variable entering	1
Sequential F-test	104.2013300
Per cent variation explained $R - Sq$	84.5780900
Standard deviation of residuals	4.0982407
Mean of the response	17.5238090
Std. dev. as per cent of response mean	23.387%
Degrees of freedom	19
Determinant value	1.0000000

ANOVA

Source	df	SS	MS	Overall F
Total	20	2069.2370000		
Regression	1	1750.1211000	1750.1211000	104.20
Residual	19	319.1159700	16.7955770	
lack of fit	13	238.4493033	18.3422541	1.36
pure error	6	80.6666667	13.4444445	

B Coefficients and Confidence Limits

Var. No.	Mean	Decoded B Coefficient	Limits Upper/Lower	Standard Error	Partial F-test
1	60.4285700	1.0203093	1.2295106 0.8111080	0.0999529	104.20

Constant term in prediction equation −44.1320220.

Squares of Partial Correlation Coefficients of Variables Not in Regression

Variables	Square of Partials
2	0.40838
3	0.03127

Residual Analysis for $\hat{X}_4 = f(X_1)$

Obs. No.	Observed Y	Predicted Y	Residual	Normal Deviate
1	42.0000000	37.4927220	4.5072780	1.0998080
2	37.0000000	37.4927220	−0.4927220	−0.1202277
3	37.0000000	32.3911750	4.6088250	1.1245862
4	28.0000000	19.1271540	8.8728460	2.1650377
5	18.0000000	19.1271540	−1.1271540	−0.2750336
6	18.0000000	19.1271540	−1.1271540	−0.2750336
7	19.0000000	19.1271540	−0.1271540	−0.0310265
8	20.0000000	19.1271540	0.8728460	0.2129807
9	15.0000000	15.0459170	−0.0459170	−0.0112041
10	14.0000000	15.0459170	−1.0459170	−0.2552112
11	14.0000000	15.0459170	−1.0459170	−0.2552112
12	13.0000000	15.0459170	−2.0459170	−0.4992184
13	11.0000000	15.0459170	−4.0459170	−0.9872326
14	12.0000000	15.0459170	−3.0459170	−0.7432255
15	8.0000000	6.8834430	1.1165570	0.2724479
16	7.0000000	6.8834430	0.1165570	0.0284407
17	8.0000000	6.8834430	1.1165570	0.2724479
18	8.0000000	6.8834430	1.1165570	0.2724479
19	9.0000000	6.8834430	2.1165570	0.5164550
20	15.0000000	13.0052980	1.9947020	0.4867215
21	15.0000000	27.2896290	−12.2896290	−2.9987572

2. *Control Information*

No. of observations	21
Response variable is no.	04
Risk level for B conf. interval	05%
List of excluded variables	01 03
Variable entering	2
Sequential F-test	62.3732090
Per cent variation explained $R - SQ$	76.6507900
Standard deviation of residuals	5.0427155
Mean of the response	17.5238090
Std. dev. as per cent of response mean	28.776%
Degrees of freedom	19
Determinant value	1.0000000

ANOVA

Source	df	SS	MS	Overall F
Total	20	2069.2370000		
Regression	1	1586.0865000	1586.0865000	62.37
Residual	19	483.1506100	25.4289790	
lack of fit	13	402.4839433	30.9603033	2.30
pure error	6	80.6666667	13.4444445	

B Coefficients and Confidence Limits

Var. No.	Mean	Decoded B Coefficient	Limits Upper/Lower	Standard Error	Partial F-test
2	21.0952370	2.8174450	3.5641096 2.0707804	0.3567438	62.37

Constant term in prediction equation −41.9108610.

Square of Partial Correlations Coefficients of Variables Not in Regression

Variables	Square of Partials
1	0.60924
3	0.01675

Residual Analysis for $\hat{X}_4 = f(X_2)$

Obs. No.	Observed Y	Predicted Y	Residual	Normal Deviate
1	42.0000000	34.1601540	7.8398460	1.5546873
2	37.0000000	34.1601540	2.8398460	0.5631581
3	37.0000000	28.5252640	8.4747360	1.6805897
4	28.0000000	25.7078190	2.2921810	0.4545529
5	18.0000000	20.0729290	−2.0729290	−0.4110740
6	18.0000000	22.8903740	−4.8903740	−0.9697898
7	19.0000000	25.7078190	−6.7078190	−1.3301997
8	20.0000000	25.7078190	−5.7078190	−1.1318939
9	15.0000000	22.8903740	−7.8903740	−1.5647073
10	14.0000000	8.8031490	5.1968510	1.0305659
11	14.0000000	8.8031490	5.1968510	1.0305659
12	13.0000000	5.9857040	7.0142960	1.3909759
13	11.0000000	8.8031490	2.1968510	0.4356484
14	12.0000000	11.6205940	0.3794060	0.0752384
15	8.0000000	8.8031490	−0.8031490	−0.1592691
16	7.0000000	8.8031490	−1.8031490	−0.3575750
17	8.0000000	11.6205940	−3.6205940	−0.7179850
18	8.0000000	11.6205940	−3.6205940	−0.7179850
19	9.0000000	14.4380390	−5.4380390	−1.0783949
20	15.0000000	14.4380390	0.5619610	0.1114402
21	15.0000000	14.4380390	0.5619610	0.1114402

3. *Control Information*

No. of observations	21
Response variable is no.	04
Risk level for B conf. interval	05%
List of excluded variables	01 02
Variable entering	3
Sequential F-test	3.6153784
Per cent variation explained $R - SQ$	15.9863400
Standard deviation of residuals	9.5654028
Mean of the response	17.5238090
Std. dev. as per cent of response mean	54.585%
Degrees of freedom	19
Determinant value	1.0000000

ANOVA

Source	df	SS	MS	Overall F
Total	20	2069.2370000		
Regression	1	330.7952600	330.7952600	3.62
Residual	19	1738.4417000	91.4969310	
lack of fit	13	1657.7750333	127.5211563	9.49
pure error	6	80.6666667	13.4444445	

B Coefficients and Confidence Limits

Var. No.	Mean	Decoded B Coefficient	Limits Upper/Lower	Standard Error	Partial F-test
3	86.2857130	0.7589552	1.5943822 −0.0764717	0.3991529	3.62

Constant term in prediction equation −47.9631840.

Squares of Partial Correlation Coefficients of Variables Not in Regression

Variables	Square of Partials
1	0.82218
2	0.72673

Residual Analysis for $\hat{X}_4 = f(X_3)$

Obs. No.	Observed Y	Predicted Y	Residual	Normal Deviate
1	42.0000000	19.5838310	22.4161690	2.3434631
2	37.0000000	18.8248760	18.1751240	1.9000897
3	37.0000000	20.3427860	16.6572140	1.7414022
4	28.0000000	18.0659210	9.9340790	1.0385426
5	18.0000000	18.0659210	−0.0659210	−0.0068916
6	18.0000000	18.0659210	−0.0659210	−0.0068916
7	19.0000000	22.6196520	−3.6196520	−0.3784108
8	20.0000000	22.6196520	−2.6196520	−0.2738674
9	15.0000000	18.0659210	−3.0659210	−0.3205219
10	14.0000000	12.7532340	1.2467660	0.1303412
11	14.0000000	19.5838310	−5.5838310	−0.5837528
12	13.0000000	18.8248760	−5.8248760	−0.6089525
13	11.0000000	14.2711440	−3.2711440	−0.3419766
14	12.0000000	22.6196520	−10.6196520	−1.1102148
15	8.0000000	19.5838310	−11.5838310	−1.2110134
16	7.0000000	17.3069650	−10.3069650	−1.0775254
17	8.0000000	6.6815920	1.3184080	0.1378309
18	8.0000000	11.9942790	−3.9942790	−0.4175756
19	9.0000000	12.7532340	−3.7532340	−0.3923759
20	15.0000000	14.2711440	0.7288560	0.0761971
21	15.0000000	21.1017410	−6.1017410	−0.6378969

4. *Control Information*

No. of observations	21	
Response variable is no.	04	
Risk level for B conf. interval	05%	
List of excluded variables	03	
Variables entering	1, 2	
Sequential F-test	12.4249510	
Percent variation explained $R - SQ$	90.8761000	
Standard deviation of residuals	3.2386150	
Mean of the response	17.5238090	
Std. dev. as per cent of response mean	18.481%	
Degrees of freedom	18	
Determinant value	0.3887071	

ANOVA

Source	df	SS	MS	Overall F
Total	20	2069.2370000		
Regression	2	1880.4418000	940.2209000	89.64
Residual	18	188.7952900	10.4886270	
lack of fit	12	108.1286233	9.0107186	<1
pure error	6	80.6666667	13.4444445	

B Coefficients and Confidence Limits

Var. No.	Mean	Decoded B Coefficient	Limits Upper/Lower	Standard Error	Partial F-test
1	60.4285700	0.6711545	0.9373324 0.4049767	0.1266910	28.06
2	21.0952370	1.2953510	2.0674378 0.5232642	0.3674854	12.42

Constant term in prediction equation −50.3588360.

Squares of Partial Correlation Coefficients of Variables Not in Regression

Variables	Square of Partials
3	0.05278

Residual Analysis for $\hat{X}_4 = f(X_1, X_2)$

Obs. No.	Observed Y	Predicted Y	Residual	Normal Deviate
1	42.0000000	38.3080040	3.6919960	1.1399922
2	37.0000000	38.3080040	−1.3080040	−0.4038776
3	37.0000000	32.3615290	4.6384710	1.4322390
4	28.0000000	22.3411690	5.6588310	1.7472996
5	18.0000000	19.7504670	−1.7504670	−0.5404986
6	18.0000000	21.0458180	−3.0458180	−0.9404693
7	19.0000000	22.3411690	−3.3411690	−1.0316660
8	20.0000000	22.3411690	−2.3411690	−0.7228920
9	15.0000000	18.3612000	−3.3612000	−1.0378510
10	14.0000000	11.8844450	2.1155550	0.6532283
11	14.0000000	11.8844450	2.1155550	0.6532283
12	13.0000000	10.5890940	2.4109060	0.7444250
13	11.0000000	11.8844450	−0.8844450	−0.2730936
14	12.0000000	13.1797960	−1.1797960	−0.3642903
15	8.0000000	6.5152090	1.4847910	0.4584648
16	7.0000000	6.5152090	0.4847910	0.1496908
17	8.0000000	7.8105600	0.1894400	0.0584941
18	8.0000000	7.8105600	0.1894400	0.0584941
19	9.0000000	9.1059110	−0.1059110	−0.0327026
20	15.0000000	13.1328380	1.8671620	0.5765310
21	15.0000000	22.5290010	−7.5290010	−2.3247595

5. *Control Information*

No. of observations	21
Response of variable is no.	04
Risk level for B conf. interval	05%
List of excluded variables	02

Variables entering	1, 3
Sequential F-test	0.5810138
Per cent variation explained $R - SQ$	85.0603200
Standard deviation of residuals	4.1441891
Mean of the response	17.5238090
Std. dev. as per cent of response mean	23.649%
Degrees of freedom	18
Determinant value	0.7498572

ANOVA

Source	df	SS	MS	Overall F
Total	20	2069.2370000		
Regression	2	1760.0996000	880.0498000	51.24
Residual	18	309.1374600	17.1743030	
lack of fit	12	228.4707933	19.0392328	1.42
pure error	6	80.6666667	13.4444445	

B Coefficients and Confidence Limits

Var. No.	Mean	Decoded B Coefficient	Limits Upper/Lower	Standard Error	Partial F-test
1	60.4285700	1.0648068	1.3100370 0.8195766	0.1167207	83.22
3	86.2857130	−0.1522227	0.2673549 −0.5718004	0.1997038	0.58

Constant term in prediction equation −33.6862970.

Squares of Partial Correlation Coefficients of Variables Not in Regression

Variables	Square of Partials
2	0.42152

Residual Analysis for $\hat{X}_4 = f(X_1, X_3)$

Obs. No.	Observed Y	Predicted Y	Residual	Normal Deviate
1	42.0000000	37.9504250	4.0495750	0.9771695
2	37.0000000	38.1026470	−1.1026470	−0.2660706
3	37.0000000	32.4741680	4.5258320	1.0920910
4	28.0000000	19.0883470	8.9116530	2.1503972
5	18.0000000	19.0883470	−1.0883470	−0.2626200
6	18.0000000	19.0883470	−1.0883470	−0.2626200
7	19.0000000	18.1750110	0.8249890	0.1990713
8	20.0000000	18.1750110	1.8249890	0.4403730
9	15.0000000	14.8291200	0.1708800	0.0412336
10	14.0000000	15.8946790	−1.8946790	−0.4571893
11	14.0000000	14.5246750	−0.5246750	−0.1266050
12	13.0000000	14.6768970	−1.6768970	−0.4046381
13	11.0000000	15.5902340	−4.5902340	−1.1076314
14	12.0000000	13.9157840	−1.9157840	−0.4622820
15	8.0000000	6.0062210	1.9937790	0.4811023
16	7.0000000	6.4628890	0.5371110	0.1296058
17	8.0000000	8.5940070	−0.5940070	−0.1433349
18	8.0000000	7.5284480	0.4715520	0.1137863
19	9.0000000	7.3762250	1.6237750	0.3918197
20	15.0000000	13.4606200	1.5393800	0.3714551
21	15.0000000	26.9979110	−11.9979110	−2.8951166

6. *Control Information*

No. of observations	21
Response variable is no.	04
Risk level for B conf. interval	05%
List of excluded variables	01

Variables entering	2, 3
Sequential F-test	0.3066295
Per cent variation explained $R - SQ$	77.0418800
Standard deviation of residuals	5.1373253
Mean of the response	17.5238090
Std. dev. as per cent of response mean	29.316%
Degrees of freedom	18
Determinant value	0.8471664

ANOVA

Source	df	SS	MS	Overall F
Total	20	2069.2370000		
Regression	2	1594.1790000	797.0895000	30.20
Residual	18	475.0580100	26.3921110	
lack of fit	12	394.3913433	32.8659453	2.44
pure error	6	80.6666667	13.4444445	

B Coefficients and Confidence Limits

Var. No.	Mean	Decoded B Coefficient	Limits Upper/Lower	Standard Error	Partial F-test
2	21.0952370	2.7319656	3.5615693 1.9023619	0.3948614	47.87
3	86.2857130	0.1289721	0.6183167 −0.3603725	0.2329103	0.31

Constant term in prediction equation −51.2360990.

Squares of Partial Correlation Coefficients of Variables Not in Regression

Variables	Square of Partials
1	0.62356

Residual Analysis for $X_4 = f(X_2, X_3)$

Obs. No.	Observed Y	Predicted Y	Residual	Normal Deviate
1	42.0000000	34.0054870	7.9945130	1.5561625
2	37.0000000	33.8765150	3.1234850	0.6079983
3	37.0000000	28.6705280	8.3294720	1.6213635
4	28.0000000	25.5516450	2.4483550	0.4765817
5	18.0000000	20.0877140	−2.0877140	−0.4063815
6	18.0000000	22.8196790	−4.8196790	−0.9381689
7	19.0000000	26.3254780	−7.3254780	−1.4259322
8	20.0000000	26.3254780	−6.3254780	−1.2312784
9	15.0000000	22.8196790	−7.8196790	−1.5221303
10	14.0000000	8.2570470	5.7429530	1.1178877
11	14.0000000	9.4177960	4.5822040	0.8919435
12	13.0000000	6.5568590	6.4431410	1.2541820
13	11.0000000	8.5149910	2.4850090	0.4837165
14	12.0000000	12.6656500	−0.6656500	−0.1295713
15	8.0000000	9.4177960	−1.4177960	−0.2759794
16	7.0000000	9.0308790	−2.0308790	−0.3953184
17	8.0000000	9.9572360	−1.9572360	−0.3809835
18	8.0000000	10.8600410	−2.8600410	−0.5567179
19	9.0000000	13.7209790	−4.7209790	−0.9189566
20	15.0000000	13.9789230	1.0210770	0.1987565
21	15.0000000	15.1396720	−0.1396720	−0.0271877

7. *Control Information*

No. of observations	21
Response variable is no.	04
Risk level for B conf. interval	05%

Variables entering	1, 2, 3
Sequential F-test	0.9473322
Per cent variation explained $R - SQ$	91.3576900
Standard deviation of residuals	3.2433636
Mean of the response	17.5238090
Std. dev. as per cent of response mean	18.508%
Degrees of freedom	17
Determinant value	0.2914747

ANOVA

Source	df	SS	MS	Overall F
Total	20	2069.2370000		
Regression	3	1890.4071000	630.1357000	59.90
Residual	17	178.8299200	10.5194070	
lack of fit	11	98.1632533	8.9239321	<1
pure error	6	80.6666667	13.4444445	

B Coefficients and Confidence Limits

Var. No.	Mean	Decoded B Coefficient	Limits Upper/Lower	Standard Error	Partial F-test
1	60.4285700	0.7156403	1.0001910 0.4310896	0.1348582	28.16
2	21.0952370	1.2952857	2.0718168 0.5187546	0.3680242	12.39
3	86.2857130	−0.1521225	0.1776579 −0.4819029	0.1562940	0.95

Constant term in prediction equation −39.9196680.

Residual Analysis for $\hat{X_4} = f(X_1, X_2, X_3)$

Obs. No.	Observed Y	Predicted Y	Residual	Normal Deviate
1	42.0000000	38.7653640	3.2346360	0.9973091
2	37.0000000	38.9174870	−1.9174870	−0.5912032
3	37.0000000	32.4444690	4.5555310	1.4045699
4	28.0000000	22.3022260	5.6977740	1.7567484
5	18.0000000	19.7116550	−1.7116550	−0.5277407
6	18.0000000	21.0069410	−3.0069410	−0.9271057
7	19.0000000	21.3894910	−2.3894910	−0.7367324
8	20.0000000	21.3894910	−1.3894910	−0.4284105
9	15.0000000	18.1443800	−3.1443800	−0.9694812
10	14.0000000	12.7328090	1.2671910	0.3907027
11	14.0000000	11.3637060	2.6362940	0.8128272
12	13.0000000	10.2205430	2.7794570	0.8569674
13	11.0000000	12.4285640	−1.4285640	−0.4404576
14	12.0000000	12.0505020	−0.0505020	−0.0155709
15	8.0000000	5.6385840	2.3614160	0.7280762
16	7.0000000	6.0949520	0.9050480	0.2790461
17	8.0000000	9.5199530	−1.5199530	−0.4686348
18	8.0000000	8.4550960	−0.4550960	−0.1403161
19	9.0000000	9.5982590	−0.5982590	−0.1844563
20	15.0000000	13.5878550	1.4121450	0.4353952
21	15.0000000	22.2377170	−7.2377170	−2.2315465

E. The demand for a consumer product is affected by many factors. In one study, measurements on the relative urbanization, educational level, and relative income of nine geographic areas were made in an attempt to determine their effect on the product usage. The data collected were as follows:

Area No.	Relative Urbanization X_1	Educational Level X_2	Relative Income X_3	Product Usage $X_4 = Y$
1	42.2	11.2	31.9	167.1
2	48.6	10.6	13.2	174.4
3	42.6	10.6	28.7	160.8
4	39.0	10.4	26.1	162.0
5	34.7	9.3	30.1	140.8
6	44.5	10.8	8.5	174.6
7	39.1	10.7	24.3	163.7
8	40.1	10.0	18.6	174.5
9	45.9	12.0	20.4	185.7
Means	41.8555	10.6222	22.4222	167.0666
Standard deviation	s_1 4.176455	s_2 0.7462871	s_3 7.927921	s_4 12.645157

The correlation matrix is

	X_1	X_2	X_3	X_4
X_1	1	0.683742	−0.615790	0.801752
X_2	0.683742	1	−0.172493	0.769950
X_3	−0.615790	−0.172493	1	−0.628746
X_4	0.801752	0.769950	−0.628746	1

Requirements

1. Follow the stepwise procedure and determine an appropriate fitted first order model using the following criteria:

$$F = 2.00 \text{ for entering variables}$$
$$F = 2.00 \text{ for rejecting variables}$$

2. Write out the analysis of variance table at each step.
3. Comment on the adequacy of the final fitted equation after examining the residuals.

CHAPTER 7

A SPECIFIC PROBLEM

7.0. Introduction

When the statistician can suggest the experimental runs to be made in a series of tests, data analysis can be quite straightforward. Often, however, an analysis must be carried out on results which have been accumulated as *part* of a test program or as a series of related test programs, and an empirical predictive model must be constructed from the results obtained. Terms for possible inclusion in the model might involve not only the principal variables but also variables such as cross-products, squares, or other combinations, or transformations, of the principal variables. If computer programs for procedures such as those described in Chapter 6 are available, the calculations can be automated. Whether or not such a program exists, the intelligent use of plots of residuals against new variable candidates can be very helpful. In this chapter we show how an empirical model can be constructed through an examination of such plots.

7.1. The Problem

A rocket engine's performance is affected by a number of environmental conditions. A standard chamber pressure is usually established which corresponds to normal performance. As the engines are made and tested, environmental conditions and chamber pressures are recorded. A typical set of such data is shown in Table 7.1. We wish to use this data to develop an empirical predictive model for chamber pressure.

7.2. Examination of the Data

As can be seen by adopting the format of Table 7.2, the data of Table 7.1 are not well balanced in the independent variables.

217

Table 7.1   Rocket Engine Data

Unit No.	Temperature of Cycle X_1	Vibration X_2	Drop (Shock) X_3	Static Fire X_4	Chamber Pressure Y
1	−75	0	0	−65	1.4
2	175	0	0	150	26.3
3	0	−75	0	150	26.5
4	0	175	0	−65	5.8
5	0	−75	0	150	23.4
6	0	175	0	−65	7.4
7	0	0	−65	150	29.4
8	0	0	165	−65	9.7
9	0	0	0	150	32.9
10	−75	−75	0	150	26.4
11	175	175	0	−65	8.4
12	0	−75	−65	150	28.8
13	0	175	165	−65	11.8
14	−75	−75	−65	150	28.4
15	175	175	165	−65	11.5
16	0	−75	0	150	26.5
17	0	175	0	−65	5.8
18	0	0	−65	−65	1.3
19	0	0	165	150	21.4
20	0	−75	−65	−65	0.4
21	0	175	165	150	22.9
22	0	−75	−65	150	26.4
23	0	175	165	−65	11.4
24	0	0	0	−65	3.7

The data (except for chamber pressures) are taken from the book *Reliability: Management, Methods, and Mathematics*, by David K. Lloyd and Myron Lipow, Prentice-Hall, 1962, page 360.

We can, nevertheless, attempt to develop an empirical prediction model from these results. If a suitable relationship can be obtained it can be used to select additional runs to check the usefulness of the fitted model or to aid in the selection of an entire, new experimental design.

There is an obvious rise in chamber pressure as static time increases and so X_4 immediately appears to be an important variable. The replication in the runs allows estimation of pure error—a great advantage in any set of data. All the replication occurs when $X_1 = 0$ but, as usual, the size of the random error is assumed to be constant throughout the experimental region. Radical departures from this assumption, if there are any, should show up in residuals plots.

Table 7.2 An Alternative Data Format

		$X_1(-75)$			$X_1(0)$			$X_1(175)$		
		$X_2(-75)$	$X_2(0)$	$X_2(175)$	$X_2(-75)$	$X_2(0)$	$X_2(175)$	$X_2(-75)$	$X_2(0)$	$X_2(175)$
$X_3(-65)$	$X_4(-65)$				0.4	1.3				
	$X_4(150)$	28.4			28.8 / 26.4	29.4				
$X_3(0)$	$X_4(-65)$		1.4			3.7	5.8 / 7.4 / 5.8			8.4
	$X_4(150)$	26.4			26.5 / 23.4 / 26.5	32.9			26.3	
$X_3(165)$	$X_4(-65)$					9.7	11.8 / 11.4			11.5
	$X_4(150)$					21.4	22.9			

7.3.   Choosing the First Variable to Enter Regression

1. Although, as we have mentioned above, X_4 is a prime candidate for the most important independent variable, we examine the correlation matrix to be sure. Since the matrix is symmetric, only the upper right portion is shown below and all results are rounded to three decimal places.

	X_1	X_2	X_3	X_4	Y
X_1	1.000	0.376	0.194	−0.157	−0.065
X_2		1.000	0.538	−0.597	−0.464
X_3			1.000	−0.225	−0.128
X_4				1.000	0.944
Y					1.000

The largest correlation, 0.944, *is* between X_4 and Y. We can now fit the model $Y = \beta_0 + \beta_4 X_4 + \epsilon$. We omit the details of the calculation. The fitted equation takes the form:

$$\hat{Y} = 12.614147 + 0.0932946 X_4$$

and the analysis of variance table is as follows.

Source	df	SS	MS	F
Total (corrected)	23	2711.60		
Regression	1	2414.02	2414.02	
Residual	22	297.58	13.53	
Lack of fit	16	286.51	17.91	9.70
Pure error	6	11.07	1.85	

The calculated value of R^2 is $2414.02/2711.60 = 0.890$, which means that this model accounts for 89% of the variation about the mean in the data. In spite of this, the lack of fit F-test value, 9.70, exceeds $F(16, 6, 0.975)$ so that the model must be regarded as inadequate. To investigate this we can examine the residuals, in particular plotting them against candidates for new variables. The original observations, fitted values, and residuals are shown in Table 7.3. Figures 7.1, 7.2, and 7.3 show plots of residuals against X_1, X_2, and X_3 respectively.

Table 7.3   Observations, Fitted Values, and Residuals

Observed Y	Predicted $\hat{Y}$ from $\hat{Y} = f(X_4)$	Residual
1.4	6.55	−5.15
26.3	26.61	−0.31
26.5	26.61	−0.11
5.8	6.55	−0.75
23.4	26.61	−3.21
7.4	6.55	0.94
29.4	26.61	2.79
9.7	6.55	3.15
32.9	26.61	6.29
26.4	26.61	−0.21
8.4	6.55	1.85
28.8	26.61	2.19
11.8	6.55	5.25
28.4	26.61	1.79
11.5	6.55	4.95
26.5	26.61	−0.11
5.8	6.55	−0.75
1.3	6.55	−5.25
21.4	26.61	−5.21
0.4	6.55	−6.15
22.9	26.61	−3.71
26.4	26.61	−0.21
11.4	6.55	4.85
3.7	6.55	−2.85

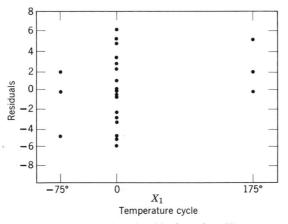

Figure 7.1   Plot of residuals against X_1.

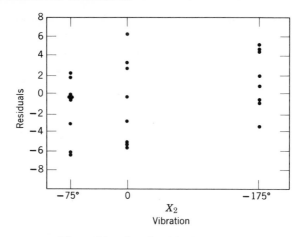

Figure 7.2 Plot of residuals against X_2.

Due to the vertical spread of the plotted values, indications of dependence of the residuals on X_1, X_2, or X_3 are not startling. (Note, however, that the variables X_1, X_2, and X_3 are not adjusted for X_4 the variable already in regression. This sometimes conceals relationships which might appear *after* such an adjustment.) Instead we turn our attention to the possibility of adding other types of variables to the model.

7.4. Construction of New Variables

We now construct the six possible cross-product variable columns of the form $X_i X_j$. Terms of this type in the model enable us to take account of the

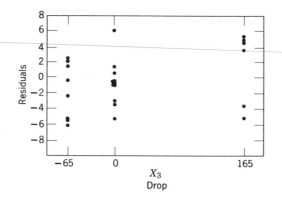

Figure 7.3 Plot of residuals against X_3.

Table 7.4   Original Data and Variable Cross-Product Columns

Unit No.	X_1	X_2	X_3	X_4	X_1X_2	X_1X_3	X_1X_4	X_2X_3	X_2X_4	X_3X_4	Y
1	-75	0	0	-65	0	0	4875	0	0	0	1.4
2	175	0	0	150	0	0	26250	0	0	0	26.3
3	0	-75	0	150	0	0	0	0	-11250	0	26.5
4	0	175	0	-65	0	0	0	0	-11375	0	5.8
5	0	-75	0	150	0	0	0	0	-11250	0	23.4
6	0	175	0	-65	0	0	0	0	-11375	0	7.4
7	0	0	-65	150	0	0	0	0	0	-9750	29.4
8	0	0	165	-65	0	0	0	0	0	-10725	9.7
9	0	0	0	150	0	0	0	0	0	0	32.9
10	-75	-75	0	150	5625	0	-11250	0	-11250	0	26.4
11	175	175	0	-65	30625	0	-11375	0	-11375	0	8.4
12	0	-75	-65	150	0	0	0	4875	-11250	-9750	28.8
13	0	175	165	-65	0	0	0	28875	-11375	-10725	11.8
14	-75	-75	-65	150	5625	4875	-11250	4875	-11250	-9750	28.4
15	175	175	165	-65	30625	28875	-11375	28875	-11375	-10725	11.5
16	0	-75	0	150	0	0	0	0	-11250	0	26.5
17	0	175	0	-65	0	0	0	0	-11375	0	5.8
18	0	0	-65	-65	0	0	0	0	0	4225	1.3
19	0	0	165	150	0	0	0	0	0	24750	21.4
20	0	-75	-65	-65	0	0	0	4875	4875	4225	0.4
21	0	175	165	150	0	0	0	28875	26250	24750	22.9
22	0	-75	-65	150	0	0	0	4875	-11250	-9750	26.4
23	0	175	165	-65	0	0	0	28875	-11375	-10725	11.4
24	0	0	0	-65	0	0	0	0	0	0	3.7

way the variables X_i and X_j interact in their effect on the response. An element of the X_iX_j column is simply the cross-product of the corresponding elements in the X_i and X_j columns. For example the first X_1X_4 element is $(-75)(-65) = 4875$. These columns are shown in Table 7.4.

7.5.   The Addition of a Cross-Product Term to the Model

A plot of the residuals from Table 7.3 against the corresponding cross-product entries X_3X_4 from Table 7.4 is shown in Figure 7.4. This indicates a greater dependence than is shown by residuals plots against other cross-product variables, which we have not reproduced. We thus now add a term $\beta_{34}X_3X_4$ to the initial model and refit. The fitted equation takes the form

$$\hat{Y} = 12.153563 + 0.979107X_4 - 0.0002650X_3X_4$$

and the analysis of variance table is as follows:

Source	df	SS	MS	F
Total (corrected)	23	2711.60		
Regression	2	2553.93	1276.97	
Residual	21	157.67	7.51	
Lack of fit	15	146.59	9.77	5.30
Pure error	6	11.07	1.85	

Now $R^2 = 2553.93/2711.60 = 0.942$ so that over 94% of the variation about the mean in the data is explained by the model. The F value for lack of fit, 5.30, exceeds $F(15, 6, 0.975)$; thus further improvement appears possible.

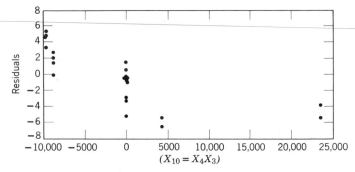

Figure 7.4   Plot of residuals against X_4X_3.

7.6. Enlarging the Model

At this stage the residuals should again be evaluated and plotted against the corresponding column entries in Table 7.4 (except for X_4 and X_3X_4 which are already in the model). We leave the details as an exercise. The plots show, however, that X_2 now appears to be the most likely candidate, now that more of the variation has been taken up by the addition of the $\beta_{34}X_3X_4$ term. Adding this variable to the model to give

$$Y = \beta_0 + \beta_4X_4 + \beta_{34}X_3X_4 + \beta_2X_2 + \epsilon$$

we can evaluate a fitted equation of the form

$$\hat{Y} = 10.713137 + 0.112340X_4 - 0.0003140X_3X_4 + 0.0233483X_2$$

and the analysis of variance table is as follows:

Source	df	SS	MS	F
Total (corrected)	23	2711.60		
Regression	3	2641.59	880.53	251.55
Residual	20	70.01	3.50	
Lack of fit	14	58.94	4.21	2.28
Pure error	6	11.07	1.85	

Lack of fit is not significant so the model is not inadequate. The fitted equation provides an $R^2 = 2641.59/2711.60 = 0.974$ or accounts for 97.4% of the variation about the mean in the data. The overall regression is significant and so are all individual coefficients.

b Coefficients and 95% Confidence Limits for the β's

Var. No.	Mean	Decoded b Coefficient	Limits Upper/Lower
X_4	42.5	0.1123395	0.1220046
			0.1026743
X_3X_4	−997.9	−0.0003140	−0.0002242
			−0.0004038
X_2	33.3	0.0233483	0.0330812
			0.0136155

We have now obtained an empirical model that can be used for predictive purposes. Created in a somewhat arbitrary fashion it makes no claim to explain why response responds to variation in the independent variables but merely provides an empirical explanation of the data that may be

Table 7.5 Observations, Predicted Values, and Residuals
from the Final Model

Obs. No.	Y	$\hat{Y}$	Residual $Y - \hat{Y}$	Normal Deviate Residuals
1	1.4	3.41	−2.01	−1.07
2	26.3	27.56	−1.26	−0.68
3	26.5	25.81	0.69	0.37
4	5.8	7.50	−1.70	−0.91
5	23.4	25.81	−2.41	−1.29
6	7.4	7.50	−0.10	−0.05
7	29.4	30.63	−1.23	−0.66
8	9.7	6.78	2.92	1.56
9	32.9	27.56	5.34	2.85
10	26.4	25.81	0.59	0.31
11	8.4	7.50	0.90	0.48
12	28.8	28.87	−0.07	−0.04
13	11.8	10.86	0.94	0.50
14	28.4	28.87	−0.47	−0.25
15	11.5	10.86	0.64	0.34
16	26.5	25.81	0.69	0.37
17	5.8	7.50	−1.70	−0.91
18	1.3	2.08	−0.78	−0.42
19	21.4	19.79	1.61	0.86
20	0.4	0.33	0.07	0.04
21	22.9	23.88	−0.98	−0.52
22	26.4	28.87	−2.47	−1.32
23	11.4	10.86	0.54	0.29
24	3.7	3.41	0.29	0.15

useful in future work. It might be wise to re-examine the current residuals at this point to see if the model performs satisfactorily over the whole region in which observations were taken. Table 7.5 contains the residuals from the final model in two forms—their original form and normal deviate form (see Section 3.1). Approximately 5% or one in twenty of the normal-deviate form residuals would be expected to fall outside the range $(-2, 2)$.

Since only one value, 2.85, from unit 9, lies outside this range there is no cause for alarm. Moreover, unit 9 was tested with $X_1 = X_2 = X_3 = 0$ and $X_4 = 150$, and the observed chamber pressure 32.9 was the highest recorded. Actually low chamber pressures are more desirable and so the empirical model fits worst in a region of little interest.

EXERCISES

A. Twenty (20) mature wood samples of slash pine cross sections were prepared, 30 μ in thickness. These sections were stained jet-black in Chlorozol E. The following determinations were made.

Data on Anatomical Factors and Wood Specific Gravity of Slash Pine

No. of Fibers/mm² in Springwood	No. of Fibers/mm² in Summerwood	Spring-wood %	Light Absorption Springwood %	Light Absorption Summerwood %	Wood Specific Gravity
573	1059	46.5	53.8	84.1	0.534
651	1356	52.7	54.5	88.7	0.535
606	1273	49.4	52.1	92.0	0.570
630	1151	48.9	50.3	87.9	0.528
547	1135	53.1	51.9	91.5	0.548
557	1236	54.9	55.2	91.4	0.555
489	1231	56.2	45.5	82.4	0.481
685	1564	56.6	44.3	91.3	0.516
536	1182	59.2	46.4	85.4	0.475
685	1564	63.1	56.4	91.4	0.486
664	1588	50.6	48.1	86.7	0.554
703	1335	51.9	48.4	81.2	0.519
653	1395	62.5	51.9	89.2	0.492
586	1114	50.5	56.5	88.9	0.517
534	1143	52.1	57.0	88.9	0.502
523	1320	50.5	61.2	91.9	0.508
580	1249	54.6	60.8	95.4	0.520
448	1028	52.2	53.4	91.8	0.506
476	1057	42.9	53.2	92.9	0.595
528	1057	42.4	56.6	90.9	0.568

Source: Anatomical Factors Influencing Wood Specific Gravity of Slash Pines and the Implications for the Development of a High-Quality Pulpwood. J. P. Van Buijtenen, *Tappi*, **47** (7), 1964, 401–404.

Requirements

Develop a prediction equation for wood specific gravity. Use the stepwise procedure and adopt critical F values of 2.00 for both accepting and rejecting variables. Examine the residuals from your model and state conclusions.

B. The following data were collected on spray congealing.

Values for the Experimental Operating Variables and
Average Particle Sizes

Run	Feed Rate per unit Whetted Wheel Periphery (gm/sec/cm) (X_1)	Peripheral Wheel Velocity (cm/sec) (X_2)	Feed Viscosity (poise) (X_3)	Mean Surface–Volume Particle Size of Product (μ) (Y)
1	0.0174	5300	0.108	25.4
2	0.0630	5400	0.107	31.6
3	0.0622	8300	0.107	25.7
4	0.0118	10800	0.106	17.4
5	0.1040	4600	0.102	38.2
6	0.0118	11300	0.105	18.2
7	0.0122	5800	0.105	26.5
8	0.0122	8000	0.100	19.3
9	0.0408	10000	0.106	22.3
10	0.0408	6600	0.105	26.4
11	0.0630	8700	0.104	25.8
12	0.0408	4400	0.104	32.2
13	0.0415	7600	0.106	25.1
14	0.1010	4800	0.106	39.7
15	0.0170	3100	0.106	35.6
16	0.0412	9300	0.105	23.5
17	0.0170	7700	0.098	22.1
18	0.0170	5300	0.099	26.5
19	0.1010	5700	0.098	39.7
20	0.0622	6200	0.102	31.5
21	0.0622	7700	0.102	26.9
22	0.0170	10200	0.100	18.1
23	0.0118	4800	0.102	28.4
24	0.0408	6600	0.102	27.3
25	0.0622	8300	0.102	25.8
26	0.0170	7700	0.102	23.1
27	0.0408	9000	0.613	23.4
28	0.0170	10100	0.619	18.1
29	0.0408	5300	0.671	30.9
30	0.0622	8000	0.624	25.7
31	0.1010	7300	0.613	29.0
32	0.0118	6400	0.328	22.0
33	0.0170	8000	0.341	18.8
34	0.0118	9700	1.845	17.9
35	0.0408	6300	1.940	28.4

Source: "Spray congealing: particle size relationships using a centrifugal wheel atomizer" by M. W. Scott, M. J. Robinson, J. F. Pauls, R. J. Lantz, *Journal of Pharmaceutical Sciences*, **53** (6), June 1964, 670–675.

A proposed model, based on theoretical considerations, is

$$Y = \alpha X_1^\beta X_2^\gamma X_3^\delta \epsilon$$

Requirements

After transformation, fit the proposed model by least squares. State which independent variable appears most important and check all coefficients for statistical significance (take $\alpha = 0.05$). Is the model a satisfactory one?

C. The data shown below which relate to a study of the quantity of vitamin B_2 in turnip green are taken from the "Annual Progress Report on the Soils-weather Project, 1948" by J. T. Wakeley, University of North Carolina (Raleigh) Institute of Statistics Mimeo Series 19 (1949). The variables are

X_1 = radiation in relative gram calories per minute during the preceding half day of sunlight (coded by dividing by 100)
X_2 = average soil moisture tension (coded by dividing by 100)
X_3 = air temperature in degrees Fahrenheit (coded by dividing by 10), and
Y = milligrams of vitamin B_2 per gram of turnip green.

These data were used by R. L. Anderson and T. A. Bancroft in *Statistical Theory in Research* (1959), McGraw-Hill, New York, on page 192, to fit the model $Y = \beta_0 + \beta_1 X_1 + \beta_2 X_2 + \beta_3 X_3 + \beta_{12} X_1 X_2 + \epsilon.$

Requirements

Develop a suitable fitted equation using these data and compare its form with the form of the one fitted by Anderson and Bancroft.

X_1	X_2	X_3	Y
1.76	0.070	7.8	110.4
1.55	0.070	8.9	102.8
2.73	0.070	8.9	101.0
2.73	0.070	7.2	108.4
2.56	0.070	8.4	100.7
2.80	0.070	8.7	100.3
2.80	0.070	7.4	102.0
1.84	0.070	8.7	93.7
2.16	0.070	8.8	98.9
1.98	0.020	7.6	96.6
0.59	0.020	6.5	99.4
0.80	0.020	6.7	96.2
0.80	0.020	6.2	99.0
1.05	0.020	7.0	88.4
1.80	0.020	7.3	75.3
1.80	0.020	6.5	92.0
1.77	0.020	7.6	82.4
2.30	0.020	8.2	77.1
2.03	0.474	7.6	74.0
1.91	0.474	8.3	65.7
1.91	0.474	8.2	56.8
1.91	0.474	6.9	62.1
0.76	0.474	7.4	61.0
2.13	0.474	7.6	53.2
2.13	0.474	6.9	59.4
1.51	0.474	7.5	58.7
2.05	0.474	7.6	58.0

D. Using the data shown in Appendix A, determine a model for predicting Monthly Steam Usage, omitting variable 6 from consideration, and ensuring that:

1. R^2 exceeds 0.8.
2. All b coefficients are statistically significant (take $\alpha = 0.05$).
3. No evidence exists of patterns in the residuals obtained from the fitted model.
4. The residual standard deviation expressed as a percentage of the mean response is less than 7%.

(If a selection procedure is used, choose a critical F value of 3.00 for both acceptance and rejection of variables.)

E. In the CAED Report 17, Iowa State University, 1963, the following data are shown for the state of Iowa, 1930–1962.

		X_1	X_2	X_3	X_4	X_5	X_6	X_7	X_8	X_9	Y
Year		Preseason Precip., in.	May Temp., °F	June Rain, in.	June Temp., °F	July Rain, in.	July Temp., °F	Aug. Rain, in.	Aug. Temp., °F	Corn Yield, bu/acre
1930	1	17.75	60.2	5.83	69.0	1.49	77.9	2.42	74.4	34.0
	2	14.76	57.5	3.83	75.0	2.72	77.2	3.30	72.6	32.9
	3	27.99	62.3	5.17	72.0	3.12	75.8	7.10	72.2	43.0
	4	16.76	60.5	1.64	77.8	3.45	76.1	3.01	70.5	40.0
	5	11.36	69.5	3.49	77.2	3.85	79.7	2.84	73.4	23.0
	6	22.71	55.0	7.00	65.9	3.35	79.4	2.42	73.6	38.4
	7	17.91	66.2	2.85	70.1	0.51	83.4	3.48	79.2	20.0
	8	23.31	61.8	3.80	69.0	2.63	75.9	3.99	77.8	44.6
	9	18.53	59.5	4.67	69.2	4.24	76.5	3.82	75.7	46.3
	10	18.56	66.4	5.32	71.4	3.15	76.2	4.72	70.7	52.2
	11	12.45	58.4	3.56	71.3	4.57	76.7	6.44	70.7	52.3
	12	16.05	66.0	6.20	70.0	2.24	75.1	1.94	75.1	51.0
	13	27.10	59.3	5.93	69.7	4.89	74.3	3.17	72.2	59.9
	14	19.05	57.5	6.16	71.6	4.56	75.4	5.07	74.0	54.7
	15	20.79	64.6	5.88	71.7	3.73	72.6	5.88	71.8	52.0
	16	21.88	55.1	4.70	64.1	2.96	72.1	3.43	72.5	43.5
	17	20.02	56.5	6.41	69.8	2.45	73.8	3.56	68.9	56.7
	18	23.17	55.6	10.39	66.3	1.72	72.8	1.49	80.6	30.5
	19	19.15	59.2	3.42	68.6	4.14	75.0	2.54	73.9	60.5
	20	18.28	63.5	5.51	72.4	3.47	76.2	2.34	73.0	46.1
	21	18.45	59.8	5.70	68.4	4.65	69.7	2.39	67.7	48.2
	22	22.00	62.2	6.11	65.2	4.45	72.1	6.21	70.5	43.1
	23	19.05	59.6	5.40	74.2	3.84	74.7	4.78	70.0	62.2
	24	15.67	60.0	5.31	73.2	3.28	74.6	2.33	73.2	52.9
	25	15.92	55.6	6.36	72.9	1.79	77.4	7.10	72.1	53.9
	26	16.75	63.6	3.07	67.2	3.29	79.8	1.79	77.2	48.4
	27	12.34	62.4	2.56	74.7	4.51	72.7	4.42	73.0	52.8
	28	15.82	59.0	4.84	68.9	3.54	77.9	3.76	72.9	62.1
	29	15.24	62.5	3.80	66.4	7.55	70.5	2.55	73.0	66.0
	30	21.72	62.8	4.11	71.5	2.29	72.3	4.92	76.3	64.2
	31	25.08	59.7	4.43	67.4	2.76	72.6	5.36	73.2	63.2
	32	17.79	57.4	3.36	69.4	5.51	72.6	3.04	72.4	75.4
1962	33	26.61	66.6	3.12	69.1	6.27	71.6	4.31	72.5	76.0
Averages:	17	19.09	60.8	4.85	70.3	3.55	75.2	3.82	73.2	50.0

Requirements

Construct a predictive model for corn yield (bu/acre). Comment on the relative importance of the independent variables concerned and suggest what further investigations could be made.

F. The density of a finished product is an important performance characteristic. It can be controlled to a large extent by four main manufacturing variables:

X_1 = the amount of water in the product mix,
X_2 = the amount of reworked material in the product mix,
X_3 = the temperature of the mix, and
X_4 = the air temperature in the drying chamber.

In addition, the raw material received from the supplier is important to the process, and a measure of its quality is X_5, the temperature rise. The following data were collected:

X_1	X_2	X_3	X_4	X_5	Y
0	800	135	578	13.195	104
0	800	135	578	13.195	102
0	800	135	578	13.195	100
0	800	135	578	13.195	96
0	800	135	578	13.195	93
0	800	135	578	13.195	103
0	800	150	585	13.180	118
0	800	150	585	13.180	113
0	800	150	585	13.180	107
0	800	150	585	13.180	114
0	800	150	585	13.180	110
0	800	150	585	13.180	114
0	1000	135	590	13.440	97
0	1000	135	590	13.440	87
0	1000	135	590	13.440	92
0	1000	135	590	13.440	85
0	1000	135	590	13.440	94
0	1000	135	590	13.440	102
0	1000	150	590	13.600	104
0	1000	150	590	13.600	102
0	1000	150	590	13.600	101
0	1000	150	590	13.600	104
0	1000	150	590	13.600	98
0	1000	150	590	13.600	101
75	800	135	550	12.745	103
75	800	135	550	12.745	111
75	800	135	550	12.745	111
75	800	135	550	12.745	107
75	800	135	550	12.745	112
75	800	135	550	12.745	106
75	800	150	595	13.885	111
75	800	150	595	13.885	107
75	800	150	595	13.885	104

Continued at top of next page.

X_1	X_2	X_3	X_4	X_5	Y
75	800	150	595	13.885	103
75	800	150	595	13.885	104
75	800	150	595	13.885	103
75	1000	135	530	11.705	116
75	1000	135	530	11.705	108
75	1000	135	530	11.705	104
75	1000	135	530	11.705	116
75	1000	135	530	11.705	116
75	1000	135	530	11.705	112
75	1000	150	590	13.835	111
75	1000	150	590	13.835	110
75	1000	150	590	13.835	115
75	1000	150	590	13.835	114
75	1000	150	590	13.835	114
75	1000	150	590	13.835	114

Requirements

1. Examine the data carefully and state preliminary conclusions.
2. Estimate the coefficients in the model

$$Y = \beta_0 + \beta_1 X_1 + \beta_2 X_2 + \beta_3 X_3 + \beta_4 X_4 + \beta_5 X_5 + \epsilon.$$

3. Is the above model adequate?
4. Would you recommend an alternative model?
5. Draw some general conclusions about this experiment.

G. An experiment was conducted to determine the effect of six factors on rate. From the data shown below, develop a prediction equation for rate, and use it to propose a set of operating values which may provide an increased rate.

			Factors			
X_1	X_2	X_3	X_4	X_5	X_6	Y (Rate)
149	66	−15	150	105	383	267
143	66	− 5	115	105	383	269
149	73	− 5	150	105	383	230
143	73	−15	115	105	383	233
149	73	−15	115	78	383	222
143	66	−15	150	78	383	267
143	73	− 5	150	78	383	231
149	66	− 5	115	78	383	260
149	73	−15	150	78	196	238
149	66	− 5	150	78	196	262
143	73	− 5	115	78	196	252
143	66	−15	115	78	196	263
143	66	− 5	150	105	196	263
149	73	− 5	115	105	196	236
149	66	−15	115	105	196	268
143	73	−15	150	105	196	242

H. The data below consist of 16 indexed observations on rubber consumptions from 1948 to 1963. Use these data to develop suitable fitted equations for predicting Y_1 and Y_2 separately in terms of the independent variables X_1, X_2, X_3, and X_4.

Obser- vation No.	Total Rubber Consump- tion Y_1	Tire Rubber Consump- tion Y_2	Car Produc- tion X_1	Gross National Product X_2	Dispos- able Personal Income X_3	Motor Fuel Consump- tion X_4
1	0.909	0.871	1.287	0.984	0.987	1.046
2	1.252	1.220	1.281	1.078	1.064	1.081
3	0.947	0.975	0.787	1.061	1.007	1.051
4	1.022	1.021	0.796	1.013	1.012	1.046
5	1.044	1.002	1.392	1.028	1.029	1.036
6	0.905	0.890	0.893	0.969	0.993	1.020
7	1.219	1.213	1.400	1.057	1.047	1.057
8	0.923	0.918	0.721	1.001	1.024	1.034
9	1.001	1.014	1.032	0.996	1.003	1.014
10	0.916	0.914	0.685	0.972	0.993	1.013
11	1.173	1.170	1.291	1.046	1.027	1.037
12	0.938	0.952	1.170	1.004	1.001	1.007
13	0.965	0.946	0.817	1.002	1.014	1.008
14	1.106	1.096	1.231	1.049	1.032	1.024
15	1.011	0.999	1.086	1.023	1.020	1.030
16	1.080	1.093	1.001	1.035	1.053	1.029

CHAPTER 8

MULTIPLE REGRESSION AND MATHEMATICAL MODEL BUILDING

8.0. Introduction

The multiple linear regression techniques we have discussed can be very useful but also very dangerous if improperly used and interpreted. Before tackling a large problem by multiple regression methods it makes sense to pre-plan the project as far ahead as possible, to specify the objectives of the work, and to provide checkpoints as the work progresses. This planning will be the subject of this chapter. First, however, we shall discuss three main types of mathematical models often used by scientists:

1. The functional model,
2. The control model,
3. The predictive model.

The Functional Model

If the true functional relationship between a response and the independent variables in a problem is known, then the experimenter is in an excellent position to be able to understand, control, and predict the response. However, there are very few situations in practice in which such models can be determined. Even in those situations the functional equations are usually very complicated, difficult to interpret and to use, and are usually of nonlinear form. (Examples of nonlinear models were mentioned in Chapter 5 and the fitting of nonlinear models will be discussed in Chapter 10.) In such situations the linear regression procedures do not apply or else linear models can be used only as approximations to the correct models in iterative estimation procedures.

The Control Model

Even if it is known completely, the functional model is not always suitable for controlling a response variable. For example, in the problem

234

on the amount of steam used in a plant, one of the most important variables is the ambient temperature, and this is not controllable in the sense that process temperature, process pressure, and other process variables are controllable. An advertising man who wishes to understand the effect of a TV commercial on sales is quite aware that his competitors' activities are very important and are a necessary element in any functional model for sales. However, these activities constitute uncontrollable variables no matter how clearly they are specified in the functional model. A model which contains variables under the control of the experimenter is essential for control of a response.

A useful control model can sometimes be constructed by multiple regression techniques, if they are used carefully. If a designed experiment using the controllable variables is feasible, then the effect of these variables on the response can be obtained from a simple application of multiple regression such as those discussed in Chapter 9. However, there are many situations where designed experiments are not feasible: for example, an experiment conducted in a manufacturing plant usually disrupts day-to-day operations, and unless the potential return from a change in the response indicated by this experiment is great enough, the experiment will *not* be performed; as another example, an experiment conducted in the market place could be well designed and handled, but the uncontrollable factors (each identifiable) would make any calculated mathematical effect of the controlled variable so confusing as to be useless. These situations lead the practitioner to the use of predictive models.

Predictive Models

When the functional model is very complex and when the ability to obtain independent estimates of the effects of the control variables is limited, one can often obtain a linear predictive model which, though it may be in some senses unrealistic, at least reproduces the main features of the behavior of the response under study. These predictive models are very useful and under certain conditions can lead to real insight into the process or problem. It is in the construction of this type of predictive model that multiple regression techniques have their greatest contribution to make. These problems are usually referred to as "problems with messy data"—that is, data in which much intercorrelation exists. The predictive model is not necessarily functional and need not be useful for control purposes. This, of course, does not make it useless, contrary to the opinion of some scientists. If nothing else, it can and does provide guidelines for further experimentation, it pinpoints important variables, and it is a very useful variable screening device.

It is necessary, however, to be very careful in using multiple regression,

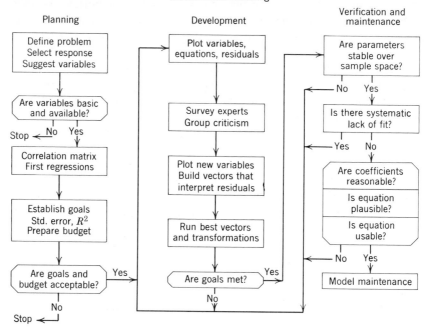

Figure 8.1   Summary of the model building procedure.

for it is easily misused and misunderstood. An organized plan for solving problems amenable to a multiple regression approach is both appropriate and necessary. This chapter is intended to be a proposal only, and anyone using this proposed scheme will find it necessary to adjust it to suit his particular situation.

While the plan below covers the development of a mathematical model for prediction purposes, it is sufficiently general in scope for use in building both functional and/or control models. The emphasis here will be on the "messy data" type problem. The plan is divided into three stages—planning, development, and maintenance. A schematic diagram of the plan is shown in Figure 8.1 and will be discussed in detail.

8.1.   Planning the Model Building Process

Define the Problem, Select the Response, Suggest Variables

The specific statement of the problem is the most important phase of any problem-solving procedure. It is necessary that the engineer, scientist, and businessman be *exact* in the specification of his problem. For example, statements like "Why does the consumer purchase my product?" or "Why isn't line no. 5 operating very well today?" are interesting, but they are not specific enough for any feasible solution. The problem definition must be to the point and both the response and predictor variables clearly identified. At the beginning of this planning phase, there should be no restraint on the scientist; he should write down every conceivable variable and response that he considers has any possible effect on the problem. This list will be large, but pursuant discussions will gradually reduce these to a reasonable number. The important point to remember is that the screening of variables should never be left to the sole discretion of any statistical procedure, including the multiple regression procedures covered in Chapter 6. Finally one arrives at a specific problem statement with a specified response or responses to be investigated by means of a specific set of potential independent variables.

Are the Proposed Variables Fundamental to the Problem and Are They Available?

Next, the list of variables obtained from the problem specification discussion must be examined carefully. Many of these independent variables may turn out to be unmeasurable; for example, the temperature drop in the process could be considered a fundamental variable but it is not measured at the present time. Either a substitute variable which is measured and is related to temperature drop would have to be used, or else new instrumentation is needed. This latter alternative will cost money, and the scientist will have to determine which of these two alternatives is better. This scientific, practical assessment of all variables must be done at this point in the planning. No data have been collected and no statistical analyses made. After a complete check on the variables has been made, a reassessment of the feasibility of the problem solution must be made.

Is the Problem Potentially Solvable?

The above screening procedure could conceivably eliminate most of the independent variables from consideration, thus reducing the chance of solving the problem at all. Generally, this will not happen, however. Nevertheless, one of three possible decisions must be made at this planning stage:
1. The project should be abandoned.
2. The project should be redefined in the light of the knowledge gained up to this point.
3. The project looks feasible, and planning should continue.

Correlation Matrix and First Regression Runs

If the project has the "go" sign, the project planner must next get some indication from data of just how difficult this problem may be analytically. Ultimately, it will be necessary to set up a timetable for the project's completion, including manpower, budgets, and so on. If some data are available, the analyst should have a correlation matrix for all independent and dependent variables computed. This matrix is examined carefully, and some simple regressions run to gain some insight into the reasonableness of an analytical solution and into the amount of time that may be necessary. For example, questions like, "Is there a dominant X-variable?" or "Are the X's essentially uncorrelated with each other?" are asked. This is the stage in planning where analytical feasibilities are considered.

Establish Goals and Prepare Budget

At this stage of the investigation, the scientist and the analyst should establish goals for the project, estimate the timetable for the problem's solution, and prepare manpower and computer requirements. Time and manpower checkpoints throughout the project timetable should be definitely agreed upon, and the whole project submitted for approval. The simple outline shown below is an example of what might conceivably be submitted for approval.

Project Form

PROJECT TITLE: Estimating Equation for the Consumption of Steam in Plant A.

PROJECT ENGINEER: Joe Doe

TYPE OF ANALYSIS TO BE USED: Multiple Regression

GOALS OF THE PROJECT:

1. The final equation should explain more than 80% of the variation ($R^2 > 0.8$).
2. The standard error of estimate should be less than 5% of the mean amount of steam used.
3. All estimated coefficients should be statistically significant with $\alpha = 0.05$.
4. There should be no discernible patterns in the residuals.

Budget	Total Dollars
Manpower	
2 man mos.	$4000
Computer	
6 hours	1800
Total	$5800

TIMETABLE:
    1. Gathering data, first runs, and preliminary analysis   2 weeks
    2. Modification and improvement   4 weeks
    3. Wrap-up and project completion   2 weeks

Are Goals and Budget Acceptable?

If the project proposal is agreed to by all concerned, the project then passes to the development stage; if not, the project is reviewed, then revived on a smaller basis, or stopped.

8.2. Development of the Mathematical Model

The project is now ready for the development of the prediction equation. The techniques used in the development of a regression model have been discussed in Chapters 1 through 7. Thus, this section contains just a single suggested outline of the various checkpoints in the process. Many variations are possible.

Plot Variables and Try Various Models

At an early stage it is helpful to plot the response against various independent variables and to examine the correlation matrix. The residuals from any fitted equations can also be plotted against potential new variables as was illustrated in Chapter 7.

Consult Experts for Criticism

As the work progresses, potential equations should be submitted to specialists in the situation under study, for their appraisal and comments.

Plot New Variables and Build Vectors that Explain Residuals

If (as usually happens) the experts suggest new variables for possible use, data on these must be obtained and examined in conjunction with corresponding response values and residuals from fitted equations.

Examine Some Fitted Regression Equations

Selected independent variables can now be tried in regression equations. Also transformed independent variables may be used. For example, a particular residuals plot might indicate that the logarithm of an independent variable should be used rather than the independent variable itself. Many transformations are suggested initially by the scientists themselves.

Are Goals Met?

Eventually, sometimes after many attempts, the investigator arrives at what he considers his best equation and must examine it in the light of the goals established in the planning stage. If it does not meet these standards, the decision must be made as to whether the project should be stopped or whether the development cycle should be repeated. Perhaps more money will be needed, or perhaps a change in the initial goals is proper. This is another checkpoint in the process.

8.3.   Verification and Maintenance of the Mathematical Model

After an equation meets the goals set in the Planning stage and is accepted as a useful predictive model, it is good practice to establish procedures for its verification and maintenance.

Are Parameters Stable over the Sample Space?

If the model has been built using observations across a long time span one can test the stability of the b coefficients by fitting the model on shorter spans and determining the pattern of successive estimates of the b's. For example if data is available monthly for the four years 1961–1964, the selected model might be fitted to data from each of the four years separately to obtain four sets of estimates of the regression coefficients. If the estimated coefficients showed trendlike patterns, for example, use of the fitted equation obtained from *all* the data for prediction purposes would be unwise. Verifications of this sort are a necessary part of the multiple regression procedure and sometimes the verification step will lead to a complete reconsideration of the whole problem.

Is There Systematic Lack of Fit?

Even when a fitted equation is acceptable with very stable parameters, variables may have been omitted. The residuals should always be examined in all possible ways to see if any patterns are discernable which indicate such omissions.

Practical Aspects of the Model

ARE THE COEFFICIENTS REASONABLE? This may seem an ususual question but it must be remembered that the model will be used by some people who are unaware of the fact that the least-squares regression coefficients are adjusted for other variables in the regression. Thus, they may attempt to predict the response by changing only one variable, using its coefficient to decide how much to change it. If all the estimated coefficients are independently estimated, this may do little harm. However, when the independent variables are highly correlated and the estimated coefficients are also correlated, reliance on individual coefficients can be dangerous. It is wise to restrict prediction to the region of the X-space from which the original data were obtained, in any event. A check can also be made to see if individual coefficients are directionally correct, for example, if X_1 is the amount of production and Y is the total yield, then the coefficient b_1 should be positive.

IS THE EQUATION PLAUSIBLE? Does the equation pass the scrutiny of experts? Are the appropriate variables in the equation, and are any obvious variables missing?

IS THE EQUATION USABLE? The final model may contain a set of variables which can be used for prediction but perhaps not for control. The following is an example of this.

> A process has a set of standard operating instructions that require the specification and setting of k variables. The fitted equation predicting the yield of the process contains only p of these variables where $p < k$. If one attempts to use the fitted equation to improve the yield, one would be ignoring the $k - p$ variables not in the equation. This happens because under standard operating conditions the $k - p$ variables are so closely tied in with the p variables in the prediction model that they are not needed for prediction. However, specification of the $k - p$ variables as well as the p variables is necessary for making the process variable settings.

The Maintenance of the Model

If all the previous criteria and checkpoints have been satisfied, it is necessary to set up a procedure for maintaining the model. Physical conditions do change, and it is necessary to be in a position to determine when the deviations of actual observations from predicted values indicate that the model is beginning to show signs of becoming obsolete. If the analyst has set up a control chart for deviations, the standard quality control chart procedures should be an adequate check on the model. One final word on maintenance: Arrange to have the model checked periodically by a statistical analyst since he will be able to detect more complicated sources of trouble. Never leave the maintenance of the model completely in the hands of the chemist or the engineer.

Summary

If the scientist desires to use multiple regression as a tool to help him solve problems, it is imperative that he follow an outline similar to the one illustrated above. Much time and effort can be wasted by trying to make sense of highly intercorrelated data; a series of planned, cost-oriented checkpoints for using multiple regression techniques is a necessity. Finally, no scientist should be persuaded to abandon his scientific insight and principles in favor of some computerized statistical screening procedure. The use of multiple regression techniques is a powerful tool only if it is applied with intelligence and caution.

CHAPTER 9

MULTIPLE REGRESSION
APPLIED TO ANALYSIS
OF VARIANCE PROBLEMS

9.0. Introduction

The procedures for multiple regression have been discussed in Chapters 1, 2, and 3. We have mentioned that in order to obtain a solution, $\mathbf{b} = (\mathbf{X}'\mathbf{X})^{-1}\mathbf{X}'\mathbf{Y}$, of the normal equations $\mathbf{X}'\mathbf{X}\mathbf{b} = \mathbf{X}'\mathbf{Y}$, it is necessary that $\mathbf{X}'\mathbf{X}$ be a nonsingular matrix. All this means in practice is that the normal equations must involve as many independent equations as there are parameters to be estimated. This is often the case. If data are obtained from a designed experiment, however, some care is needed to check that all the normal equations are independent, or if they are not, to take steps to obtain estimates just the same.

One frequently used method of analysis on data from designed experiments is the analysis of variance technique. It is usually treated as being something foreign to and quite different from general regression, and special analysis of variance programs exist at most computer centers. Some workers are unaware that *any* "fixed-effects" (sometimes called Model I) analysis of variance situation can be handled by a general regression routine, if the model is correctly identified and if precautions are taken to achieve independent normal equations. (Actually models other than fixed-effects models can also be handled but we limit our discussion to Model I examples.)

We are *not* recommending that fixed-effects analysis of variance problems be handled by general regression methods. We are pointing out that they *can* be, if the correct steps are taken in handling the problem and that it is valuable to realize that this is possible. One thought behind this comment is the fact that the question "What model are you considering?" is often met with "I am not considering one—I am using analysis of variance."

The realization that a model exists for *all* analysis of variance situations, and that it and it alone is the basis for the construction of an analysis of variance table, might be aided by knowing that analysis of variance is, practically, equivalent to a regression analysis. *Everyone* knows a model is necessary for regression analysis! We shall now discuss two basic analysis of variance situations and show how they can be handled through general regression methods. We assume knowledge of analysis of variance and thus do not discuss it in detail.

9.1. The One-Way Classification

Suppose we have data in I groups, $i = 1, 2, \ldots, I$, with J_i observations in each group as given below.

$$\begin{array}{llll} \text{Group 1} & Y_{11}, Y_{12}, \ldots, Y_{1J_1}, & \text{mean } \overline{Y}_1 \\ \text{Group 2} & Y_{21}, Y_{22}, \ldots, Y_{2J_2}, & \text{mean } \overline{Y}_2 \\ \quad \cdots \\ \text{Group } I & Y_{I1}, Y_{I2}, \ldots, Y_{IJI}, & \text{mean } \overline{Y}_I \end{array}$$

The usual fixed effects analysis of variance model for such a situation is

$$E(Y_{ij}) = \mu + t_i, \qquad i = 1, 2, \ldots, I \tag{9.1.1}$$
$$j = 1, 2, \ldots, J_i$$

where $t_1, t_2, \ldots, t_I$ are parameters such that

$$J_1 t_1 + J_2 t_2 + \cdots + J_I t_I = 0. \tag{9.1.2}$$

The restriction of Eq. (9.1.2) is necessary since Eq. (9.1.1) contains more parameters than are really needed, and it is usual to regard μ as the overall mean level and t_i as difference between the ith group mean and the overall mean level. Thus the total of all differences between groups and the overall level is zero, as Eq. (9.1.2) expresses. The usual analysis of variance table is (where $\overline{Y} = \sum_{i=1}^{I} J_i \overline{Y}_i / \sum_{i=1}^{I} J_i$):

Source	Sum of Squares	Degrees of Freedom	Mean Square
Between Groups	$\displaystyle\sum_{i=1}^{I} J_i(\overline{Y}_i - \overline{Y})^2$	$I - 1$	s_B^2
Within groups	$\displaystyle\sum_{i=1}^{I}\sum_{j=1}^{J}(Y_{ij} - \overline{Y}_i)^2$	$\displaystyle\sum_{i=1}^{I}(J_i - 1) = n - I$	s_W^2
Mean	$n\,\overline{Y}^2$	1	
Total	$\displaystyle\sum_{i=1}^{I}\sum_{j=1}^{J} Y_{ij}^2$	$\displaystyle\sum_{i=1}^{I} J_i = n$	

It is usual to test the hypothesis that there are no differences between means of groups; that is, $H_0: t_1 = t_2 = \cdots = t_I = 0$ by comparing the ratio $F = s_B{}^2/s_W{}^2$ to a suitable percentage point of the $F(I - 1, \sum_{i=1}^{I} (J_i - 1))$ distribution.

9.2. Regression Treatment of the One-Way Classification Using the Original Model

Instead of Eq. (9.1.1), write

$$E(Y) = \mu X_0 + t_1 X_1 + t_2 X_2 + \cdots + t_I X_I. \qquad (9.2.1)$$

We want this to express the fact that if we consider an observation Y_{ij} from the ith group it must have expectation $\mu + t_i$. We define

$$\mathbf{Y}' = (Y_{11}, Y_{12}, \ldots, Y_{1J_1}; Y_{21}, Y_{22}, \ldots, Y_{2J_2}; \ldots; Y_{I1}, Y_{I2}, \ldots, Y_{IJ_I})$$

and

$$\mathbf{X} = \begin{array}{cccccc} X_0 & X_1 & X_2 & X_3 & \cdots & X_I \\ \left[\begin{array}{cccccc} 1 & 1 & 0 & 0 & \cdots & 0 \\ 1 & 1 & 0 & 0 & \cdots & 0 \\ \cdots & \cdots & \cdots & \cdots & \cdots & \cdots \\ 1 & 1 & 0 & 0 & \cdots & 0 \\ \hline 1 & 0 & 1 & 0 & \cdots & 0 \\ 1 & 0 & 1 & 0 & \cdots & 0 \\ \cdots & \cdots & \cdots & \cdots & \cdots & \cdots \\ 1 & 0 & 1 & 0 & \cdots & 0 \\ \hline \cdots & \cdots & \cdots & \cdots & \cdots & \cdots \\ \hline 1 & 0 & 0 & 0 & \cdots & 1 \\ 1 & 0 & 0 & 0 & \cdots & 1 \\ \cdots & \cdots & \cdots & \cdots & \cdots & \cdots \\ 1 & 0 & 0 & 0 & \cdots & 1 \end{array}\right] \end{array}$$

where the dashes divide the matrix into sets of rows, there being J_1,

$J_2, \ldots, J_I$ rows in successive sets. The headings show to which X the columns relate. Furthermore, define

$$\boldsymbol{\beta}' = (\mu; t_1, t_2, \ldots, t_I)$$

then

$$E(\mathbf{Y}) = \mathbf{X}\,\boldsymbol{\beta}$$

expresses Eq. (9.1.1) in matrix notation. Now

$$\mathbf{X'X} = \begin{bmatrix} n & J_1 & J_2 & \cdots & J_I \\ J_1 & J_1 & 0 & \cdots & 0 \\ J_2 & 0 & J_2 & \cdots & 0 \\ \cdots & & & \cdots & \\ J_I & 0 & 0 & \cdots & J_I \end{bmatrix}, \quad \mathbf{X'Y} = \begin{bmatrix} n\,\bar{Y} \\ J_1\bar{Y}_1 \\ J_2\bar{Y}_2 \\ \cdots \\ J_I\bar{Y}_I \end{bmatrix} \qquad (9.2.2)$$

If we write b_0, b_i for the least squares estimates of μ and t_i, we can write the normal equations $(\mathbf{X'X})\mathbf{b} = \mathbf{X'Y}$ as

$$\begin{aligned}
nb_0 + J_1 b_1 + J_2 b_2 + \cdots + J_I b_I &= n\,\bar{Y} \\
J_1 b_0 + J_1 b_1 \qquad\qquad\qquad\quad &= J_1 \bar{Y}_1 \\
J_2 b_0 \qquad\quad + J_2 b_2 \qquad\qquad &= J_2 \bar{Y}_2 \qquad (9.2.3) \\
\cdots \qquad\qquad\qquad \cdots \\
J_I b_0 \qquad\qquad\qquad + J_I b_I &= J_I \bar{Y}_I
\end{aligned}$$

Here the $(\mathbf{X'X})^{-1}$ matrix does not exist since $\mathbf{X'X}$ is singular. This is due to the fact that Eqs. (9.2.3) are *not* independent because the first equation is the sum of the other I equations. There are in fact only I equations in the $(I + 1)$ unknowns b_0, b_1, $\ldots$, b_I, because the original model (9.1.1) contained more parameters than were actually necessary. This "singularity" of the $\mathbf{X'X}$ matrix is also clear from examination of the $\mathbf{X}$ matrix where the X_0 column is equal to the sum of the X_1, X_2, $\ldots$, X_I columns, a dependence which becomes transmitted into the normal equations of (9.2.3) as we have noted. How can we proceed then? A condition we have not so far taken into account is Eq. (9.1.2) which, if true of the parameters, must also hold for the estimates of the parameters. Hence

$$J_1 b_1 + J_2 b_2 + \cdots + J_I b_I = 0. \qquad (9.2.4)$$

This provides the additional independent equation we require. We now take *any* I equations from (9.2.3) together with Eq. (9.2.4) and use these as the normal equations. It is most convenient to drop the first equation of (9.2.3) since it has the most terms. This leaves, as the equations to solve,

$$J_1 b_1 + J_2 b_2 + \cdots + J_I b_I = 0$$

$$
\begin{aligned}
J_1 b_0 + J_1 b_1 \qquad\qquad &= J_1 \overline{Y}_1 \\
J_2 b_0 \qquad\quad + J_2 b_2 \qquad &= J_2 \overline{Y}_2 \qquad\qquad (9.2.5) \\
\cdots \qquad\qquad \cdots \qquad\qquad & \\
J_I b_0 \qquad\qquad\qquad + J_I b_I &= J_I \overline{Y}_I
\end{aligned}
$$

To maintain symmetry we have not divided through the second to $(I + 1)$st equations by their common factors. In matrix form we can write Eq. (9.2.5) as

$$
\begin{bmatrix}
0 & J_1 & J_2 & \cdots & J_I \\
J_1 & J_1 & 0 & \cdots & 0 \\
J_2 & 0 & J_2 & \cdots & \\
\cdots & & \cdots & & \\
J_I & 0 & 0 & \cdots & J_I
\end{bmatrix}
\begin{bmatrix}
b_0 \\
b_1 \\
b_2 \\
\cdots \\
b_I
\end{bmatrix}
=
\begin{bmatrix}
0 \\
J_1 \overline{Y}_1 \\
J_2 \overline{Y}_2 \\
\cdots \\
J_I \overline{Y}_I
\end{bmatrix}
\qquad (9.2.6)
$$

Since we cannot express these in the form $(\mathbf{X'X})\mathbf{b} = \mathbf{X'Y}$ it will usually be impractical to use this procedure when the work is done by a computer routine which requires this form.

From Eq. (9.2.5),

$$b_i = \overline{Y}_i - b_0, \qquad i = 1, 2, \ldots, I.$$

Substituting in the first equation

$$
0 = \sum_{i=1}^{I} J_i b_i = \sum_{i=1}^{I} J_i (\overline{Y}_i - b_0)
$$

$$
= \sum_{i=1}^{I} J_i \overline{Y}_i - b_0 \sum_{i=1}^{I} J_i
$$

$$
= n\overline{Y} - n b_0
$$

Thus

$$b_0 = \overline{Y}$$

and

$$b_i = \bar{Y}_i - \bar{Y}$$

The sum of squares due to a vector of estimates **b** determined from equations $\mathbf{X'Xb} = \mathbf{X'Y}$ is defined by $\mathbf{b'X'Y}$ even if $\mathbf{X'X}$ is singular and cannot be inverted and extra conditions of the form $\mathbf{Qb} = 0$ are needed to give a unique solution. No matter what form $\mathbf{Q}$ takes (here it is a vector $(0, 1, 1, \ldots, 1)$), $\mathbf{b'X'Y}$ is invariant. This is so since if $\mathbf{b}_1, \mathbf{b}_2$ are two solutions arising from different "extra conditions"

$$\mathbf{b}_1'(\mathbf{X'Y}) = \mathbf{b}_1'(\mathbf{X'Xb}_2)$$

$$= (\mathbf{X'Xb}_1)'\mathbf{b}_2,$$

[regrouping and using the matrix theory fact that $(\mathbf{AB})' = \mathbf{B'A'}$],

$$= (\mathbf{X'Y})'\mathbf{b}_2$$

$$= \mathbf{b}_2'\mathbf{X'Y}.$$

Thus the sum of squares due to regression is

$$\mathbf{b'X'Y} = n\,\bar{Y}^2 + \sum_{i=1}^{I} J_i\,\bar{Y}_i(\bar{Y}_i - \bar{Y})$$

$$= n\,\bar{Y}^2 + \sum_{i=1}^{I} J_i(\bar{Y}_i - \bar{Y})^2$$

with I degrees of freedom, since the additional term,

$$\sum_{i=1}^{I} (-\bar{Y})J_i(\bar{Y}_i - \bar{Y}),$$

added to the right-hand side, is zero by definition of the means. If the model had only a term μ in it we should have

$$\mathrm{SS}\,(b_0) = n\bar{Y}^2$$

with one degree of freedom. Thus

$$\mathrm{SS}(b_1, b_2, \ldots, b_I \mid b_0) = \mathbf{b'X'Y} - n\,\bar{Y}^2$$

$$= \sum_{i=1}^{I} J_i(\bar{Y}_i - \bar{Y})^2,$$

with $I - 1$ degrees of freedom.

These provide the sums of squares due to the "mean" and "between groups," of the analysis of variance table in Section (9.1). The "within groups" sum of squares is found by the difference $\mathbf{Y'Y} - \mathbf{b'X'Y}$ as usual and is the same as the form given in Section (9.1) if evaluated. The test for $H_0: t_1 = t_2 = \cdots = t_I = 0$ is made exactly as in the analysis of variance case.

We have shown here that the one-way analysis of variance problem can be done formally by regression using the original model. However, to perform the calculations on a computer, it is probably best to remove the singularity in the problem by choosing the model more carefully prior to computation.

Note. The work above shows how we must proceed in general in a regression problem, if there are more parameters to estimate than there are independent normal equations. If no natural restrictions are available as in the analysis of variance case we must make restrictions in an arbitrary fashion. While the choice of restrictions will influence the actual values of the regression coefficients, it will not affect the sum of squares due to regression. Usually we would select restrictions which would make the normal equations easier to solve.

Example.   Suppose the normal equations were

$$22b_1 + 10b_2 + 12b_3 + 5b_4 + 8b_5 + 9b_6 = 34.37$$

$$10b_1 + 10b_2 \qquad + 3b_4 + 4b_5 + 3b_6 = 21.21$$

$$12b_1 \qquad + 12b_3 + 2b_4 + 4b_5 + 6b_6 = 13.16$$

$$5b_1 + 3b_2 + 2b_3 + 5b_4 \qquad\qquad = 10.28$$

$$8b_1 + 4b_2 + 4b_3 \qquad + 8b_5 \qquad = 14.23$$

$$9b_1 + 3b_2 + 6b_3 \qquad\qquad + 9b_6 = 9.86$$

(These equations appear in *Regression Analysis* by R. L. Plackett (1960, p. 44). They arise from a two-way classification with unequal numbers of observations in the cells. Such data might also arise from an intended "equal cells" analysis if several observations were missing. We discuss the "equal cells" analysis in the next section.)

Only four of the six equations are independent since the second and third equation add to give the first as do the fourth, fifth, and sixth. Thus two additional equations are required to give six equations in six unknowns. We must add two independent restrictions on $b_1, b_2, \ldots, b_6$, which must not be linear combinations of the existing equations.

Since there are actually only four independent normal equations we can drop two dependent ones, say the first and sixth. The remaining four can be written in matrix form as follows:

$$
\begin{array}{cccccc}
\mathbf{1} & \mathbf{2} & \mathbf{3} & \mathbf{4} & \mathbf{5} & \mathbf{6}
\end{array}
$$

$$
\begin{bmatrix}
10 & 10 & 0 & 3 & 4 & 3 \\
12 & 0 & 12 & 2 & 4 & 6 \\
5 & 3 & 2 & 5 & 0 & 0 \\
8 & 4 & 4 & 0 & 8 & 0
\end{bmatrix}
\begin{bmatrix}
b_1 \\ b_2 \\ b_3 \\ b_4 \\ b_5 \\ b_6
\end{bmatrix}
=
\begin{bmatrix}
21.21 \\ 13.16 \\ 10.28 \\ 14.23
\end{bmatrix}
$$

Since the original matrix was symmetric, the row (or equation) dependence noticed earlier is also reflected in the fact that the first column is the sum of the second and third, and also the sum of the fourth, fifth, and sixth. When adding the two restrictions on the b's we must be careful on two counts. The restrictions, when placed below the four selected equations, will contribute two additional rows to the matrix and two zeros to the right-hand side vector (we would usually take the restrictions to be of the form $\sum c_i b_i = 0$). The final matrix must be such that there is no dependence between either rows *or* columns, if a unique solution is required. (There are more elegant matrix ways of expressing this in general; see, for example, Plackett (1960), but we do not provide them in our more elementary development.) For example, we *cannot* take restrictions

$$7b_1 + 6b_2 + b_3 + b_4 + b_5 + 5b_6 = 0$$
$$11b_1 + 9b_2 + 2b_3 + 4b_4 + 4b_5 + 3b_6 = 0$$

since the original column dependency is preserved. Even if only one column dependency is preserved, for example, by restrictions

$$3b_4 + 4b_5 + 3b_6 = 0$$
$$9b_1 + 5b_2 + 4b_3 \qquad\qquad\quad = 0$$

which allow column one to be the sum of columns two and three, the restrictions are useless. However,

$$3b_4 + 4b_5 + 3b_6 = 0$$
$$b_2 + b_3 \qquad\qquad\quad = 0$$

would be usable, since no dependence will occur, and we shall have six equations in six unknowns as required. (Different restrictions were used by Plackett, whose book should be consulted for the subsequent solution.)

The idea of adding arbitrary restrictions may seem somewhat peculiar at first. One must remember that it is required whenever more parameters are used than are actually necessary to express the model. At some stage this "looseness" must be removed and that is the purpose of the added restrictions.

9.3. Regression Treatment of the One-Way Classification: Independent Normal Equations

The analysis of variance model is

$$E(Y_{ij}) = \mu + t_i. \tag{9.3.1}$$

Let us write

$$\beta_i = \mu + t_i \qquad i = 1, 2, \ldots, I.$$

Then, in regression terms, we can write

$$E(\mathbf{Y}) = \mathbf{X}\boldsymbol{\beta} \tag{9.3.2}$$

where, comparing with the definitions in Section (9.1), $\mathbf{Y}$ is as before, $\mathbf{X}$ is the matrix formed by dropping the X_0 column of the previous $\mathbf{X}$ matrix, and $\boldsymbol{\beta}' = (\beta_1, \beta_2, \ldots, \beta_I)$. Let $\mathbf{b}' = (b_1, b_2, \ldots, b_I)$. Then

$$\mathbf{X'X} = \begin{bmatrix} J_1 & & & \\ & J_2 & & \mathbf{0} \\ & & \cdot & \\ & & & \cdot \\ \mathbf{0} & & & \cdot \\ & & & J_I \end{bmatrix}, \qquad \mathbf{X'Y} = \begin{bmatrix} J_1 \bar{Y}_1 \\ J_2 \bar{Y}_2 \\ \cdots \\ J_I \bar{Y}_I \end{bmatrix}. \tag{9.3.3}$$

Since $\mathbf{X'X}$ is a diagonal matrix with J_i in the ith diagonal position and zero elsewhere, its inverse is a diagonal matrix with $1/J_i$ in the ith diagonal position. From this it is easy to see that

$$b_i = \bar{Y}_i. \tag{9.3.4}$$

The sum of squares due to $\mathbf{b}$ is

$$\mathbf{b'X'Y} = \sum_{i=1}^{I} \bar{Y}_i(J_i \bar{Y}_i) = \sum_{i=1}^{I} J_i \bar{Y}_i^{\,2} \tag{9.3.5}$$

and the residual sum of squares is

$$
\begin{aligned}
\mathbf{Y'Y} - \mathbf{b'X'Y} &= \sum_{i=1}^{I} \sum_{j=1}^{J_i} Y_{ij}^2 - \sum_{i=1}^{I} J_i \bar{Y}_i^{\,2} \\
&= \sum_{i=1}^{I} \left\{ \sum_{j=1}^{J_i} Y_{ij}^2 - J_i \bar{Y}_i^{\,2} \right\} \\
&= \sum_{i=1}^{I} \sum_{j=1}^{J_i} (Y_{ij} - \bar{Y}_i)^2,
\end{aligned}
\tag{9.3.6}
$$

with $n - I$ degrees of freedom.

The hypothesis $H_0: t_1 = t_2 = \cdots = t_I = 0$ is expressed in our present notation as

$$H_0: \beta_1 = \beta_2 = \cdots = \beta_I = \mu.$$

If H_0 were true, the model would be

$$E(Y_{ij}) = \mu$$

or

$$E(\mathbf{Y}) = \mathbf{j}\mu \qquad (9.3.7)$$

where $\mathbf{j}$ is a vector of ones of the same length as $\mathbf{Y}$. The (single) normal equation would be

$$n\mu = \mathbf{j}'\mathbf{j}\mu = \mathbf{j}'\mathbf{Y} = \sum_{i=1}^{I} \sum_{j=1}^{J} Y_{ij} = n\bar{Y}.$$

Thus the estimate of μ would be

$$b_0 = \bar{Y} \qquad (9.3.8)$$

and

$$SS(b_0) = n\bar{Y}^2 \qquad (9.3.9)$$

providing a residual sum of squares of

$$\sum_{i=1}^{I} \sum_{j=1}^{J_i} Y_{ij}^2 - n\bar{Y}^2 = \sum_{i=1}^{I} \sum_{j=1}^{J_i} (Y_{ij} - \bar{Y})^2 \qquad (9.3.10)$$

as can be shown, with $n - 1$ degrees of freedom.

The sum of squares due to H_0 is the difference between Eqs. (9.3.10) and (9.3.6), namely

$$SS\,(H_0) = \sum_{i=1}^{I} \sum_{j=1}^{J_i} \{(Y_{ij} - \bar{Y})^2 - (Y_{ij} - \bar{Y}_i)^2\}$$

$$= \sum_{i=1}^{I} J_i (\bar{Y}_i - \bar{Y})^2$$

after reduction, with $(n - 1) - (n - I) = I - 1$ degrees of freedom. The test statistic for H_0 is thus

$$F = \frac{SS\,(H_0)}{I - 1} \Bigg/ \frac{\sum_{i=1}^{I} \sum_{j=1}^{J_i} (Y_{ij} - \bar{Y}_i)^2}{n - I}$$

which is exactly what we obtain from the analysis of variance. Thus if we express the model for the one-way classification as $E(Y_{ij}) = \beta_i$ and test the hypothesis $H_0: \beta_1 = \beta_2 = \cdots = \beta_I = \mu$, we can reproduce the analysis of variance through regression analysis using standard programs. We can obtain estimates of the parameters t_i from $b_i - b_0$.

9.4. The Two-Way Classification with Equal Numbers of Observations in the Cells

Suppose we have a two-way classification with I rows and J columns, with K observations Y_{ijk}, $k = 1, 2, \ldots, K$ in the cells so defined. The usual fixed-effects analysis of variance model is

$$E(Y_{ijk}) = \mu + \alpha_i + \beta_j + \gamma_{ij} \qquad i = 1, 2, \ldots, I \qquad (9.4.1)$$

$$j = 1, 2, \ldots, J$$

$$k = 1, 2, \ldots, K$$

subject to the restrictions

$$\sum_{i=1}^{I} \alpha_i = \sum_{j=1}^{J} \beta_j = \sum_{i=1}^{I} \gamma_{ij} \text{ (all } j) = \sum_{j=1}^{J} \gamma_{ij} \text{ (all } i) = 0. \qquad (9.4.2)$$

These restrictions enable μ to be regarded as the overall mean level while the α_i, β_i, and γ_{ij} are differences between row level and overall mean, column level and overall mean, and cell level and the joint row plus column level, respectively. The usual analysis of variance table is

Source	Sum of Squares	Degrees of Freedom	Mean Square
Rows	$JK \sum_{i=1}^{I} (\bar{Y}_{i..} - \bar{Y})^2$	$I - 1$	s_r^2
Columns	$IK \sum_{j=1}^{J} (\bar{Y}_{.j.} - \bar{Y})^2$	$J - 1$	s_c^2
Cells (interaction)	$K \sum_{i=1}^{I} \sum_{j=1}^{J} (\bar{Y}_{ij.} - \bar{Y}_{i..} - \bar{Y}_{.j.} + \bar{Y})^2$	$(I - 1)(J - 1)$	s_{rc}^2
Residual	By subtraction	$IJ(K - 1)$	s^2
Mean	$IJK \bar{Y}^2$	1	
Total	$\sum_{i=1}^{I} \sum_{j=1}^{J} \sum_{k=1}^{K} Y_{ijk}^2$	IJK	

where $\bar{Y}_{i..}$ is the mean of all observations in row i,

$\bar{Y}_{.j.}$ is the mean of all observations in column j,

$\bar{Y}_{ij.}$ is the mean of all observations in the cell (i, j), and

$\bar{Y}$ is the mean of all observations.

The usual tests made are

H_0: all $\alpha_i = 0$ $F = s_r^2/s^2$ is compared with $F[(I-1), IJ(K-1)]$

H_0: all $\beta_j = 0$ $F = s_c^2/s^2$ is compared with $F[(J-1), IJ(K-1)]$

H_0: all $\gamma_{ij} = 0$ $F = s_{rc}^2/s^2$ is compared with $F[(I-1)(J-1), IJ(K-1)]$

9.5. Regression Treatment of the Two-Way Classification with Equal Numbers of Observations in the Cells

We could, if desired, handle this problem in a way similar to that given in Section 9.2, by writing down the *dependent* normal equations in the parameters μ, α_i, β_j, γ_{ij}, adding the equations given by the restrictions of Eq. (9.4.2), and solving a selected set of $(I+1)(J+1)$ independent equations. We shall instead deal with the situation in another way, so that nonsingular $\mathbf{X'X}$ matrices will always occur in the calculations.

There are $1 + I + J + IJ = (I+1)(J+1)$ parameters in all, but these are dependent through the $1 + 1 + J + I - 1 = I + J + 1$ restrictions defined by Eq. (9.4.2). We have to take one off the (at first sight) apparent number of restrictions to allow for the fact that if we know all the *row* sums of the γ_{ij} are zero, so is the total sum of all γ_{ij}. This means that it is necessary to specify only that $J - 1$ of the column sums of the γ_{ij} be zero, in order to achieve zero for the final column sum. Thus we actually need only IJ parameters to describe the model and we can define these as

$$\delta_{ij} = \mu + \alpha_i + \beta_j + \gamma_{ij} \qquad \begin{array}{l} i = 1, 2, \ldots, I, \\ j = 1, 2, \ldots, J. \end{array}$$

Consider the following models:

(a) $E(Y_{ijk}) = \delta_{ij}$

(b) $E(Y_{ijk}) = \delta_{.j}$ independent of i

(c) $E(Y_{ijk}) = \delta_{i.}$ independent of j

(d) $E(Y_{ijk}) = \delta$ independent of i and j.

We can represent all these in matrix form. Let

$$\mathbf{Y} = (Y_{111}, Y_{112}, \ldots, Y_{11K}; Y_{121}, Y_{122}, \ldots, Y_{12K}; \ldots; Y_{IJ1}, Y_{IJ2}, \ldots, Y_{IJK})$$

where we order the cells in the sequence

$$(11), (12), \ldots, (1J); (21), (22), \ldots, (2J); \ldots; (I1), (I2), \ldots, (IJ)$$

and order the observations within the cells in the numerical order of the third subscript. Then, with the matrices indicated below, we can write all the models (a), (b), (c), and (d) in the form $E(\mathbf{Y}) = \mathbf{X}\boldsymbol{\beta}$.

Model (*a*)

$$
\begin{array}{cccccc}
\delta_{11} & \delta_{12}\cdots\delta_{1J} & \delta_{21} & \delta_{22}\cdots\delta_{2J}\cdots\delta_{I1} & \delta_{I2}\cdots\delta_{IJ}
\end{array}
$$

$$
\mathbf{X} =
\left[
\begin{array}{ccccc}
1 & 0\cdots 0 & 0\cdots 0 & 0\cdots 0 & 0\cdots 0 \\
1 & 0\cdots 0 & 0\cdots 0 & 0\cdots 0 & 0\cdots 0 \\
\cdots & \cdots\ \cdots & \cdots\ \cdots & \cdots\ \cdots & \cdots\ \cdots \\
1 & 0\cdots 0 & 0\cdots 0 & 0\cdots 0 & 0\cdots 0 \\
\hline
0 & 1\cdots 0 & 0\cdots 0 & 0\cdots 0 & 0\cdots 0 \\
0 & 1\cdots 0 & 0\cdots 0 & 0\cdots 0 & 0\cdots 0 \\
\cdots & \cdots\ \cdots & \cdots\ \cdots & \cdots\ \cdots & \cdots\ \cdots \\
0 & 1\cdots 0 & 0\cdots 0 & 0\cdots 0 & 0\cdots 0 \\
\hline
\cdots & \cdots\ \cdots & \cdots\ \cdots & \cdots\ \cdots & \cdots\ \cdots \\
\hline
0 & 0\cdots 0 & 0\cdots 0 & 0\cdots 0 & 0\cdots 1 \\
0 & 0\cdots 0 & 0\cdots 0 & 0\cdots 0 & 0\cdots 1 \\
\cdots & \cdots\ \cdots & \cdots\ \cdots & \cdots\ \cdots & \cdots\ \cdots \\
0 & 0\cdots 0 & 0\cdots 0 & 0\cdots 0 & 0\cdots 1
\end{array}
\right]
$$

where each row segment is of depth K

$$\boldsymbol{\beta}' = (\delta_{11}, \delta_{12}, \ldots, \delta_{1J}; \delta_{21}, \delta_{22}, \ldots, \delta_{2J}; \ldots; \delta_{I1}, \delta_{I2}, \ldots, \delta_{IJ}).$$

Model (*b*)

$$
\begin{array}{ccccc}
\delta_{\cdot 1} & \delta_{\cdot 2} & \delta_{\cdot 3} & \cdots & \delta_{\cdot J}
\end{array}
$$

$$
\mathbf{X} =
\left[
\begin{array}{ccccc}
\mathbf{j} & \mathbf{0} & \mathbf{0} & \cdots & \mathbf{0} \\
\mathbf{0} & \mathbf{j} & \mathbf{0} & \cdots & \mathbf{0} \\
\mathbf{0} & \mathbf{0} & \mathbf{j} & \cdots & \mathbf{0} \\
\cdots & \cdots & \cdots & \cdots & \cdots \\
\mathbf{0} & \mathbf{0} & \mathbf{0} & \cdots & \mathbf{j} \\
\hline
\multicolumn{5}{c}{(I-1)\ \text{more blocks}} \\
\multicolumn{5}{c}{\text{exactly as above}}
\end{array}
\right]
$$

where $\mathbf{j}$ denotes a $K \times 1$ vector of unities

$$\boldsymbol{\beta}' = (\delta_{.1}, \delta_{.2}, \ldots, \delta_{.J}).$$

Model (c)

$$
\begin{array}{cccc}
\delta_{1.} & \delta_{2.} & \delta_{3.} \cdots \delta_{I.} \\
\end{array}
$$

$$
\mathbf{X} = \begin{bmatrix}
\mathbf{j} & 0 & 0 \cdots 0 \\
0 & \mathbf{j} & 0 \cdots 0 \\
0 & 0 & \mathbf{j} \cdots 0 \\
\cdot & \cdot & \cdot \cdot \cdot \cdot \cdot \\
0 & 0 & 0 \cdots \mathbf{j}
\end{bmatrix}
$$

where $\mathbf{j}$ here denotes a $JK \times 1$ vector of unities

$$\boldsymbol{\beta}' = (\delta_{1.}, \delta_{2.}, \ldots, \delta_{I.}).$$

Model (d)

$$\mathbf{X} = \mathbf{j}$$

where $\mathbf{j}$ here denotes a $IJK \times 1$ vector of unities

$$\boldsymbol{\beta} = \delta, \text{ a scalar.}$$

We can construct the standard analysis of variance table through regression methods as follows. Denote by S_a, S_b, S_c, and S_d, the regression sums of squares, which arise from the four regression models given above and let $S = \sum_{i=1}^{I} \sum_{j=1}^{J} \sum_{k=1}^{K} Y_{ijk}^2$. Then using the "extra sum of squares" principle given in Section 2.7, we can construct the table below.

Source	Mean Square	Degrees of Freedom
Rows	$S_c - S_d$	$I - 1$
Columns	$S_b - S_d$	$J - 1$
Interaction	$S_a - S_b - S_c + S_d$	$(I - 1)(J - 1)$
Residual	$S - S_a$	$IJ(K - 1)$
Mean	S_d	1
Total	S	IJK

(The interaction sum of squares is actually obtained from

$$S_a - (S_b - S_d) - (S_c - S_d) - S_d$$

which reduces to the form given in the table.)

The equivalence between these sum of squares and the ones given in the analysis of variance procedure can be easily demonstrated mathematically but we shall not do this.

We are usually interested in obtaining estimates m, a_i, b_j, c_{ij} of the original parameters μ, α_i, β_j, γ_{ij} in the analysis of variance model. These estimates can be obtained from the estimates d_{ij}, $d_{.j}$, $d_{i.}$ d of the regression coefficients δ_{ij}, $\delta_{.j}$, $\delta_{i.}$, δ in the four models. They are

$$m = d$$
$$a_i = d_{i.} - d$$
$$b_j = d_{.j} - d$$
$$c_{ij} = d_{ij} - d_{i.} - d_{.j} + d$$

An Alternative Method

Our suggested method of dealing with the two-way analysis of variance classification has involved four symmetric regression analyses and the use of the extra sum of squares principle. To deal with this situation in a single regression analysis we must write an unsymmetric model which omits some of the dependent parameters of the standard analysis of variance model. We illustrate this with an example. Consider the two-way classification below, which has two observations in each cell.

	Column $j = 1$	$j = 2$
Row $i = 1$	Y_1, Y_2	Y_3, Y_4
$i = 2$	Y_5, Y_6	Y_7, Y_8
$i = 3$	Y_9, Y_{10}	Y_{11}, Y_{12}

The standard analysis of variance model is

$$E(Y_{ij}) = \mu + \alpha_i + \beta_j + \gamma_{ij}$$

where
$$\alpha_1 + \alpha_2 + \alpha_3 = 0 \qquad \gamma_{11} + \gamma_{12} = 0$$
$$\beta_1 + \beta_2 = 0 \qquad \gamma_{21} + \gamma_{22} = 0$$
$$\gamma_{31} + \gamma_{33} = 0$$
$$\gamma_{11} + \gamma_{21} + \gamma_{31} = 0$$
$$\gamma_{12} + \gamma_{22} + \gamma_{32} = 0$$

Thus if (for example) μ, α_1, α_2, β_1, γ_{11}, and γ_{21} are known or estimated, all other parameters or their estimates, respectively, can be found from the restrictions. We can thus write the regression model

$$E(Y_{ij}) = \mu + \alpha_1 X_1 + \alpha_2 X_2 + \beta_1 X_3 + \gamma_{11} X_4 + \gamma_{21} X_5$$

or

$$E(\mathbf{Y}) = \mathbf{X}\boldsymbol{\beta}$$

where

$$
\mathbf{Y} =
\begin{bmatrix}
Y_1 \\
Y_2 \\
Y_3 \\
Y_4 \\
Y_5 \\
Y_6 \\
Y_7 \\
Y_8 \\
Y_9 \\
Y_{10} \\
Y_{11} \\
Y_{12}
\end{bmatrix}
\qquad
\mathbf{X} =
\begin{array}{cccccc}
\mu & \alpha_1 & \alpha_2 & \beta_1 & \gamma_{11} & \gamma_{21} \\
\begin{bmatrix}
1 & 1 & 0 & 1 & 1 & 0 \\
1 & 1 & 0 & 1 & 1 & 0 \\
1 & 1 & 0 & -1 & -1 & 0 \\
1 & 1 & 0 & -1 & -1 & 0 \\
1 & 0 & 1 & 1 & 0 & 1 \\
1 & 0 & 1 & 1 & 0 & 1 \\
1 & 0 & 1 & -1 & 0 & -1 \\
1 & 0 & 1 & -1 & 0 & -1 \\
1 & -1 & -1 & 1 & -1 & -1 \\
1 & -1 & -1 & 1 & -1 & -1 \\
1 & -1 & -1 & -1 & 1 & 1 \\
1 & -1 & -1 & -1 & 1 & 1
\end{bmatrix}
\end{array}
$$

$$\boldsymbol{\beta}' = (\mu, \alpha_1, \alpha_2, \beta_1, \gamma_{11}, \gamma_{21})$$

Note. The elements of the γ_{ij} column are obtained as the product of corresponding elements of the α_i and β_j columns.

Any independent subset of the parameters can be used for such a model and there are many possible alternative forms. From this point, the usual regression methods are used to estimate $\boldsymbol{\beta}$. Because of the orthogonality in $\mathbf{X}$, we can obtain separate, orthogonal sums of squares for the estimates of (1) μ, (2) α_1 and α_2, (3) β_1, (4) γ_{11} and γ_{21}. These will be the usual sums of squares for (1) mean, (2) rows, (3) columns, (4) interaction, in the standard analysis of variance setup.

9.6.   Example: The Two-Way Classification

The data in the two-way classification below appears in *Statistical Theory and Methodology in Science and Engineering* by K. A. Brownlee (Second edition), John Wiley and Sons, 1965, page 475. Descriptive details are omitted.

	Column 1	Column 2	Column 3
Row 1	17, 21, 49, 54	64, 48, 34, 63	62, 72, 61, 91
Row 2	33, 37, 40, 16	41, 64, 34, 64	56, 62, 57, 72

Following the procedure given in Section 9.5 we can calculate the quantities below through regression methods.

(a) $S_a = 65,863$ $\begin{pmatrix} d_{11} & d_{12} & d_{13} \\ d_{21} & d_{22} & d_{23} \end{pmatrix} = \begin{pmatrix} 35.25 & 52.25 & 71.50 \\ 31.50 & 50.75 & 61.75 \end{pmatrix}$

(b) $S_b = 65,640.25$ $(d_{.1}, d_{.2}, d_{.3}) = (33.375, 51.5, 66.625)$

(c) $S_c = 61,356$ $\begin{pmatrix} d_{1.} \\ d_{2.} \end{pmatrix} = \begin{pmatrix} 53 \\ 48 \end{pmatrix}$

(d) $S_d = 61,206$ $d = 50.5$

Using the formulae given in the previous section we obtain an analysis of variance table as follows:

Source	Sum of Squares	Degrees of Freedom
Rows	150.00	1
Columns	4,434.25	2
Interaction	72.75	2
Residual	3,495.00	18
Mean	61,206.00	1
Total (uncorrected)	69,358.00	24

The same table was obtained in Brownlee (1965) through the usual analysis of variance calculations. The estimates of the usual analysis of variance parameters are given by

$$m = 50.5$$

$$a_1 = 53 - 50.5 = 2.5, \qquad a_2 = 48 - 50.5 = -2.5.$$

Note. $a_1 + a_2 = 0$, as it should.

$$b_1 = 33.375 - 50.5 = -17.125, \qquad b_2 = 51.5 - 50.5 = 1.0,$$
$$b_3 = 66.625 - 50.5 = 16.125.$$

Note. $b_1 + b_2 + b_3 = 0$ as it should.

$$c_{11} = 35.25 - 53 - 33.375 + 50.5 = -0.625$$
$$c_{12} = 52.25 - 53 - 51.5 + 50.5 = -1.750$$
$$c_{13} = 71.50 - 53 - 66.625 + 50.5 = 2.375$$
$$c_{21} = 31.50 - 48 - 33.375 + 50.5 = 0.625$$
$$c_{22} = 50.75 - 48 - 51.5 + 50.5 = 1.750$$
$$c_{23} = 61.75 - 48 - 66.625 + 50.5 = -2.375$$

Note. $\sum_{i=1}^{2} c_{ij} = \sum_{j=1}^{3} c_{ij} = 0$, as they should.

The residuals from this analysis of variance model would be

$$
\begin{aligned}
\bar{Y}_{ijk} &= Y_{ijk} - m - a_i - b_j - c_{ij} \\
&= Y_{ijk} - d - (d_{i.} - d) - (d_{.j} - d) - (d_{ij} - d_{i.} - d_{.j} + d) \\
&= Y_{ijk} - d_{ij}
\end{aligned}
$$

which are the residuals from the regression analysis using Model (*a*). These would be examined as described in Chapter 3. Also plots of residuals for each row and column can be examined.

9.7.   Comments

We have seen, in the specific cases discussed, that analysis of variance can, if necessary, be conducted by standard regression techniques. If the model is examined carefully and reparameterized properly, the analysis of variance table for other models could also be obtained in a similar way. Since additional work is required it is usually better to use an appropriate analysis of variance computation method, or computer routine, where it exists. Nevertheless it is useful to appreciate the connection between the two methods of analysis for several reasons:

1. It focuses attention on the fact that a model is necessary in analysis of variance problems.
2. It points up the fact that the residuals in analysis of variance models play the same role as residuals in regression models and *must* be examined for the information they contain on the possible inadequacy of the model under consideration. (There seems to be a tacit assumption in most variance analysis that the model is correct.)
3. When observations are missing from analysis of variance data, they can often be "estimated" by standard formulae. If this is inconvenient, or too many observations are missing, the data can usually be analyzed by a regression routine setting up models as illustrated above but deleting rows of the **X** matrices for which no observations are available. (The word estimated is placed in quotes because no real estimation takes place. The "estimates" are simply numbers inserted for calculation purposes which help provide the same estimates of parameters which would otherwise have been obtained from the incomplete data via an unsymmetrical, often difficult, analysis.)

EXERCISES

A. Perform the following analysis on each of the two-way classifications given on the following page.

1. Analyze these data using the method described in Section 9.5 and illustrated in Section 9.6.
2. Evaluate the fitted values and residuals and examine the residuals in all reasonable ways. State any defects you find.
3. Confirm that the alternative regression method given at the end of Section 9.5 leads to the same results.

   a. An experiment was conducted to determine the effect of steam pressure and blowing time on the percentage of foreign matter left in filter earth. The data are as follows:

		Blowing Time		
		1 hr	2 hr	3 hr
	10 lb	45.2, 46.0	40.0, 39.0	35.9, 34.1
Steam	20 lb	41.8, 20.6	27.8, 19.0	22.5, 17.7
pressure	30 lb	23.5, 33.1	44.6, 52.2	42.7, 48.6

   b. An experiment was conducted on the effect of premixing speed and finish mixer speed on the center heights of cakes. Three different levels of speed were chosen for each of the two variables. The data collected were as follows:

(Premix speed − 5)	(Finish mix speed − 3.5) × 2	(Center height − 2) × 100
X_1	X_2	Y
−1	−1	4, −3
−1	0	3, 2
−1	1	−1, −5
0	−1	3, 10
0	0	2, 2
0	1	0, 0
1	−1	−1, −10
1	0	1, 2
1	1	7, 9

B. A chemical experiment was performed to investigate the effect of extrusion temperature X_1 and cooling temperature X_2 on the compressibility of a finished product. Knowledge of the process suggested that a model of the form

$$Y = \beta_0 + \beta_1 X_1 + \beta_2 X_2 + \beta_{12} X_1 X_2 + \epsilon$$

would satisfactorily explain the variation observed. Two levels of extrusion temperature and two levels of cooling temperature were chosen and all four of the combinations were performed. Each of the four experiments was carried out four times and the data yielded the following information:

Analysis of Variance

Source of Variation	Degrees of Freedom	Sum of Squares	Mean Square
Total	16	921.0000	
Due to regression		881.2500	
b_0	1	798.0625	
b_1	1	18.0625	
b_2			
b_{12}		5.0625	
Residual			

1. a. Complete the above ANOVA table.
   b. Using $\alpha = 0.05$, examine the following questions:
      (1) Is the overall regression equation statistically significant?
      (2) Are all the b-coefficients significant?
   c. Calculate the square of the multiple correlation coefficient, R^2.
2. Given the following additional information,

$$\sum x_{0i} y_i = 113, \quad \sum x_{1i} y_i = 17, \quad \sum x_{2i} y_i = 31, \quad \text{and} \sum x_{1i} x_{2i} y_i = -9,$$

   where $x_{ji} = X_{ji} - \bar{X}_j, j = 0, 1, 2$, and $y_i = Y_i - \bar{Y}$,
   a. Determine b_0, b_1, b_2, and b_{12}, and write out the prediction equation.
   b. The predicted value of $\hat{Y}$ at $X_1 = 70°$ and $X_2 = 150°$ is 54. The variance of this predicted value is 0.6875. What is the variance of a single predicted observation at the point $X_1 = 70$, $X_2 = 150$?
   c. Place 95% confidence limits on the true mean value of Y at the point $X_1 = 70$, $X_2 = 150$.
3. What conclusions can be drawn from your analysis?
C. Analyze the data given at the bottom of page 258 by the alternative method of pages 257–258.

CHAPTER 10

AN INTRODUCTION TO
NONLINEAR ESTIMATION

10.1. Introduction

This chapter is a brief introduction to the problems of nonlinear estimation. Since a knowledge of the geometrical aspects of least squares helps
toward an understanding of problems in the nonlinear case, a brief discussion of this geometry is found in Sections 10.6 and 10.7. The bibliography at the end of this chapter contains many of the important papers on
nonlinear estimation, and publications suggested for initial reading are
denoted by an asterisk.

In previous chapters we have fitted, by least squares, models which were
linear in the parameters and were of the type

$$Y = \beta_0 + \beta_1 Z_1 + \beta_2 Z_2 + \cdots + \beta_p Z_p + \epsilon \qquad (10.1.1)$$

where the Z_i can represent any functions of the basic independent
variables $X_1, X_2, \ldots, X_k$. While Eq. (10.1.1) can represent a wide variety
of relationships (see Chapter 5) there are many situations in which a
model of this form is not appropriate; for example, when definite information is available about the form of the relationship between the
response and the independent variables. Such information might involve
direct knowledge of the actual form of the true model or might be represented by a set of differential equations which the model must satisfy.
Sometimes the information leads to several alternative models (in which
case methods for discriminating between them will be of interest). When we
are led to a model of nonlinear form, we would usually prefer to fit such a
model whenever possible, rather than to fit an alternative, perhaps less
realistic, linear model.

Any model which is not of the form (10.1.1) will be called a *nonlinear
model*, that is nonlinear in the *parameters*. Two examples of such models

are

$$Y = \exp(\theta_1 + \theta_2 t^2 + \epsilon) \qquad (10.1.2)$$

$$Y = \frac{\theta_1}{\theta_1 - \theta_2} [e^{-\theta_2 t} - e^{-\theta_1 t}] + \epsilon. \qquad (10.1.3)$$

In these examples the parameters to be estimated are denoted by θ's rather than β's as used previously, t is the single independent variable, and ϵ is a random error term with $E(\epsilon) = 0$, $V(\epsilon) = \sigma^2$. (We could also write these models without ϵ and replacing Y by η. Then, the models would show how *true* values of the response, η, depend on t. Here, we wish to be specific about how the error enters the model to permit the discussion which follows.)

The models (10.1.2) and (10.1.3) are both nonlinear in the sense that they involve θ_1 and θ_2 in a nonlinear way but they are of essentially different characters. The model (10.1.2) can be transformed, by taking logarithms to the base e, into the form

$$\ln Y = \theta_1 + \theta_2 t^2 + \epsilon, \qquad (10.1.4)$$

which is the form of (10.1.1) and is *linear* in the parameters. We can thus say that the model (10.1.2) is *intrinsically linear* since it can be transformed into linear form. (Some writers use the phrase *nonintrinsically nonlinear*, but we shall not.)

However, it is impossible to convert the model (10.1.3) into a form linear in the parameters. Such a model is said to be *intrinsically nonlinear*. While, at times, it may be useful to transform a model of this type so that it can be more easily fitted, it will remain a nonlinear model, whatever the transformation applied. Unless specifically noted, all models mentioned in this chapter will be intrinsically nonlinear.

Note. In models in which the error is additive, an intrinsically linear model is one which can be made linear by a transformation of parameters; for example, $Y = e^\theta X + \epsilon$ is of this type since, if we transform by $\beta = e^\theta$, the model becomes $Y = \beta X + \epsilon$. Other authors use the words *intrinsically linear* in this sense only.

10.2. Least Squares in the Nonlinear Case

Suppose the postulated model is of the form

$$Y = f(\xi_1, \xi_2, \ldots, \xi_k; \theta_1, \theta_2, \ldots, \theta_p) + \epsilon. \qquad (10.2.1)$$

If we write

$$\boldsymbol{\xi} = (\xi_1, \xi_2, \ldots, \xi_k)'$$
$$\boldsymbol{\theta} = (\theta_1, \theta_2, \ldots, \theta_p)'$$

we can shorten Eq. (10.2.1) to

$$Y = f(\boldsymbol{\xi}, \boldsymbol{\theta}) + \epsilon$$

or   (10.2.2)

$$E(Y) = f(\boldsymbol{\xi}, \boldsymbol{\theta})$$

if we assume that $E(\epsilon) = 0$. We shall also assume that errors are uncorrelated, that $V(\epsilon) = \sigma^2$ and, usually, that $\epsilon \sim N(0, \sigma^2)$ so that errors are independent.

When there are n observations of the form

$$Y_u, \xi_{1u}, \xi_{2u}, \ldots, \xi_{ku}$$

for $u = 1, 2, \ldots, n$, available, we can write the model in the alternative form

$$Y_u = f(\xi_{1u}, \xi_{2u}, \ldots, \xi_{ku}; \theta_1, \theta_2, \ldots, \theta_p) + \epsilon_u \qquad (10.2.3)$$

where ϵ_u is the uth error, $u = 1, 2, \ldots, n$. This can be abbreviated to

$$Y_u = f(\boldsymbol{\xi}_u, \boldsymbol{\theta}) + \epsilon_u \qquad (10.2.4)$$

where $\boldsymbol{\xi}_u = (\xi_{1u}, \xi_{2u}, \ldots, \xi_{ku})'$. The assumption of normality and independence of the errors can now be written as $\boldsymbol{\epsilon} \sim N(\mathbf{0}, \mathbf{I}\sigma^2)$ where $\boldsymbol{\epsilon} = (\epsilon_1, \epsilon_2, \ldots, \epsilon_n)'$, and as usual $\mathbf{0}$ is a vector of zeros and $\mathbf{I}$ is a unit matrix, both of appropriate sizes. We define the *error sum of squares* for the nonlinear model and the given data as

$$S(\boldsymbol{\theta}) = \sum_{u=1}^{n} \{Y_u - f(\boldsymbol{\xi}_u, \boldsymbol{\theta})\}^2. \qquad (10.2.5)$$

Note that since Y_u and $\boldsymbol{\xi}_u$ are fixed observations, the sum of squares is a function of $\boldsymbol{\theta}$. We shall denote by $\hat{\boldsymbol{\theta}}$, a *least squares estimate* of $\boldsymbol{\theta}$, that is a value of $\boldsymbol{\theta}$ which minimizes $S(\boldsymbol{\theta})$. (It can be shown that, if $\boldsymbol{\epsilon} \sim N(\mathbf{0}, \mathbf{I}\sigma^2)$, the least squares estimate of $\boldsymbol{\theta}$ is also the maximum likelihood estimate of $\boldsymbol{\theta}$. This is because the likelihood function for this problem can be written

$$l(\boldsymbol{\theta}, \sigma^2) = (2\pi\sigma^2)^{-n/2} e^{-S(\boldsymbol{\theta})/2\sigma^2}$$

so that if σ^2 is known, maximizing $l(\boldsymbol{\theta}, \sigma^2)$ with respect to $\boldsymbol{\theta}$ is equivalent to minimizing $S(\boldsymbol{\theta})$ with respect to $\boldsymbol{\theta}$.)

To find the least squares estimate $\hat{\boldsymbol{\theta}}$ we need to differentiate Eq. (10.2.5) with respect to $\boldsymbol{\theta}$. This provides the p *normal equations*, which must be solved for $\hat{\boldsymbol{\theta}}$. The normal equations take the form

$$\sum_{u=1}^{n} \{Y_u - f(\boldsymbol{\xi}_u, \hat{\boldsymbol{\theta}})\} \left[\frac{\partial f(\boldsymbol{\xi}_u, \boldsymbol{\theta})}{\partial \theta_i}\right]_{\theta = \hat{\theta}} = 0 \qquad (10.2.6)$$

for $i = 1, 2, \ldots, p$, where the quantity denoted by brackets is the derivative of $f(\boldsymbol{\xi}_u, \boldsymbol{\theta})$ with respect to θ_i with all θ's replaced by the corresponding $\hat{\theta}$'s, which have the same subscript. We recall that when the function $f(\boldsymbol{\xi}_u, \boldsymbol{\theta})$ was linear this quantity was a function of the $\boldsymbol{\xi}_u$ only and did not involve the θ's at all. For example, if

$$f(\boldsymbol{\xi}_u, \boldsymbol{\theta}) = \theta_1 \xi_{1u} + \theta_2 \xi_{2u} + \cdots + \theta_p \xi_{pu},$$

then

$$\frac{\partial f}{\partial \theta_i} = \xi_{iu}, \qquad i = 1, 2, \ldots, p$$

and is independent of $\boldsymbol{\theta}$. This leaves the normal equations in the form of linear equations in $\theta_1, \theta_2, \ldots, \theta_p$ as we saw in previous chapters. When the model is nonlinear in the θ's, so will be the normal equations. We now illustrate this with a simple example involving the estimation of a single parameter θ in a nonlinear model.

Example.   Suppose we wish to find the normal equation for obtaining the least squares estimate $\hat{\theta}$ of θ for the model $Y = f(\theta, t) + \epsilon$ where $f(\theta, t) = e^{-\theta t}$, and where n pairs of observations $(Y_1, t_1), (Y_2, t_2), \ldots, (Y_n, t_n)$ are available. We find that

$$\frac{\partial f}{\partial \theta} = -t e^{-\theta t}.$$

Applying Eq. (10.2.6) leads to the single normal equation

$$\sum_{u=1}^{n} [Y_u - e^{-\theta t_u}][-t_u e^{-\theta t_u}] = 0$$

or

$$\sum_{u=1}^{n} Y_u t_u e^{-\theta t_u} - \sum_{u=1}^{n} t_u e^{-2\theta t_u} = 0.$$

We see that even with one parameter and a comparatively simple nonlinear model, finding $\hat{\theta}$ by solving the (only) normal equation is not easy. When more parameters are involved and the model is more complicated, the solution of the normal equations can be extremely difficult to obtain, and iterative methods must be employed in nearly all cases. To compound the difficulties it may happen that multiple solutions exist, corresponding to multiple stationary values of the function $S(\hat{\boldsymbol{\theta}})$. We now discuss methods which have been used to estimate the parameters in nonlinear systems.

10.3.   Estimating the Parameters of a Nonlinear System

In some nonlinear problems it is most convenient to write down the normal equations (10.2.6) and develop an iterative technique for solving them. Whether this works satisfactorily or not depends on the form of the equations and the iterative method used. In addition to this approach there are several currently employed methods available for obtaining the parameter estimates by a routine computer calculation. We shall mention three of these: (1) linearization, (2) steepest descent, and (3) Marquardt's compromise.

The linearization (or Taylor series) method uses the results of linear least squares in a succession of stages. Suppose the postulated model is of the form (10.2.4). Let $\theta_{10}, \theta_{20}, \ldots, \theta_{p0}$ be initial values for the parameters $\theta_1, \theta_2, \ldots, \theta_p$. These initial values may be intelligent guesses or preliminary estimates based on whatever information is available. (They may, e.g., be values suggested by the information gained in fitting a similar equation in a different laboratory or suggested as "about right" by the experimenter based on his experience and knowledge.) These initial values will, hopefully, be improved upon in the successive iterations to be described below. If we carry out a Taylor series expansion of $f(\boldsymbol{\xi}_u, \boldsymbol{\theta})$ about the point $\boldsymbol{\theta}_0$ where $\boldsymbol{\theta}_0 = (\theta_{10}, \theta_{20}, \ldots, \theta_{p0})'$ and curtail the expansion at the first derivatives, we can say that, approximately, when $\boldsymbol{\theta}$ is close to $\boldsymbol{\theta}_0$,

$$f(\boldsymbol{\xi}_u, \boldsymbol{\theta}) = f(\boldsymbol{\xi}_u, \boldsymbol{\theta}_0) + \sum_{i=1}^{p} \left[\frac{\partial f(\boldsymbol{\xi}_u, \boldsymbol{\theta})}{\partial \theta_i} \right]_{\theta=\theta_0} (\theta_i - \theta_{i0}). \quad (10.3.1)$$

If we set

$$f_u^{\,0} = f(\boldsymbol{\xi}_u, \boldsymbol{\theta}_0)$$
$$\beta_i^{\,0} = \theta_i - \theta_{i0}$$
$$Z_{iu}^0 = \left[\frac{\partial f(\boldsymbol{\xi}_u, \boldsymbol{\theta})}{\partial \theta_i} \right]_{\theta=\theta_0} \quad (10.3.2)$$

we can see that the model (10.2.4) is of the form, approximately,

$$Y_u - f_u^{\,0} = \sum_{i=1}^{p} \beta_i^{\,0} Z_{iu}^0 + \epsilon_u; \quad (10.3.3)$$

in other words is of the linear form (10.1.1), to the selected order of approximation. We can now estimate the parameters $\beta_i^{\,0}$, $i = 1, 2, \ldots, p$ by applying linear least squares theory. If we write

$$\mathbf{Z}_0 = \begin{bmatrix} Z_{11}^0 & Z_{21}^0 & \cdots & Z_{p1}^0 \\ Z_{12}^0 & Z_{22}^0 & \cdots & Z_{p2}^0 \\ \cdots & & & \\ Z_{1u}^0 & Z_{2u}^0 & \cdots & Z_{pu}^0 \\ \cdots & & & \\ Z_{1n}^0 & Z_{2n}^0 & \cdots & Z_{pn}^0 \end{bmatrix} = \{Z_{iu}^0\}, \; n \times p \qquad (10.3.4)$$

$$\mathbf{b}_0 = \begin{bmatrix} b_1{}^0 \\ b_2{}^0 \\ \cdots \\ b_p{}^0 \end{bmatrix} \quad \text{and} \quad \mathbf{y}_0 = \begin{bmatrix} Y_1 - f_1{}^0 \\ Y_2 - f_2{}^0 \\ \cdots \\ Y_u - f_u{}^0 \\ \cdots \\ Y_n - f_n{}^0 \end{bmatrix} = \mathbf{Y} - \mathbf{f}^0, \qquad (10.3.5)$$

say, with an obvious notation, then the estimate of $\boldsymbol{\beta}_0 = (\beta_1{}^0, \beta_2{}^0, \ldots, \beta_p{}^0)'$ is given by

$$\mathbf{b}_0 = (\mathbf{Z}_0{}'\mathbf{Z}_0)^{-1}\mathbf{Z}_0{}'(\mathbf{Y} - \mathbf{f}^0). \qquad (10.3.6)$$

The vector $\mathbf{b}_0$ will therefore minimize the sum of squares

$$\text{SS}(\boldsymbol{\theta}) \equiv \sum_{u=1}^{n} \left\{ Y_u - f(\boldsymbol{\xi}_u, \boldsymbol{\theta}_0) - \sum_{i=1}^{p} \beta_i{}^0 Z_{iu}^0 \right\}^2 \qquad (10.3.7)$$

with respect to the $\beta_i{}^0$, $i = 1, 2, \ldots, p$, where $\beta_i{}^0 = \theta_i - \theta_{i0}$. Let us write $b_i{}^0 = \theta_{i1} - \theta_{i0}$. Then the θ_{i1}, $i = 1, 2, \ldots, p$ can be thought of as the revised best estimates of $\boldsymbol{\theta}$.

Note carefully the difference between the sum of squares $S(\boldsymbol{\theta})$ in Eq. (10.2.5) where the appropriate *nonlinear* model is used, and the sum of squares $\text{SS}(\boldsymbol{\theta})$ in Eq. (10.3.7) where the *approximating linear expansion* of the model is employed.

We can now place the values θ_{i1}, the revised estimates, in the same roles as were played above by the values θ_{i0} and go through exactly the same procedure described above by Eqs. (10.3.1) through (10.3.7), but replacing all the zero subscripts by ones. This will lead to another set of revised estimates θ_{i2}, and so on. In vector form, extending the previous notation in an obvious way, we can write

$$\boldsymbol{\theta}_{j+1} = \boldsymbol{\theta}_j + \mathbf{b}_j$$
$$\boldsymbol{\theta}_{j+1} = \boldsymbol{\theta}_j + (\mathbf{Z}_j{}'\mathbf{Z}_j)^{-1}\mathbf{Z}_j{}'(\mathbf{Y} - \mathbf{f}^j) \qquad (10.3.8)$$

where $\mathbf{Z}_j = \{Z_{iu}^j\}$,

$$\mathbf{f}^j = (f_1^j, f_2^j, \ldots, f_n^j)',$$ (10.3.9)

$$\boldsymbol{\theta}_j = (\theta_{1j}, \theta_{2j}, \ldots, \theta_{pj})'.$$

This iterative process is continued until the solution converges, that is, until in successive iterations j, $(j + 1)$,

$$|\{\theta_{i(j+1)} - \theta_{ij}\}/\theta_{ij}| < \delta, \quad i = 1, 2, \ldots, p,$$

where δ is some prespecified amount [e.g., 0.000001]. At each stage of the iterative procedure, $S(\boldsymbol{\theta}_j)$ can be evaluated to see if a reduction in its value has actually been achieved.

The linearization procedure has possible drawbacks for some problems in that

1. It may converge very slowly; that is, a very large number of iterations may be required before the solution stabilizes even though the sum of squares $S(\boldsymbol{\theta}_j)$ may decrease consistently as j increases. This sort of behavior is not common but can occur.

2. It may oscillate widely, continually reversing direction, and often increasing, as well as decreasing the sum of squares. Nevertheless the solution may stabilize eventually.

3. It may not converge at all, and even diverge, so that the sum of squares increases iteration after iteration without bound.

To combat these deficiencies, a program written by G. W. Booth and T. I. Peterson (1958) under the direction of G. E. P. Box amends the correction vector $\mathbf{b}_j$ in Eq. (10.3.8) by halving it if

$$S(\boldsymbol{\theta}_{j+1}) > S(\boldsymbol{\theta}_j)$$

or doubling it if

$$S(\boldsymbol{\theta}_{j+1}) < S(\boldsymbol{\theta}_j).$$

This halving and/or doubling process is continued until three points between $\boldsymbol{\theta}_j$ and $\boldsymbol{\theta}_{j+1}$ are found which include between them a local minimum of $S(\boldsymbol{\theta})$. A quadratic interpolation is used to locate the minimum, and the iterative cycle begins again.

Although in theory this method always converges (see "The modified Gauss-Newton method for fitting of non-linear regression functions by least squares, " by H. O. Hartley, *Technometrics*, **3**, 1961, 269–280), in practice difficulties may occur. In a case considered by N. H. Smith ("Transient operation of continuous stirred tank reactors," Ph.D. Thesis, University of Wisconsin, 1963) nonlinearity of the model caused tremendous "overshoot," and although the correction vector was halved ten times, no reduction in $S(\boldsymbol{\theta})$ was achieved over the original value. Although the original value was less than 10 units, the calculation produced values of $S(\boldsymbol{\theta})$ so great (exceeding 10^{308}) that they "overflowed" the machine capacity. This unsuccessful attempt to obtain estimates was not reported in the thesis, incidentally; a random search technique was used instead. We mention the example only to point out that such difficulties can occur. The linearization method is, in general, a useful one and will successfully solve many nonlinear problems. Where it does not, consideration should be given to reparameterization of the model (see Section 10.5) or to the use of Marquardt's compromise.

A REMARK ON DERIVATIVES. Many computer programs which use a method needing the values of the derivatives of a function at certain points do not use the functional values of the derivatives at all. Instead they compute ratios such as

$$\{f(\boldsymbol{\xi}_u, \theta_{10}, \theta_{20}, \ldots, \theta_{i0} + h_i, \ldots, \theta_{p0}) - f(\boldsymbol{\xi}_u, \theta_{10}, \theta_{20}, \ldots, \theta_{p0})\}/h_i,$$

$$i = 1, 2, \ldots p,$$

where h_i is a selected small increment. The ratio given above is an approximation to the expression

$$\left[\frac{\partial f(\boldsymbol{\xi}_u, \boldsymbol{\theta})}{\partial \theta_i}\right]_{\theta=\theta_0}$$

since, if h_i tends to zero, the limit of the ratio is this expression by definition.

Steepest Descent

The steepest descent method involves concentration on the sum of squares function, $S(\boldsymbol{\theta})$ as defined by Eq. (10.2.5) and use of an iterative process to find the minimum of this function. The basic idea is to move, from an initial point $\boldsymbol{\theta}_0$, along the vector with components

$$-\frac{\partial S(\boldsymbol{\theta})}{\partial \theta_1}, -\frac{\partial S(\boldsymbol{\theta})}{\partial \theta_2}, \ldots, -\frac{\partial S(\boldsymbol{\theta})}{\partial \theta_p},$$

whose values change continuously as the path is followed. One way of achieving this in practice, without evaluating functional derivatives, is to estimate the vector slope components at various places on the surface $S(\boldsymbol{\theta})$ by fitting planar approximating functions. This is a technique of great value in experimental work for finding stationary values of response surfaces. A full description of the method is given in *Design and Analysis of Industrial Experiments*, edited by O. L. Davies, Oliver and Boyd, Edinburgh, Scotland, 1954, and we shall discuss it only briefly here.

The procedure is as follows. Starting in one particular region of the $\boldsymbol{\theta}$ space (or the *parameter space* as we shall call it) several *runs* are made, by selecting n (say) combinations of levels of $\theta_1, \theta_2, \ldots, \theta_p$ and evaluating

$S(\boldsymbol{\theta})$ at those combinations of levels. The runs are usually chosen in the pattern of a two level factorial design. Using the evaluated $S(\boldsymbol{\theta})$ values as observations of a dependent variable and the combinations of levels of $\theta_1, \theta_2, \ldots, \theta_p$ as the observations of corresponding independent variables, we fit the model

$$\text{"Observed } S(\boldsymbol{\theta})\text{"} = \beta_0 + \sum_{i=1}^{p} \beta_i(\theta_i - \bar{\theta}_i)/s_i + \epsilon$$

by standard least squares. Here $\bar{\theta}_i$ denotes the mean of the levels $\theta_{iu}, u = 1, 2, \ldots n,$ of θ_i used in the runs and s_i is a scaling factor chosen so that $\sum_{u=1}^{n}(\theta_{iu} - \bar{\theta}_i)^2/s_i^2 = $ constant. This implies that we believe the true surface defined by $S(\boldsymbol{\theta})$ can be approximately represented by a plane in the region of the parameter space in which we made our runs. The estimated coefficients

$$b_1, b_2, \ldots, b_p$$

indicate the direction of steepest ascent so the negatives of these, namely

$$-b_1, -b_2, \ldots, -b_p$$

indicate the direction of steepest descent. This means that as long as the linear approximation is realistic the maximum decrease in $S(\boldsymbol{\theta})$ will be obtained by moving along the line which contains points such that

$$(\theta_i - \bar{\theta}_i)/s_i \propto -b_i.$$

Denoting the proportionality factor by λ, the path of steepest descent contains points $(\theta_1, \theta_2, \ldots, \theta_p)$ such that

$$\frac{(\theta_i - \bar{\theta}_i)}{s_i} = -\lambda b_i,$$

where $\lambda > 0$, or

$$\theta_i = \bar{\theta}_i - \lambda b_i s_i.$$

By giving λ selected values the path of steepest descent can be followed. A number of values of λ are selected and the path of steepest descent is followed as long as $S(\boldsymbol{\theta})$ decreases. When it does not, another experimental design is set down and the process is continued until it converges to the value $\hat{\boldsymbol{\theta}}$ which minimizes $S(\boldsymbol{\theta})$.

While, theoretically, the steepest descent method will converge, it may do so in practice with agonizing slowness after some rapid initial progress. Slow convergence is particularly likely when the $S(\boldsymbol{\theta})$ contours are attenuated and banana-shaped (as they often are in practice), and it happens when the path of steepest descent zigzags slowly up a narrow ridge, each iteration bringing only a slight reduction in $S(\boldsymbol{\theta})$. (This is less of a problem in laboratory-type investigations where human intervention can be permitted at each stage of calculation since then the experimental design can be revised, the scales of the independent variables can be changed, and so on.) This difficulty has led to modifications of the basic steepest descent procedure when used for nonlinear fitting. For some references see "A review of minimization techniques for nonlinear functions," by H. A. Spang, *Society for Industrial and Applied Mathematics Review*, **4**, 1962, 343–365. (One possible modification is to use a second-order approximating function rather than a first-order or planar approximation. While this provides better graduation of the true surface, it also requires additional computation in the iterative procedures.)

A further disadvantage of the steepest descent method is that it is not scale invariant. The indicated direction of movement changes if the scales s_i of the variables are changed, unless all are changed by the same factor. The steepest descent method is, on the whole, slightly less favored than the linearization method but will work satisfactorily for many nonlinear problems, especially if modifications are made to the basic technique.

Marquardt's Compromise

A method developed by D. W. Marquardt ("An algorithm for least squares estimation of nonlinear parameters," *Journal of the Society for Industrial & Applied Mathematics*, **2**, 1963, 431–441) appears to enlarge considerably the number of practical problems that can be tackled by nonlinear estimation. Marquardt's method represents a compromise between the linearization (or Taylor series) method and the steepest descent method and appears to combine the best features of both while avoiding their most serious limitations. It is good in that it almost always converges and does not "slow down" as the steepest descent method often does. However, as we again emphasize, the other methods will work perfectly well on many practical problems which do not violate the limitations of the methods. (In general, we must keep in mind that, given a particular method, a problem can usually be constructed to defeat it. Alternatively, given a particular problem and a suggested method, ad hoc modifications can often provide quicker convergence than an alternative method. The Marquardt method is one which appears to work well in many circumstances and thus is a sensible practical choice. For the reasons stated above, no method can be called "best" for all nonlinear problems.)

The idea of Marquardt's method can be explained briefly as follows. Suppose we start from a certain point in the parameter space, $\boldsymbol{\theta}$. If the method of steepest descent is applied, a certain vector direction, $\boldsymbol{\delta}_g$, where g stands for gradient, is obtained for movement away from the initial point. Because of attenuation in the $S(\boldsymbol{\theta})$ contours this may be the best *local* direction in which to move to attain smaller values of $S(\boldsymbol{\theta})$ but may not be the best *overall* direction. However the best direction must be within 90° of $\boldsymbol{\delta}_g$ or else $S(\boldsymbol{\theta})$ will get larger locally. The linearization (or Taylor series) method leads to another correction vector $\boldsymbol{\delta}$ given by a formula like (10.3.6). Marquardt found that for a number of practical problems he studied, the angle, ϕ say, between $\boldsymbol{\delta}_g$ and $\boldsymbol{\delta}$, fell in the range $80° < \phi < 90°$. In other words, the two directions were almost at right angles! The Marquardt algorithm provides a method for interpolating between the vectors $\boldsymbol{\delta}_g$ and $\boldsymbol{\delta}$ and for obtaining a suitable step size as well.

We shall not go into the detail of the method here. The basic algorithm is given in the quoted reference and a discussion of the method is contained in "Problems in the analysis of nonlinear models by least squares," by D. A. Meeter, University of Wisconsin, Ph.D. Thesis, 1964. The program references are as follows:

Marquardt, D. W. (1966). Least squares estimation of nonlinear parameters, a computer program in FORTRAN IV language. IBM SHARE Library, Distribution Number 309401, August, 1966. (Successor to Distribution Number 1428 and 3094.)
Marquardt, D. W. and Stanley, R. M. (1964). NLIN 2—Least squares estimation of nonlinear parameters, supplement to S.D.A. 3093 (NLIN). Mimeo manuscript, available from the authors.
Note. Programs are frequently rewritten to incorporate refinements and new developments; more recent versions may now be available.

Confidence Contours

Some idea of the nonlinearity in the model under study can be obtained, after the estimation of $\boldsymbol{\theta}$, by evaluating the ellipsoidal confidence region obtained on the assumption that the linearized form of the model is valid around $\hat{\boldsymbol{\theta}}$, the final estimate of $\boldsymbol{\theta}$. This is given by the formula

$$(\boldsymbol{\theta} - \hat{\boldsymbol{\theta}})'\hat{\mathbf{Z}}'\hat{\mathbf{Z}}(\boldsymbol{\theta} - \hat{\boldsymbol{\theta}}) \leq ps^2 F(p, n - p, 1 - \alpha)$$

where $\hat{\mathbf{Z}}$ denotes a matrix of the form shown in Eq. (10.3.4) but with $\hat{\boldsymbol{\theta}}$ substituted into the elements in place of $\boldsymbol{\theta}_0$ everywhere, and where

$$s^2 = S(\hat{\boldsymbol{\theta}})/(n - p).$$

Note that when the difference between successive values $\boldsymbol{\theta}_{j+1}$ and $\boldsymbol{\theta}_j$ is sufficiently small so that the linearization procedure terminates with $\boldsymbol{\theta}_{j+1} = \hat{\boldsymbol{\theta}}$ ($= \boldsymbol{\theta}_j$ for practical purposes), then $S(\hat{\boldsymbol{\theta}})$ is a minimum value of $S(\boldsymbol{\theta})$ in Eq. (10.2.5) to the accuracy imposed by the termination procedure selected. This can be seen by examining Eq. (10.3.7) with $\hat{\boldsymbol{\theta}}$, β_i^{j+1}, and Z_{iu}^{j+1} replacing $\boldsymbol{\theta}_0$, β_i^0 and Z_{iu}^0 respectively and remembering that, to the order of accuracy imposed by the termination procedure, $\mathbf{b}_{j+1} = \mathbf{0}$. The ellipsoid above will *not* be a true confidence region when the model is nonlinear. We can however determine the end points on the major axes of this ellipsoid by canonical reduction (see, e.g., *Design and Analysis of Industrial Experiments* edited by O. L. Davies, published by Oliver and Boyd, Edinburgh, Scotland, 1954); the *actual* values of $S(\theta)$ can be evaluated at these points and compared with each other. Under linear theory the values would all be the same.

An exact confidence contour is defined by taking $S(\theta) = $ constant, but since we do not know the correct distribution properties in the general nonlinear case, we are unable to obtain a specified probability level. However, we can, for example, choose the contour such that

$$S(\theta) = S(\hat{\theta})\left\{1 + \frac{p}{n-p} F(p, n-p, 1-\alpha)\right\}$$

which, if the model is linear, provides an *exact*, ellipsoidal $100(1-\alpha)\%$ boundary (see Section 2.6 or 10.6) and label it as an approximate $100(1-\alpha)\%$ confidence contour in the nonlinear case. Note that the contour so determined *will be a proper correct confidence contour in this case* (and will not be elliptical in general), *and it is only the probability level which is approximate*. When only two parameters are involved the confidence contour can be drawn. For more parameters, sectional drawings can be constructed if desired.

In general when a linearized form of a nonlinear model is used, all the usual formulae and analyses of linear regression theory can be applied. Any results obtained are, however, only valid to the extent that the linearized form provides a good approximation to the true model.

Grids and Plots

Two obvious ways of examining the sum of squares surface $S(\theta)$ are often overlooked; they can be particularly useful when an iterative procedure, beginning from chosen initial values, does not satisfactorily converge.

The first of these is to select a grid of points, that is, a factorial design, in the space of the parameters $(\theta_1, \theta_2, \ldots, \theta_p)$ and to evaluate (usually on a computer) the sum of squares function at every point of the grid. These values will provide some idea of the form of the sum of squares surface and may reveal, for example, that multiple minima are possible. In any case, the grid point at which the smallest sum of squares is found can be used as the starting point of an iterative parameter estimation procedure, or a reduced grid can be examined in the best neighbourhood, to obtain a better starting point. The simplest type of grid available is that in which *two* levels of every parameter are selected. In this case, the grid points are those of a 2^p factorial design, and it is possible to use standard methods to evaluate the factorial effects and interactions and so provide information on the effects of changes in the parameters on the sum of squares function $S(\theta)$.

The second possibility is to draw sum of squares contours in any particular region of the parameter space in which difficulty in convergence occurs or in which additional information would be helpful. This is usually straightforward when only one or two parameters are involved. When there are more than two parameters, two-dimensional slices of the contours can be obtained for selected values of all but two of the parameters, and a composite picture can be built up.

The Importance of Good Starting Values

All iterative procedures require initial values $\theta_{10}, \theta_{20}, \ldots, \theta_{p0}$, of the parameters $\theta_1, \theta_2, \ldots, \theta_p$, to be selected. All available prior information should be used to make these starting values as reliable as they possibly can be. Good starting values will often allow an iterative technique to converge to a solution much faster than would otherwise be possible. Also if multiple minima exist or if there are several local minima in addition to an absolute minimum, poor starting values may result in convergence to an unwanted stationary point of the sum of squares surface. This unwanted point may have parameter values which are physically impossible or which do not provide the true minimum value of $S(\theta)$. As suggested above, a preliminary evaluation of $S(\theta)$ at a number of grid points in the parameter space is often useful.

10.4. An Example

The example which follows is taken from an investigation performed at Procter and Gamble and reported by H. Smith and S. D. Dubey in "Some reliability problems in the chemical industry," *Industrial Quality Control*, **21**, 1964, no. 2, 64–70. We shall use this example to illustrate how a solution can be obtained for a nonlinear estimation problem by solving the normal equations directly, or alternatively by the linearization method. We shall not provide an example of the use of steepest descent (but see, e.g., the paper "Application of digital computers in the exploration of functional relationships," by G. E. P. Box and G. A. Coutie, *Proceedings of the Institution of Electrical Engineers*, **103**, Part B, Supplement no. 1 1956, 100–107) nor an example of Marquardt's compromise procedure. The investigation involved a product A which must have a fraction 0.50 of Available Chlorine at the time of manufacture. The fraction of Available Chlorine in the product decreases with time; this is known. In the eight weeks before the product reaches the consumer a decline to a level 0.49 occurs but since many uncontrolled factors then arise (such as warehousing environments, handling facilities), theoretical calculations are not reliable for making extended predictions of the Available Chlorine fraction present at later times. To assist management in decisions such as (1) When should warehouse material be scrapped? (2) When should store stocks be replaced? cartons of the product were analyzed over a period to provide the data of Table 10.1 (Note that the product is made only every other week and code-dated only by the week of the year. The predicted values shown in the table are obtained from the fitted equation to be found in what follows.)

Table 10.1   Per Cent of Available Chlorine in a Unit of Product

Length of Time since Produced (weeks) X	Available Chlorine Y	Average Available Chlorine $\bar{Y}$	Predicted Y, Using the Model $\hat{Y}$
8	0.49, 0.49	0.490	0.490
10	0.48, 0.47, 0.48, 0.47	0.475	0.472
12	0.46, 0.46, 0.45, 0.43	0.450	0.457
14	0.45, 0.43, 0.43	0.437	0.445
16	0.44, 0.43, 0.43	0.433	0.435
18	0.46, 0.45	0.455	0.427
20	0.42, 0.42, 0.43	0.423	0.420
22	0.41, 0.41, 0.40	0.407	0.415
24	0.42, 0.40, 0.40	0.407	0.410
26	0.41, 0.40, 0.41	0.407	0.407
28	0.41, 0.40	0.405	0.404
30	0.40, 0.40, 0.38	0.393	0.401
32	0.41, 0.40	0.405	0.399
34	0.40	0.400	0.397
36	0.41, 0.38	0.395	0.396
38	0.40, 0.40	0.400	0.395
40	0.39	0.390	0.394
42	0.39	0.390	0.393

It was postulated that a nonlinear model of the form

$$Y = \alpha + (0.49 - \alpha)e^{-\beta(X-8)} + \epsilon \qquad (10.4.1)$$

would suitably account for the variation observed in the data, for $X \geq 8$. This model provides a true level, without error, of $\eta = 0.49$ when $X = 8$, and it exhibits the proper sort of decay. An additional point of information, agreed upon by knowledgable chemists, was that an equilibrium, asymptotic level of Available Chlorine somewhere above 0.30 should be expected. The problem is to estimate the parameters α and β of the nonlinear model

(10.4.1) using the data given in the table. The residual sum of squares for this model can be written as

$$S(\alpha, \beta) = \sum_{u=1}^{n} [Y_u - \alpha - (0.49 - \alpha)e^{-\beta(X_u-8)}]^2 \qquad (10.4.2)$$

where (X_u, Y_u), $u = 1, 2, 3, \ldots, 44$ are the corresponding pairs of observations from the table (e.g., $X_1 = 8$, $Y_1 = 0.49, \ldots, X_{44} = 42$, $Y_{44} = 0.39$).

A Solution Through the Normal Equations

Differentiating Eq. (10.4.2) first with respect to α, and then with respect to β, and setting the results equal to zero provides two normal equations. After removal of a factor of 2 from the first equation and a factor of $2(0.49 - \alpha)$ from the second equation and some rearrangement, the equations reduce to

$$\alpha = \frac{\sum Y_u - \sum Y_u e^{-\beta t_u} - 0.49 \sum e^{-\beta t_u} + 0.49 \sum e^{-2\beta t_u}}{n - 2 \sum e^{-\beta t_u} + \sum e^{-2\beta t_u}} \qquad (10.4.3)$$

and

$$\alpha = \frac{0.49 \sum t_u e^{-2\beta t_u} - \sum Y_u t_u e^{-\beta t_u}}{\sum t_u e^{-2\beta t_u} - \sum t_u e^{-\beta t_u}} \qquad (10.4.4)$$

where all summations are from $u = 1$ to $u = 44$ and $t_u = X_u - 8$. We see that these normal equations have a particular simplification in that the parameter α can be eliminated by subtracting one equation from the other. If this is done, a single nonlinear equation of the form $f(\beta) = 0$ in β results. This can be solved by applying the Newton–Raphson technique, first guessing a value for β, call it β_0, and then "correcting" it by h_0 obtained as follows. If the root of $f(\beta) = 0$ is at $(\beta_0 + h_0)$, then

$$0 = f(\beta_0 + h_0) = f(\beta_0) + h_0 \left[\frac{df(\beta)}{d\beta}\right]_{\beta=\beta_0} \qquad (10.4.5)$$

approximately, or

$$h_0 = -f(\beta_0) \bigg/ \left[\frac{df(\beta)}{d\beta}\right]_{\beta=\beta_0} \qquad (10.4.6)$$

approximately. We can now use $\beta_1 = \beta_0 + h_0$ instead of β_0 and repeat the correction procedure to find h_1 and so $\beta_2 = \beta_1 + h_1$. This process can be continued until it converges to a value $\hat{\beta}$ which is then the least squares estimate of β. The value of $\hat{\alpha}$, the least squares estimate of α, can be obtained from Eqs. (10.4.3) and (10.4.4) by substituting $\hat{\beta}$ in the right-hand sides. As a check, the same value should be obtained from both equations.

We can guess β_0 initially by noting, for example, that when $X_{44} = 42$, $Y_{44} = 0.39$. If Y_{44} contained no error then we should have

$$0.39 = \alpha + (0.49 - \alpha)e^{-34\beta_0}$$

whereas if we assume that Y tends to 0.30 as X tends to infinity (on the basis of the prior information given by the chemists) then we can make an initial guess of $\alpha_0 = 0.30$ for α, from which it follows that

$$0.39 = 0.30 + (0.49 - 0.30)e^{-34\beta_0}$$

or

$$e^{-34\beta_0} = \frac{0.09}{0.19}$$

so that

$$\beta_0 = \frac{-[\ln (0.09/0.19)]}{34} = 0.02$$

approximately. (Note that we must have $\beta > 0$ or else a decay cannot be represented by the function.)

If we denote Eqs. (10.4.3) and (10.4.4) by

$$\alpha = f_1(\beta)$$
$$\alpha = f_2(\beta) \qquad\qquad (10.4.7)$$

respectively, then

$$f(\beta) \equiv f_1(\beta) - f_2(\beta) = 0 \qquad\qquad (10.4.8)$$

and

$$\frac{\partial f(\beta)}{\partial \beta} \equiv \frac{\partial f_1(\beta)}{\partial \beta} - \frac{\partial f_2(\beta)}{\partial \beta}. \qquad\qquad (10.4.9)$$

Rather than write down the rather lengthy expressions which result from the differentiation of $f_1(\beta)$ and $f_2(\beta)$ we shall adopt a simpler method of finding $\hat\alpha$, $\hat\beta$.

An alternative procedure for estimating α and β in this case is to plot the functions (10.4.7) over a reasonable range of β and note where the two curves intersect. This provides both $\hat\alpha$ and $\hat\beta$ immediately since the point of intersection is at $(\hat\beta, \hat\alpha)$. Some values of $f_1(\beta)$ and $f_2(\beta)$ for a range of values of β are shown in Table 10.2 and the resulting plot is shown in Figure 10.1. The estimates can be read off squared paper to sufficient accuracy as $\hat\alpha = 0.39$, $\hat\beta = 0.10$. We can see from the figure that the two curves plotted are quite close together for a comparatively large range of β and for a somewhat smaller range of α. This indicates that β is somewhat less well determined than α. For example, $|f_1(\beta) - f_2(\beta)| < 0.0025$ is achieved by a range of values of β between about 0.07 and 0.12 and a range of values of α between about 0.37 and 0.40. Many pairs of values

Table 10.2 Points on the Curves $f_i(\beta)$

β	$f_1(\beta)$	$f_2(\beta)$
0.06	0.3656	0.3627
0.07	0.3743	0.3720
0.08	0.3808	0.3791
0.09	0.3857	0.3847
0.10	0.3896	0.3894
0.11	0.3927	0.3935
0.12	0.3953	0.3970
0.13	0.3975	0.4002
0.14	0.3993	0.4031
0.15	0.4009	0.4057
0.16	0.4023	0.4082

of (β, α) such as, for example, $(0.09, 0.385)$, $(0.11, 0.393)$ bring the two curves of the figure almost together, and thus do not appear unreasonable estimates for (β, α) in the light of the data even though they do not actually minimize $S(\alpha, \beta)$. We shall see that these comments are substantiated by the confidence regions for the true (β, α) which can be constructed for this problem (see Figure 10.3). The fitted equation now takes the form

$$\hat{Y} = 0.39 + 0.10e^{-0.10(X-8)} \qquad (10.4.10)$$

The observed values of X can be inserted in this equation to give the fitted values shown in Table 10.1. The fitted curve and the observations are

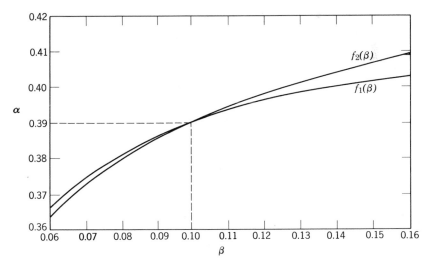

Figure 10.1 Finding $(\hat{\beta}, \hat{\alpha})$ as the intersection of $f_1(\beta)$ and $f_2(\beta)$.

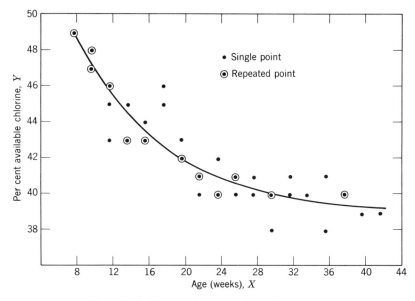

Figure 10.2   The fitted curve and the observations.

shown in Figure 10.2. The usual analysis of residuals could now be carried out as discussed in Chapter 3. (The large residuals at $X = 18$ strike the eye immediately. According to the authors no assignable cause could be found for these.)

A Solution through the Linearization Technique

To linearize the model into the form (10.3.1) we need to evaluate the first derivatives of

$$f(\boldsymbol{\xi}_u, \boldsymbol{\theta}) = f(X_u; \alpha, \beta)$$
$$= \alpha + (0.49 - \alpha)e^{-\beta(X_u - 8)} \qquad (10.4.11)$$

namely,

$$\frac{\partial f}{\partial \alpha} = 1 - e^{-\beta(X_u - 8)},$$
$$\qquad\qquad\qquad\qquad\qquad (10.4.12)$$
$$\frac{\partial f}{\partial \beta} = -(0.49 - \alpha)(X_u - 8)e^{-\beta(X_u - 8)}.$$

Thus if $\alpha = \alpha_j$, $\beta = \beta_j$ are the values inserted at the jth stage, as described in Section 10.3, we have, in the notation implied in that section, a model of form (at the jth stage):

$$Y_u - f_u{}^j = [1 - e^{-\beta_j(X_u - 8)}](\alpha - \alpha_j)$$
$$+ [-(0.49 - \alpha_j)(X_u - 8)e^{-\beta_j(X_u - 8)}](\beta - \beta_j) + \epsilon_u$$

or in matrix form

$$\mathbf{Y} - \mathbf{f}^j = \mathbf{Z}_j \begin{bmatrix} \alpha - \alpha_j \\ \beta - \beta_j \end{bmatrix} + \boldsymbol{\epsilon}$$

where

$$f_u^{\ j} = \alpha_j + (0.49 - \alpha_j)e^{-\beta_j(X_u - 8)} \tag{10.4.13}$$

$$\mathbf{Z}_j = \begin{bmatrix} 1 - e^{-\beta_j(X_1 - 8)} & -(0.49 - \alpha_j)(X_1 - 8)e^{-\beta_j(X_1 - 8)} \\ \cdots & \\ 1 - e^{-\beta_j(X_u - 8)} & -(0.49 - \alpha_j)(X_u - 8)e^{-\beta_j(X_u - 8)} \\ \cdots & \\ 1 - e^{-\beta_j(X_n - 8)} & -(0.49 - \alpha_j)(X_n - 8)e^{-\beta_j(X_n - 8)} \end{bmatrix} \tag{10.4.14}$$

and the vector of quantities to be estimated is

$$\begin{bmatrix} \alpha - \alpha_j \\ \beta - \beta_j \end{bmatrix} \tag{10.4.15}$$

with estimate given by

$$\begin{bmatrix} \alpha_{j+1} - \alpha_j \\ \beta_{j+1} - \beta_j \end{bmatrix} = (\mathbf{Z}_j'\mathbf{Z}_j)^{-1}\mathbf{Z}_j' \begin{bmatrix} Y_1 - f_1^{\ j} \\ Y_2 - f_2^{\ j} \\ \cdots \\ Y_n - f_n^{\ j} \end{bmatrix} \tag{10.4.16}$$

If we begin the iterations with initial guesses $\alpha_0 = 0.30$ and $\beta_0 = 0.02$ as before, and apply Eq. (10.4.16) iteratively, we obtain estimates as follows

Iteration	α_j	β_j	$S(\alpha_j, \beta_j)$
0	0.30	0.02	0.0263
1	0.8416	0.1007	4.4881
2	0.3901	0.1004	0.0050
3	0.3901	0.1016	0.0050
4	0.3901	0.1016	0.0050

Note. These figures were rounded from the end results of computer calculations which carried more significant figures. Numerical differences might occur if a parallel calculation was made on a desk calculator.

This process converges to the same least squares estimates as before, namely $\hat{\alpha} = 0.39$, $\hat{\beta} = 0.10$ to give the fitted model (10.4.10). Note that this happens in spite of the rather alarming fact that, after the first stage, $S(\alpha_1, \beta_1) = 4.4881$, which is about 170 times the initial $S(\alpha_0, \beta_0) = 0.0263$.

The reduction in the next iteration is dramatic and practically final, the subsequent reduction in $S(\alpha, \beta)$ being in the sixth place of decimals, which is not shown. In some nonlinear problems no correction, dramatic or otherwise, occurs and the process diverges providing larger and larger values for $S(\theta)$. (For a possible reason for this sort of behavior, see Section 10.7.)

Further Analysis

The usual tests which are appropriate in the linear model case are, in general, *not* appropriate when the model is nonlinear. As a practical procedure we can compare the unexplained variation with an estimate of $V(Y_u) = \sigma^2$ but cannot use the F-statistic to obtain conclusions at any stated level. The unexplained variation is $S(\hat{\alpha}, \hat{\beta}) = 0.0050$. In the absence of exact results for the nonlinear case, we can regard this sum of squares as being based on approximately $44 - 2 = 42$ degrees of freedom (since two parameters have been estimated). In the nonlinear case this does not, in general, lead to an unbiased estimate of σ^2 as in the linear case, even when the model is correct.

A pure error estimate of σ^2 (see Section 1.5) can be obtained from the repeat observations. This provides a sum of squares $S_{pe} = 0.0024$ with 26 degrees of freedom.

An approximate idea of possible lack of fit can be obtained by evaluating

$$S(\hat{\alpha}, \hat{\beta}) - S_{pe} = 0.0026 \quad \text{with} \quad 42 - 26 = 16 \text{ degrees of freedom}$$

and comparing the mean squares

$$\frac{(S(\hat{\alpha}, \hat{\beta}) - S_{pe})}{16} = 0.00016$$

$$\frac{S_{pe}}{26} = 0.00009.$$

An F-test is *not* applicable here but we can use the value of $F(16, 26, 0.95) = 2.08$ as a measure of comparison. We see that $16/9 = 1.8$, which would make us tentatively feel that the model does not fit badly.

Confidence Regions

We can calculate approximate $100(1 - q)\%$ confidence contours (described in Section 10.3) by finding points (α, β) which satisfy

$$S(\alpha, \beta) = S(\hat{\alpha}, \hat{\beta})\left[1 + \frac{p}{n - p} F(p, n - p, 1 - q)\right]$$

$$= 0.0050[1 + F(2, 42, 1 - q)/21]$$

$$= S_q,$$

From Eq. (10.4.2) we can write this as

$$\sum_{u=1}^{n} \{(Y_u - 0.49e^{-\beta(X_u-8)}) + \alpha(e^{-\beta(X_u-8)} - 1)\}^2 = S_q$$

or

$$A\alpha^2 + 2B\alpha + C - S_q = 0$$

where

$$A = \sum_{u=1}^{n} (e^{-\beta(X_u-8)} - 1)^2$$

$$B = \sum_{u=1}^{n} (Y_u - 0.49e^{-\beta(X_u-8)})(e^{-\beta(X_u-8)} - 1)$$

$$C = \sum_{u=1}^{n} (Y_u - 0.49e^{-\beta(X_u-8)})^2$$

are all functions of β alone. We can thus select a value for q and then evaluate

$$\alpha = \frac{\{-B \pm [B^2 - A(C - S_q)]^{1/2}\}}{A}$$

for a range of β to obtain points on the boundaries of confidence regions with approximately $100(1 - q)\%$ confidence coefficients. The 75%, 95%, 99%, and 99.5% regions, obtained from $q = 0.25, 0.05, 0.01$, and 0.005, respectively, are shown in Figure 10.3. The dot denotes the point $(\hat{\beta}, \hat{\alpha})$.

Points (β, α) that lie within the contour marked (say) 95% are considered, by the data, as not unreasonable for the true values of (β, α), at an approximate 95% level of confidence. The orientation and shape of the contours

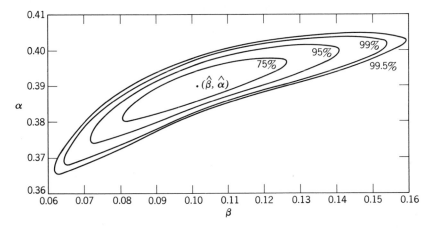

Figure 10.3 Confidence regions for $(\hat{\beta}, \hat{\alpha})$. The regions are exact, the confidence levels approximate.

indicate that $\hat{\beta}$ is less well determined than $\hat{\alpha}$. (For discussion on this point in the linear case, see Section 10.6; see also Section 10.5.)

10.5. A Note on Reparameterization of the Model

When the sum of squares surface, defined by Eq. (10.2.5), is attenuated and contains long ridges, slow convergence of any iterative estimation procedure is likely. As a simple example of this in the *linear* model case consider the model $Y_u = \theta_0 + \theta_1 X_u + \epsilon_u$ and suppose we have three observations Y_1, Y_2, and Y_3 at $X = 9$, 10, and 11. Then

$$S(\boldsymbol{\theta}) = (Y_1 - \theta_0 - 9\theta_1)^2 + (Y_2 - \theta_0 - 10\theta_1)^2 + (Y_3 - \theta_0 - 11\theta_1)^2$$

$$= \sum_{u=1}^{3} Y_u^2 - 2\theta_0 \sum_{u=1}^{3} Y_u - 2\theta_1(9Y_1 + 10Y_2 + 11Y_3)$$

$$+ 3\theta_0^2 + 302\theta_1^2 + 60\theta_0\theta_1.$$

In coordinates (θ_0, θ_1), the contours of $S(\boldsymbol{\theta}) = $ constant are long thin ellipses. Such a sum of squares surface can be called *poorly conditioned* and usually results in extremely slow convergence. However, if we rewrite the model as

$$Y_u = (\theta_0 + \theta_1 \bar{X}) + \theta_1(X_u - \bar{X}) + \epsilon_u$$

$$= \phi_0 + \phi_1 x_u + \epsilon_u$$

where $\phi_0 = \theta_0 + \theta_1 \bar{X}$, $\phi_1 = \theta_1$, $x_u = X_u - \bar{X} (= -1, 0, 1$ for $u = 1, 2, 3)$ we obtain a sum of squares in terms of $\boldsymbol{\phi} = (\phi_1, \phi_2)'$ given by

$$S(\boldsymbol{\phi}) = \sum (Y_u - \phi_0 - \phi_1 x_u)^2$$

$$= (Y_1 - \phi_0 + \phi_1)^2 + (Y_2 - \phi_0)^2 + (Y_3 - \phi_0 - \phi_1)^2$$

$$= \sum_{u=1}^{3} Y_u^2 - 2\phi_0 \sum_{u=1}^{3} Y_u + 2\phi_1(Y_1 - Y_3) + 3\phi_0^2 + 2\phi_1^2.$$

In coordinates (ϕ_0, ϕ_1) these contours are "well-rounded" ellipses—the surface is said to be *well-conditioned*—and almost any system of numerical minimization would converge rapidly.

A similar sort of ill-conditioning can occur in nonlinear models of the form

$$Y_u = \theta_0 e^{\theta_1 X_u} + \epsilon_u$$

if the mean of the X_u, $\bar{X}$, is not close to zero. When expressions of this type occur it is sometimes better to consider the model in the alternative form

$$Y_u = (\theta_0 e^{\theta_1 \bar{X}})(e^{\theta_1(X_u - \bar{X})}) + \epsilon_u$$

$$= \phi_0 e^{\phi_1 x_u} + \epsilon_u$$

where $\phi_0 = \theta_0 e^{\theta_1 \bar{X}}$, $\phi_1 = \theta_1$, and $x_u = X_u - \bar{X}$.

Note. In our example in Section 10.4 we did not do this. There it would have complicated the model since α occurred in two places and more than a simple replacement of one parameter by another is involved.

Suitable reparameterizations which will improve the conditioning of a sum of squares surface in a general case are not always apparent. Simple transformations which permit a "centering" of some variables, as in the examples above, may often be beneficial, however, and are often, at worst, harmless. For additional comments on reparameterization see "Some notes on nonlinear estimation," by G. E. P. Box, Statistics Department Technical Report no. 25, University of Wisconsin, Madison, Wisconsin, 1964.

10.6. The Geometry of Linear Least Squares

To understand why iterative methods applied to nonlinear problems are not always successful, it is helpful to consider the geometrical interpretation of *linear* least squares first of all. In the linear case, in the notation of this chapter we can write the model as

$$Y = f(\xi, \theta) + \epsilon$$
$$= \theta_1 X_1 + \theta_2 X_2 + \cdots + \theta_p X_p + \epsilon$$

where the X_i are functions of ξ. If we have observations Y_u containing errors ϵ_u when the X_i take the values $X_{1u}, X_{2u}, \ldots, X_{pu}$, for $u = 1, 2, \ldots, n$, then we can write the model in the alternative form:

$$Y = X\theta + \epsilon$$

where

$$Y = \begin{bmatrix} Y_1 \\ Y_2 \\ \cdots \\ Y_n \end{bmatrix}, \quad X = \begin{bmatrix} X_{11} & X_{21} & \cdots & X_{p1} \\ X_{12} & X_{22} & \cdots & X_{p2} \\ \cdots & & & \\ X_{1n} & X_{2n} & \cdots & X_{pn} \end{bmatrix}, \quad \theta = \begin{bmatrix} \theta_1 \\ \theta_2 \\ \cdots \\ \theta_p \end{bmatrix}, \quad \epsilon = \begin{bmatrix} \epsilon_1 \\ \epsilon_2 \\ \cdots \\ \epsilon_n \end{bmatrix}$$

(Note that we can obtain a "β_0 term" in the model in this form by taking $X_{1u} = 1$ for $u = 1, 2, \ldots, n$.) The sum of squares surface in Eq. (10.2.5) can be written as

$$S(\theta) = \sum_{u=1}^{n} \left[Y_u - \sum_{i=1}^{p} \theta_i X_{iu} \right]^2$$
$$= (Y - X\theta)'(Y - X\theta)$$
$$= Y'Y - 2\theta'X'Y + \theta'X'X\theta.$$

If we differentiate this expression with respect to $\boldsymbol{\theta}$, set the result equal to $\mathbf{0}$, and write $\hat{\boldsymbol{\theta}}$ for $\boldsymbol{\theta}$, we obtain the normal *equations*

$$\mathbf{X'X}\hat{\boldsymbol{\theta}} = \mathbf{X'Y},$$

with solution, if $\mathbf{X'X}$ is nonsingular, given by

$$\hat{\boldsymbol{\theta}} = (\mathbf{X'X})^{-1}\mathbf{X'Y}.$$

We recall that the regression sum of squares is $\hat{\boldsymbol{\theta}}'\mathbf{X'Y}$ and the residual sum of squares is $\mathbf{Y'Y} - \hat{\boldsymbol{\theta}}'\mathbf{X'Y}$. Now

$$\begin{aligned}
S(\hat{\boldsymbol{\theta}}) &= \mathbf{Y'Y} - 2\hat{\boldsymbol{\theta}}'\mathbf{X'Y} + \hat{\boldsymbol{\theta}}'\mathbf{X'X}\hat{\boldsymbol{\theta}} \\
&= \mathbf{Y'Y} - \hat{\boldsymbol{\theta}}'\mathbf{X'Y} - \hat{\boldsymbol{\theta}}'(\mathbf{X'Y} - \mathbf{X'X}\hat{\boldsymbol{\theta}}) \\
&= \mathbf{Y'Y} - \hat{\boldsymbol{\theta}}'\mathbf{X'Y}
\end{aligned}$$

since $\hat{\boldsymbol{\theta}}$ satisfies the normal equations. Thus $S(\hat{\boldsymbol{\theta}})$, the smallest value of $S(\boldsymbol{\theta})$, is equal to the residual sum of squares in the analysis of variance table. We can also write

$$\begin{aligned}
S(\boldsymbol{\theta}) - S(\hat{\boldsymbol{\theta}}) &= \boldsymbol{\theta}'\mathbf{X'X}\boldsymbol{\theta} - 2\boldsymbol{\theta}'\mathbf{X'Y} + \hat{\boldsymbol{\theta}}'\mathbf{X'X}\hat{\boldsymbol{\theta}} \\
&= (\boldsymbol{\theta} - \hat{\boldsymbol{\theta}})'\mathbf{X'X}(\boldsymbol{\theta} - \hat{\boldsymbol{\theta}}).
\end{aligned}$$

If the errors ϵ_u are independent and each follow the distribution $N(0, \sigma^2)$; that is, if $\boldsymbol{\epsilon} \sim N(\mathbf{0}, \mathbf{I}\sigma^2)$, then it can be shown that, if the model is correct, the following results are true:

(1) $\hat{\boldsymbol{\theta}} \sim N[\boldsymbol{\theta}, (\mathbf{X'X})^{-1}\sigma^2]$
(2) $S(\hat{\boldsymbol{\theta}}) \sim \sigma^2 \chi^2_{n-p}$
(3) $S(\boldsymbol{\theta}) - S(\hat{\boldsymbol{\theta}}) \sim \sigma^2 \chi_p^2$
(4) $S(\boldsymbol{\theta}) - S(\hat{\boldsymbol{\theta}})$ and $S(\hat{\boldsymbol{\theta}})$ are distributed independently so that the ratio

$$\frac{[S(\boldsymbol{\theta}) - S(\hat{\boldsymbol{\theta}})]/p}{S(\hat{\boldsymbol{\theta}})/(n-p)} \sim F(p, n-p).$$

The contours defined by $S(\boldsymbol{\theta}) = $ constant can be examined in two different but related ways. We can examine them in the *sample space* (in which the mechanism of linear least squares can be best understood) or in the *parameter space* (in which we concentrate on the contours of $S(\boldsymbol{\theta})$ alone). We shall now discuss these two representations.

The Sample Space

The sample space is an n-dimensional space. The vector of observations $\mathbf{Y} = (Y_1, Y_2, \ldots, Y_n)'$ defines a vector $\overrightarrow{OY}$ from the origin O to the point Y with coordinates $(Y_1, Y_2, \ldots, Y_n)$. The $\mathbf{X}$ matrix has p column vectors, each containing n elements. The elements of the jth column define the coordinates $(X_{j1}, X_{j2}, \ldots, X_{jn})$ of a point X_j in the sample space and the jth column vector of $\mathbf{X}$ defines the vector $\overrightarrow{OX_j}$ in the sample space. The p vectors $\overrightarrow{OX_1}, \overrightarrow{OX_2}, \ldots, \overrightarrow{OX_p}$ define a subspace of p dimensions, called the *estimation space*, which is contained within the sample space. Any point of this subspace can be represented by the end point of a vector which is a linear combination of the vectors defining the space—that is, which is a linear combination of the columns of $\mathbf{X}$, such as, for example, $\mathbf{X\theta}$ where $\mathbf{\theta} = (\theta_1, \theta_2, \ldots, \theta_p)'$ is a $p \times 1$ vector. Suppose the vector $\mathbf{X\theta}$ defines the point T. Then the squared distance YT^2 is given by

$$(\mathbf{Y} - \mathbf{X\theta})'(\mathbf{Y} - \mathbf{X\theta}) = S(\mathbf{\theta})$$

as defined earlier. Thus the sum of squares $S(\mathbf{\theta})$ represents, in the sample space, the squared distance of Y from a general point T of the estimation space. Minimization of $S(\mathbf{\theta})$ with respect to $\mathbf{\theta}$ implies finding that value of $\mathbf{\theta}$, $\hat{\mathbf{\theta}}$, say, which provides a point P (defined by the vector $\hat{\mathbf{Y}} = \mathbf{X}\hat{\mathbf{\theta}}$) of the estimation space closest to the point Y. Geometrically, then, P must be the foot of the perpendicular from Y to the estimation space, that is, the foot of a line passing through Y and orthogonal to all the columns of the $\mathbf{X}$ matrix. In terms of vectors from the origin, we can write

$$\mathbf{Y} = \hat{\mathbf{Y}} + (\mathbf{Y} - \hat{\mathbf{Y}})$$
$$= \hat{\mathbf{Y}} + \mathbf{e}$$

where $\mathbf{e}$ is the vector of *residuals*. The vector $\mathbf{Y}$ is thus divided into two orthogonal components: (1) $\hat{\mathbf{Y}}$, which lies entirely in the estimation space, and (2) $\mathbf{Y} - \hat{\mathbf{Y}} = \mathbf{e}$, the vector of residuals, which lies in what is called the *error space*. The error space is defined as the $(n - p)$-dimensional subspace which remains of the full n-dimensional space, after the p-dimensional estimation space has been defined. The estimation and error spaces are thus orthogonal. We can confirm algebraically that $\hat{\mathbf{Y}}$ and $\mathbf{e}$ are orthogonal as follows:

$$\hat{\mathbf{Y}}'\mathbf{e} = (\mathbf{X}\hat{\mathbf{\theta}})'(\mathbf{Y} - \mathbf{X}\hat{\mathbf{\theta}})$$
$$= \hat{\mathbf{\theta}}'\mathbf{X}'(\mathbf{Y} - \mathbf{X}\hat{\mathbf{\theta}})$$
$$= \hat{\mathbf{\theta}}'(\mathbf{X}'\mathbf{Y} - \mathbf{X}'\mathbf{X}\hat{\mathbf{\theta}})$$
$$= 0$$

since $\hat{\mathbf{\theta}}$ satisfies the normal equations, thus causing the parenthesis to vanish. The vector $\mathbf{e}$ is a vector $\overrightarrow{OR}$ say from the origin O, with length $OR = YP$, and with OR parallel to PY.

If T is a general point of the estimation space and YP is orthogonal to the space, then

$$YT^2 = YP^2 + PT^2$$

or

$$S(\boldsymbol{\theta}) = S(\hat{\boldsymbol{\theta}}) + PT^2.$$

Thus the contours for which $S(\boldsymbol{\theta}) = $ constant must be such that

$$PT^2 = S(\boldsymbol{\theta}) - S(\hat{\boldsymbol{\theta}}) = \text{a constant}.$$

In the sample space, then, the contours defined by $S(\boldsymbol{\theta}) = $ constant consist of all points T such that $PT^2 = $ constant, that is, points in the estimation space and of the form $\mathbf{X}\boldsymbol{\theta}$ which lie on a p-dimensional sphere centered at the point P defined by $\mathbf{X}\hat{\boldsymbol{\theta}}$. The radius of this sphere is $[S(\boldsymbol{\theta}) - S(\hat{\boldsymbol{\theta}})]^{1/2}$. By using the fact, given earlier, that

$$\frac{[S(\boldsymbol{\theta}) - S(\hat{\boldsymbol{\theta}})]/p}{S(\hat{\boldsymbol{\theta}})/(n-p)} \sim F(p, n-p),$$

we can define the boundary of a $100(1 - \alpha)\%$ confidence region for the point $\mathbf{X}\boldsymbol{\theta}$, which arises from the true (but unknown) value of $\boldsymbol{\theta}$, by

$$\frac{[S(\boldsymbol{\theta}) - S(\hat{\boldsymbol{\theta}})]/p}{S(\hat{\boldsymbol{\theta}})/(n-p)} = F(p, n-p, 1-\alpha)$$

that is, by

$$S(\boldsymbol{\theta}) = S(\hat{\boldsymbol{\theta}})\left[1 + \frac{p}{n-p} F(p, n-p, 1-\alpha)\right]$$

which is of the sensible form $S(\hat{\boldsymbol{\theta}})(1 + q^2)$ indicating values of $S(\boldsymbol{\theta})$ somewhat greater than the minimum value $S(\hat{\boldsymbol{\theta}})$. The confidence region will thus consist of the inside of a sphere in the estimation space centered at P and with radius

$$[S(\boldsymbol{\theta}) - S(\hat{\boldsymbol{\theta}})]^{1/2} = \left[S(\hat{\boldsymbol{\theta}}) \frac{p}{n-p} F(p, n-p, 1-\alpha)\right]^{1/2}.$$

The Sample Space When $n = 3$, $p = 2$

In order to illustrate the foregoing remarks by a diagram, we shall suppose that $n = 3$. When $n > 3$ the complete situation cannot be drawn but the mental extension to higher dimensions is not difficult.

Figure 10.4 shows the sample space when $n = 3$, the coordinate axes being labeled 1, 2, and 3 to correspond to the three components (Y_1, Y_2, Y_3) of the vector $\mathbf{Y}'$. We shall suppose that there are $p = 2$ parameters θ_1 and

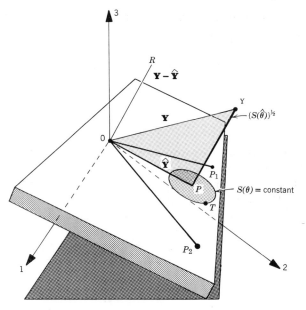

Figure 10.4   The sample space when $n = 3, p = 2$.

θ_2 so that $\mathbf{X}$ is a three by two matrix of form

$$\mathbf{X} = \begin{bmatrix} X_{11} & X_{21} \\ X_{12} & X_{22} \\ X_{13} & X_{23} \end{bmatrix}.$$

The columns of $\mathbf{X}$ define two points P_1 and P_2 with coordinates $(X_{11},$ $X_{12}, X_{13})$ and (X_{21}, X_{22}, X_{23}), respectively, and the vectors $\overrightarrow{OP_1}$ and $\overrightarrow{OP_2}$ define a plane which represents the two-dimensional estimation space in which the vector $\hat{\mathbf{Y}} = \mathbf{X}\hat{\boldsymbol{\theta}}$ must lie. The point Y lies above this plane and the perpendicular YP from Y to the plane OP_1P_2 hits the plane at P. Thus YP is the shortest distance from Y to any point in the estimation space, P is defined by $\hat{\mathbf{Y}} = \mathbf{X}\hat{\boldsymbol{\theta}}$ and $S(\hat{\boldsymbol{\theta}}) = YP^2$. In addition, since $OY^2 = \mathbf{Y}'\mathbf{Y}$, the standard analysis of variance breakup

$$\mathbf{Y}'\mathbf{Y} = \hat{\boldsymbol{\theta}}'\mathbf{X}'\mathbf{Y} + (\mathbf{Y}'\mathbf{Y} - \hat{\boldsymbol{\theta}}'\mathbf{X}'\mathbf{Y})$$

or

$$\mathbf{Y}'\mathbf{Y} = \hat{\boldsymbol{\theta}}'\mathbf{X}'\mathbf{Y} + S(\hat{\boldsymbol{\theta}})$$

is equivalent to the Pythagoras result:

$$OY^2 = OP^2 + YP^2.$$

If we draw a line OR through O, equal in length (so that $OR^2 = S(\hat{\theta})$) and parallel to PY, then $\overrightarrow{OR}$ represents the vector of residuals $\mathbf{e} = \mathbf{Y} - \hat{\mathbf{Y}}$. The vector $\overrightarrow{OP}$ is $\hat{\mathbf{Y}}$ so we have the vector equation

$$\overrightarrow{OY} = \overrightarrow{OP} + \overrightarrow{OR}$$

or

$$\mathbf{Y} = \hat{\mathbf{Y}} + (\mathbf{Y} - \hat{\mathbf{Y}}).$$

We recall that, in general, contours of constant $S(\theta)$ are represented by p-dimensional spheres in the estimation space. Here, then, the contours must be circles on the plane OP_1P_2. This is easy to see, for if T is a general point $\mathbf{X}\theta$ on the plane, $S(\theta) = $ constant means $YT^2 = $ constant, so that $PT^2 = YT^2 - YP^2 = $ constant. We thus obtain circles about P. One such circle is shown on the figure. The circle which provides a $100(1 - \alpha)\%$ confidence interval for the true point $\mathbf{X}\theta$ has radius given by

$$[2S(\hat{\theta})F(2, 1, 1 - \alpha)]^{1/2},$$

obtained by putting $n = 3$, $p = 2$ in the general formula.

The Sample Space Geometry When the Model is Wrong

Suppose $\mathbf{Y} = \mathbf{X}\theta + \boldsymbol{\epsilon}$ is the postulated linear model containing p parameters but that the true linear model is

$$\mathbf{Y} = \mathbf{X}\theta + \mathbf{X}_2\theta_2 + \boldsymbol{\epsilon}$$

and contains additional terms $\mathbf{X}_2\theta_2$ not considered. Then since the estimation space consists only of points of the form $\mathbf{X}\theta$, the true point $\eta = \mathbf{X}\theta + \mathbf{X}_2\theta_2$ cannot lie in the estimation space. In this case the perpendicular YP from Y onto the estimation space (whose foot P is given by $\hat{\mathbf{Y}} = \mathbf{X}\hat{\theta}$) will be longer than it would have been if the correct model and estimation space had been used. To illustrate this point we shall give, in Figure 10.5, a diagram for the case $n = 3$, $p = 1$ where the true model contains two parameters θ and θ_2. The true model takes the form

$$\begin{bmatrix} Y_1 \\ Y_2 \\ Y_3 \end{bmatrix} = \begin{bmatrix} X_{11} \\ X_{12} \\ X_{13} \end{bmatrix} \theta + \begin{bmatrix} X_{21} \\ X_{22} \\ X_{23} \end{bmatrix} \theta_2 + \begin{bmatrix} \epsilon_1 \\ \epsilon_2 \\ \epsilon_3 \end{bmatrix}$$

$$= \mathbf{X}\theta \quad + \mathbf{X}_2\theta_2 \quad + \boldsymbol{\epsilon}$$

and the postulated model is given when $\theta_2 = 0$. The single column of the $\mathbf{X}$ matrix defines the point P_1 and the line OP_1 which is the estimation space for the postulated model. The line YP is the perpendicular from Y

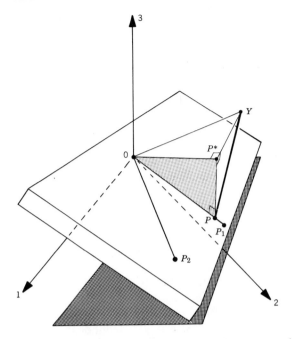

Figure 10.5   The sample space for $n = 3$, $p = 1$; wrong model.

onto OP_1 and P is the point $\hat{\mathbf{Y}} = \mathbf{X}\hat{\theta}$. Therefore, the shortest squared distance $S(\hat{\theta})$, of all the squared distances $S(\theta)$ from the point Y to points $\mathbf{X}\theta$ on the line OP_1, is represented by the square of the length of YP. The true value of θ defines an unknown point $\mathbf{X}\theta$ on the line OP_1. A confidence interval for the true value $\mathbf{X}\theta$ can be constructed on OP_1 and around the point P.

Now the second vector $\mathbf{X}_2$ in the true model defines a line OP_2 and the lines OP_1 and OP_2 define a plane in which the true point $\mathbf{X}\theta + \mathbf{X}_2\theta_2$ lies. Suppose YP^* is the perpendicular from Y to the *correct* estimation space given by the plane OP_1P_2. Then P^* represents the point that would have given the correct fitted value $\hat{\mathbf{Y}}^*$, say, *if* the correct model had been used. This is always of length less than or equal to YP since a perpendicular to a space (a plane here, OP_1P_2) cannot be longer than the perpendicular to an included space (here the line OP_1). If the model is incorrect, therefore, $S(\hat{\theta}) = YP^2$ will be too long, if anything. (Note that it *could* happen that P and P^* coincide, so that the same minimum value $S(\hat{\theta})$ would occur, whichever model was used. This would be very unusual, of course.)

When the postulated model is correct, in the general case, $S(\hat{\theta})$ has expected or mean value $(n - p)\sigma^2$. If a pure error or prior estimate of σ^2 is available we know how big, roughly, the quantity YP^2 should be.

However, if the postulated model is inadequate, YP^2 will probably be too long. The standard lack of fit test given in Eq. (2.6.12) is thus examining the question, "Is the squared length YP^2 greater than we should expect on the basis of the good information we have about the size of the random error?" How much greater is *too great* is determined through the distribution properties involved, as formalized earlier.

The Parameter Space

The parameter space is a p-dimensional space in which a set of values $(\theta_1, \theta_2, \ldots, \theta_p)$ of the parameters defines a point. The minimum value of $S(\boldsymbol{\theta})$ is attained at the point $\hat{\boldsymbol{\theta}} = (\hat{\theta}_1, \hat{\theta}_2, \ldots, \hat{\theta}_p)$. We recall that

$$S(\boldsymbol{\theta}) - S(\hat{\boldsymbol{\theta}}) = (\boldsymbol{\theta} - \hat{\boldsymbol{\theta}})'\mathbf{X}'\mathbf{X}(\boldsymbol{\theta} - \hat{\boldsymbol{\theta}}).$$

All values of $\boldsymbol{\theta}$ which satisfy $S(\boldsymbol{\theta}) = \text{constant} = K$ are given by

$$(\boldsymbol{\theta} - \hat{\boldsymbol{\theta}})'\mathbf{X}'\mathbf{X}(\boldsymbol{\theta} - \hat{\boldsymbol{\theta}}) = K - S(\hat{\boldsymbol{\theta}})$$

and it can be shown that this is the equation of a closed ellipsoidal contour surrounding the point $\hat{\boldsymbol{\theta}}$. When $K_1 > K_2$ the contour $S(\boldsymbol{\theta}) = K_1$ completely encloses the contour $S(\boldsymbol{\theta}) = K_2$ and $\hat{\boldsymbol{\theta}}$ lies in the center of these nested p-dimensional "eggs." A $100(1 - \alpha)\%$ confidence region for the true (but unknown) value of $\boldsymbol{\theta}$ is enclosed by the contour which is such that

$$\frac{[S(\boldsymbol{\theta}) - S(\hat{\boldsymbol{\theta}})]/p}{S(\hat{\boldsymbol{\theta}})/(n - p)} = F(p, n - p, 1 - \alpha)$$

if the errors are normally distributed, i.e., $\boldsymbol{\epsilon} \sim N(\mathbf{0}, \mathbf{I}\sigma^2)$. This can be rearranged as

$$S(\boldsymbol{\theta}) = S(\hat{\boldsymbol{\theta}})\left\{1 + \frac{p}{n - p} F(p, n - p, 1 - \alpha)\right\}$$

in which the expression on the right-hand side is the constant value that defines the contour.

The Parameter Space When p = 2

We again use a simple case to illustrate the situation. Figure 10.6 shows some possible contours of the form $S(\boldsymbol{\theta}) = \text{constant}$ for three values of the constant when $p = 2$. The outer contour is labeled as a $100(1 - \alpha)\%$ confidence contour, defined as above. In the two-dimensional space of (θ_1, θ_2), the contours are concentric ellipses about the point $(\hat{\theta}_1, \hat{\theta}_2)$. Note that contours of this type are obtained no matter what the value of n (the number of observations) may be, since the dimension of the parameter space depends on p alone.

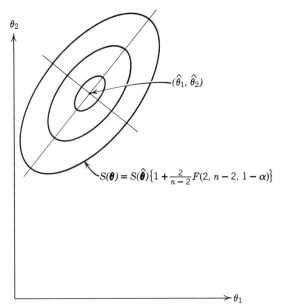

Figure 10.6 Contours of $S(\boldsymbol{\theta})$ in the parameter space when $p = 2$.

In general, the orientation and the shape of the ellipses are both of importance. If the axes of the ellipses are parallel to the θ_1 and θ_2 axes, then the value $\hat{\theta}_1$ which makes $S(\theta_1, \theta_2)$ a minimum has no dependence on θ_2; that is, if we fix θ_2 at any value the same value of $\theta_1 = \hat{\theta}_1$ minimizes $S(\theta_1, \theta_2 \mid \theta_2$ fixed$)$. This means that specific information about θ_2 which fixed its value would not alter the least squares estimate $\hat{\theta}_1$. This situation occurs when the expression for $S(\theta_1, \theta_2)$ can be written without a cross product term in $\theta_1\theta_2$. The model when $p = 2$ can be written as

$$Y_u = \theta_1 X_{1u} + \theta_2 X_{2u} + \epsilon_u, \qquad u = 1, 2, \ldots, n.$$

Thus

$$S(\boldsymbol{\theta}) = S(\theta_1, \theta_2) = \sum_{u=1}^{n} (Y_u - \theta_1 X_{1u} - \theta_2 X_{2u})^2$$

$$= \sum Y_u^2 - \theta_1 2 \sum X_{1u} Y_u - \theta_2 2 \sum X_{2u} Y_u$$
$$+ \theta_1^2 \sum X_{1u}^2 + \theta_2^2 \sum X_{2u}^2 + \theta_1 \theta_2 2 \sum X_{1u} X_{2u}$$

where all summations are over $u = 1, 2, \ldots, n$. It is clear from this that the minimizing value of θ_1, namely $\hat{\theta}_1$, which satisfies $\partial S(\boldsymbol{\theta})/\partial \theta_1 = 0$ will not depend on θ_2 (and vice versa) if the coefficient of $\theta_1 \theta_2$ vanishes; that is, if $\sum X_{1u} X_{2u} = 0$, or if the columns of the $\mathbf{X}$ matrix are orthogonal.

When the X_1 and X_2 columns of the $\mathbf{X}$ matrix are not orthogonal, a $\theta_1 \theta_2$ term occurs in $S(\theta_1, \theta_2)$, and the ellipses are obliquely oriented with respect to the θ_1 and θ_2 axes.

Figure 10.7   The interpretation of some possible 95% confidence regions for parameters (θ_1, θ_2).

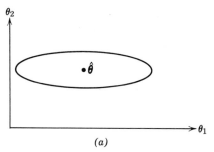

(a)

Figure 10.7(a) $\hat{\theta}_1$ not well determined; $\hat{\theta}_2$ well determined; no dependence between $\hat{\theta}_1$ and $\hat{\theta}_2$.

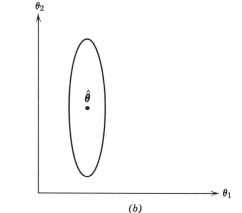

(b)

Figure 10.7(b) $\hat{\theta}_1$ well determined; $\hat{\theta}_2$ not well determined; no dependence between $\hat{\theta}_1$ and $\hat{\theta}_2$.

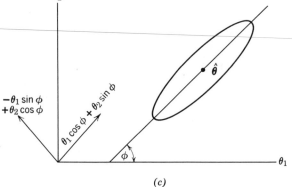

(c)

Figure 10.7(c) $\hat{\theta}_1 \cos \phi + \hat{\theta}_2 \sin \phi$ not well determined; $-\hat{\theta}_1 \sin \phi + \hat{\theta}_2 \cos \phi$ well determined; dependence between $\hat{\theta}_1$ and $\hat{\theta}_2$.

The shape of the $S(\theta_1, \theta_2)$ contours shows the relative precisions with which the estimates $\hat{\theta}_1$ and $\hat{\theta}_2$ are determined. Figure 10.7 illustrates some of the possibilities. The single contour shown is intended to represent the 95% confidence region boundary, and the point $\hat{\theta}$ with coordinates $(\hat{\theta}_1, \hat{\theta}_2)$ is the least squares estimate of θ, in each case.

10.7. The Geometry of Nonlinear Least Squares

The Sample Space

When the model is nonlinear rather than linear there is no **X** matrix in the linear model sense. While there is still an estimation space it is not one defined by a set of vectors and may be very complex. The estimation space (also called the *solution locus*) consists of all points with coordinates expressible as

$$\{f(\xi_1, \theta), f(\xi_2, \theta), \ldots, f(\xi_n, \theta)\}.$$

Since the sum of squares function $S(\theta)$ still represents the square of the distance from the point $(Y_1, Y_2, \ldots, Y_n)$ to a point of the estimation space, minimization of $S(\theta)$ still corresponds geometrically to finding a point P of the estimation space which is nearest to Y. The sample space for a very simple nonlinear example involving only $n = 2$ observations Y_1 and Y_2 taken at $\xi = \xi_1$ and $\xi = \xi_2$, respectively, and a single parameter θ, is shown in Figure 10.8. The estimation space consists of the curved line which contains points

$$\{f(\xi_1, \theta), f(\xi_2, \theta)\}$$

as θ varies, where ξ_1, ξ_2 are fixed. Y has coordinates (Y_1, Y_2), and P is the point of the estimation space nearest to Y.

Figure 10.9 shows the sample space for an example involving $n = 3$ observations Y_1, Y_2, and Y_3 taken at $\xi = \xi_1, \xi_2$, and ξ_3, respectively, and two parameters θ_1 and θ_2. The curved lines indicate the coordinate systems of the parameters on the estimation space or solution locus, which consists of all points of the form

$$\{f(\xi_1, \theta_1, \theta_2), f(\xi_2, \theta_1, \theta_2), f(\xi_3, \theta_1, \theta_2)\}$$

as θ_1 and θ_2 vary, where ξ_1, ξ_2, and ξ_3 are fixed. Y has coordinates (Y_1, Y_2, Y_3) and P is the point of the estimation space nearest to Y. When we apply the linearization technique to nonlinear problems we are selecting a point of the estimation space, θ_0 say, as new origin, defining a linearized estimation space in the form of the tangent space at θ_0 and solving the linearized least-squares problem so defined. The solution to this (which will be given

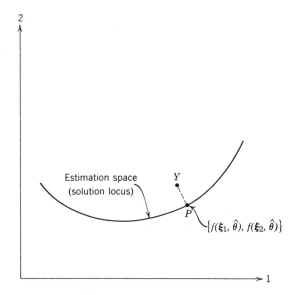

Figure 10.8   The sample space when $n = 2$, $f(\xi, \theta)$ nonlinear.

in units of rate of change of $\boldsymbol{\theta}$ which are appropriate at $\boldsymbol{\theta}_0$ only) are applied to the nonlinear problem, where they may not be correct, and another iteration is attempted. For a nonlinear problem involving only two observations and one parameter the effect will be as shown in Figure 10.10. Figure 10.10 shows the estimation space or solution locus with units of θ shown upon it. We assume here that $\theta_0 = 0$ and that the point marked $\theta = 1$ is the point of the estimation space attained where $\theta = 1$, and so on. Note that the markings for θ are *not* equally spaced due to the nonlinearity and the nonuniformity of the coordinate system. The line tangent to the estimation space curve at $\theta = \theta_0 = 0$ is shown, graduated with units $\theta = 0, 1, 2, \ldots$ which are obtained from the rate of change found at θ_0. These units are equally spaced. We now find the least squares estimate of θ based on the linear assumption.

Geometrically, this means finding the point Q_0 so that YQ_0 is perpendicular to the tangent line. We see that, in the linearized units, a value of θ of about 3.2 (at Q_0) is indicated. In the next iteration of the linearization procedure we thus use the tangent line at the point where $\theta = 3.2$ on the *estimation space curve*, that is, at the point Q.

It is easy to see from this one reason why the linearization procedure sometimes fails. If the rate of change of $f(\xi, \boldsymbol{\theta})$ is small at $\boldsymbol{\theta}_0$, but increases rapidly, the units on the tangent line may be quite unrealistic. For example, in Figure 10.11, the rate of change at $\theta_0 = 0$ is small and so the linearized

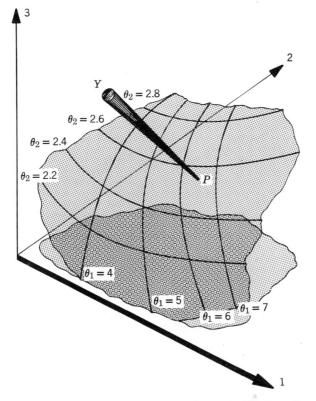

Figure 10.9   The sample space when $n = 3$, $p = 2$, $f(\xi, \theta)$ nonlinear.

units of θ are small. The actual units increase sharply, however. Thus, if we begin a further iteration using the indicated value of θ of about 26 at Q_0, our starting point on the estimation space will be farther from the best point P than was our original guess $\theta = \theta_0 = 0$. The situation may or may

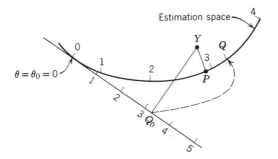

Figure 10.10   Geometrical interpretation of linearization method ($n = 2$, $p = 1$).

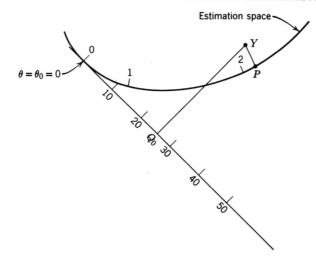

Figure 10.11   The effect on the linearization method of gross inequities in the systems of units ($n = 2, p = 1$).

not be corrected in successive iterations. (Although we have used $\theta_0 = 0$ and units 1, 2, . . . , for simplicity, similar remarks apply in general whatever the value of the initial guess θ_0 and whatever the system of units near θ_0 may be.)

When there are more observations than two and more than one parameter the same ideas hold but the situation is more complicated and is difficult or impossible to draw.

When the model is linear, contours of constant $S(\theta)$ in the sample space consist of spheres. In nonlinear problems this is no longer true and quite irregular contours may arise consisting of all points of the estimation space equidistant, a selected distance, from the point Y: $(Y_1, Y_2, . . . , Y_n)$.

The Parameter Space

In the linear model case, contours of constant $S(\theta)$ in the parameter space, or θ-space, consist of concentric ellipses. When the model is nonlinear the contours are sometimes banana-shaped, often elongated. Sometimes the contours stretch to infinity and do not even close, or they may have multiple loops surrounding a number of stationary values. When several stationary values exist they may have various levels or provide alternative minima for $S(\theta)$. Consider, for example, the model

$$Y = \frac{(\theta_1 e^{-\theta_2 t} - \theta_2 e^{-\theta_1 t})}{(\theta_2 - \theta_1)} + \epsilon.$$

Interchange of θ_1 and θ_2 leaves the model unaltered. Thus if the minimum $S(\boldsymbol{\theta})$ is attained at $(\theta_1, \theta_2) = (\hat{\theta}_1, \hat{\theta}_2)$, the same minimum value is given at $(\theta_1, \theta_2) = (\hat{\theta}_2, \hat{\theta}_1)$, so that a double solution exists. Multiple solutions are not easy to spot in this way in general. An example of banana-shaped contours is given in Section 10.4.

Confidence Contours in the Nonlinear Case

When the model is nonlinear a number of results which are true for the linear case no longer apply. When the error ϵ of the nonlinear model (10.2.1) is assumed to be normally distributed, $\hat{\boldsymbol{\theta}}$ is no longer normally distributed, $s^2 = S(\hat{\boldsymbol{\theta}})/(n - p)$ is no longer an unbiased estimate of σ^2, and there is no variance-covariance matrix of form $(\mathbf{X'X})^{-1}\sigma^2$ in general.

Although confidence regions can still be *defined* by the expression

$$S(\boldsymbol{\theta}) = S(\hat{\boldsymbol{\theta}})\left\{1 + \frac{p}{n - p} F(p, n - p, 1 - \alpha)\right\}$$

which provides a $100(1 - \alpha)\%$ confidence region in the linear model, normal error situation, the confidence coefficient will not be $1 - \alpha$, in the nonlinear case. We do not know in general what the confidence will be but we can call such regions *approximate* $100(1 - \alpha)\%$ *confidence regions* for $\boldsymbol{\theta}$. The banana-shaped regions for the example in Section 10.4 were obtained in this manner. While suitable comparisons of mean squares can still be made visually, the usual F-tests for regression and lack of fit are not valid, in general, in the nonlinear case. Measures of nonlinearity suggested by E. M. L. Beale could be used to help decide when linearized results provide acceptable approximations, however. For a discussion see "On Beale's measures of nonlinearity" by I. Guttman and D. A. Meeter, *Technometrics*, **7**, 1965, 623–637.

EXERCISES

A. Estimate the parameter θ in the nonlinear model

$$Y = e^{-\theta t} + \epsilon$$

from the following observations:

t	Y
1	0.80
4	0.45
16	0.04

Construct an approximate 95% confidence interval for θ.

B. Estimate the parameter θ in the nonlinear model

$$Y = e^{-\theta t} + \epsilon$$

from the following observations:

t	Y
0.5	0.96, 0.91
1	0.86, 0.79
2	0.63, 0.62
4	0.48, 0.42
8	0.17, 0.21
16	0.03, 0.05

Construct an approximate 95% confidence interval for θ.

C. Estimate the parameters α, β in the nonlinear model

$$Y = \alpha + (0.49 - \alpha)e^{-\beta(X-8)} + \epsilon$$

from the following observations

X	Y
10	0.48
20	0.42
30	0.40
40	0.39

Construct an approximate 95% confidence region for (α, β).

D. The relationship between the yield of a crop, Y, and the amounts of fertilizer, X, applied to that crop has been formulated as $Y = \alpha - \beta \rho^X + \epsilon$, where $0 < \rho < 1$. Given:

X	Y
0	44.4
1	54.6
2	63.8
3	65.7
4	68.9

obtain estimates of α, β, and ρ. Also construct an approximate 95% confidence region for (α, β, ρ).

E. The relationship between pressure and temperature in saturated steam can be written as

$$Y = \alpha(10)^{\beta t/(\gamma+t)} + \epsilon$$

where Y = pressure
       t = temperature
       α, β, γ = unknown constants.

The following data were collected:

$t(°C)$	Y(Pressure)
0	4.14
10	8.52
20	16.31
30	32.18
40	64.62
50	98.76
60	151.13
70	224.74
80	341.35
85	423.36
90	522.78
95	674.32
100	782.04
105	920.01

Obtain estimates of α, β, and γ. Also construct an approximate 95% confidence region for (α, β, γ).

F. Consider the model

$$Y = \theta + \alpha X_1 X_3 + \beta X_2 X_3 + \alpha\gamma X_1 + \beta\gamma X_2 + \epsilon.$$

Is this model nonlinear? How can estimates of the parameters be obtained by using only linear regression methods, if data $(Y_u, X_{1u}, X_{2u}, X_{3u})$ are available?

Nonlinear Bibliography

Note: Publications marked with an asterisk comprise a suggested short-list for initial reading or investigation.

Barnes, J. G. P. "An algorithm for solving non-linear equations based on the secant method," *The Computer Journal*, **8**, 66–72 (1965).

Beale, E. M. L. "Confidence regions in non-linear estimation," *J. Roy. Statis. Soc.* **B-22**, 41–76 (1960).

Beale, E. M. L. "The computation of measures of non-linearity in nonlinear estimation," Report N1/M2-64, Admiralty Research Laboratory, Teddington, England (1960).

Behnken, D. W. "Estimation of copolymer reactivity ratios: an example of non-linear estimation," *J. Polymer Sci.*, **A-2**, 645–668 (1964).

Blakemore, J. W., and A. E. Hoerl. "Fitting non-linear reaction rate equations to data," *Chem. Eng. Progr. Symp. Ser.*, **59**, 42 14–27 (1963).

*Booth, G. W., and T. L. Peterson. "Non-linear estimation," I.B.M. Share Program Pa., no. 687 WLNL1 (1958).

Box, G. E. P. "Some notes on non-linear estimation," unpublished, reissued as Technical, Report no. 25, Department of Statistics, University of Wisconsin, Madison Wisconsin (1964) (originally 1956).

Box, G. E. P. "Use of statistical methods in the elucidation of basic mechanism," *Bull. Inst. Intern. Statist.*, **36**, 215–225 (1958).

Box, G. E. P. "Discussion on paper by Satterthwaite and Budne on random balance," *Technometrics*, **1**, 174–180 (1959).

Box, G. E. P. "Fitting empirical data," *Annals N.Y. Acad. Sci.*, **86**, 792–816 (1960).

Box, G. E. P. "The effects of errors in the factor levels and experimental design," *Bull. Intern. Statist. Inst.*, **38**, 339–355 (1961). Reprinted in *Technometrics*, **5**, 247–262 (1963).

*Box, G. E. P., and G. A. Coutie. "Application of digital computers in the exploration of functional relationships," *Proc. Inst. Elec. Engs.*, **103**, Part B, Supplement no. 1, 100–107 (1956).

Box, G. E. P., and D. R. Cox. "An analysis of transformations," *J. Roy. Statist. Soc.*, **B-26**, 211–243, discussion 244–252 (1964).

*Box, G. E. P., and N. R. Draper. "The Bayesian estimation of common parameters from several responses," *Biometrika*, **52**, 355–365 (1965).

*Box, G. E. P., and H. L. Lucas. "Design of experiments in non-linear situations," *Biometrika*, **46**, 77–90 (1959).

*Box, G. E. P., and W. G. Hunter. "A useful method for model building," *Technometrics*, **4**, 301–318 (1962).

*Box, G. E. P., and W. G. Hunter. "The experimental study of physical mechanisms," *Technometrics*, **7**, 23–42 (1965).

*Box, G. E. P., and P. W. Tidwell. "Transformation of the independent variables," *Technometrics*, **4**, 531–550 (1962).

Box, G. E. P., and K. B. Wilson. "On the experimental attainment of optimum conditions," *J. Roy. Statist. Soc.*, **B-13**, 1–45 (1951).

*Box, G. E. P., and P. V. Youle. "The exploration and exploitation of response surfaces: an example of the link between the fitted surface and the basic mechanism of the system," *Biometrics*, **11**, 287–323 (1955).

Box, M. J. "A new method of constrained optimization and a comparison with other methods," *Computer J.*, **8**, 42–52 (1965).

Carroll, C. W. "The created response surface technique for optimizing non-linear restrained systems," *Operations Res.*, **9**, 169–185 (1961).

Cornell, R. G. "A method for fitting linear combinations of exponentials," *Biometrics*, **18**, 104–113 (1962).

Cornell, R. G. "Spearman estimation for a simple exponential model," *Biometrics*, **21**, 858–864 (1965).

Cornell, R. G., and J. A. Speckman. "Estimation for a one parameter exponential model," *J. Am. Statist. Assoc.*, **60**, 560–572 (1965).

Crockett, J. B., and H. Chernoff. "Gradient methods of maximization," *Pacific J. Math.*, **5**, 33–59 (1955).

Curry, H. B. "The method of steepest descent for non-linear minimization problems," *Quart. Appl. Math.*, **2**, 258–261 (1944).

Davidon, W. C. "Variable metric method for minimization," (British) Atomic Energy Commission Research and Development Report ANL-5990 (Rev.), 1959.

Ferguson, R. C., and D. W. Marquardt. "Computer analysis of NMR spectra: magnetic equivalence factoring," *J. Chem. Phys.*, **41**, 2087–2095 (1964).

Fisher, R. A. *The Design of Experiments*, Oliver and Boyd, Edinburgh, 1935.

Fisher, R. A. *Statistical Methods and Scientific Inference*, Oliver and Boyd, Edinburgh, 1956.

Fletcher, R., and M. J. D. Powell. "A rapidly convergent descent method for minimization," *Computer J.*, **6**, 163–168 (1963).

Fletcher, R., and C. M. Reeves. "Function minimization by conjugate gradients," *Computer J.*, **7**, 2, 149–153 (1964).

Franklin, N. L., P. H. Pinchbeck, and F. Popper. "A statistical approach to catalyst development. Part I: the effect of process variables on the vapor phase oxidation of naphthalene," *Trans. Inst. Chem. Engrs.*, **34**, 280–293 (1956).

Gauss, C. F. "Theory of least squares," English translation by Hale F. Trotter, Princeton University, Statistical Techniques Research Group, Technical Report no. 5 (1957) (original 1821).

Goldfeld, S. M., R. E. Quandt, and H. F. Trotter. "Maximization by quadratic hill-climbing," Princeton University Econometric Research Program Memorandum no. 72, January 19, 1965.

*Guttman, I., and D. A. Meeter. "Use of transformations on parameters in non-linear theory, I. Transformations to accelerate convergence in non-linear least squares," Technical Report no. 37, Department of Statistics, University of Wisconsin, Madison, Wisconsin (1964).

*Guttman, I., and D. A. Meeter. "On Beale's measures of nonlinearity," *Technometrics*, **7**, 623–637 (1965).

Halperin, M. "Confidence interval estimation in non-linear regression," Program 360, Applied Mathematics Department SRRC-RR-62-28, Sperry-Rand Research Center, 1962.

Hartley, H. O. "The estimation of non-linear parameters by 'internal least squares'," *Biometrika*, **35**, 32–45 (1948).

Hartley, H. O. "The modified Gauss-Newton method for the fitting of non-linear regression functions by least squares," *Technometrics*, **3**, 269–280 (1961).

Hartley, H. O. "Exact confidence regions for the parameters in non-linear regression laws," *Biometrika*, **51**, 347–353 (1964).

Hartley, H. O., and A. Booker. "Non-linear least squares estimation," *Annals Math. Statist.*, **36**, 638–650 (1965).

Hunter, W. G. "Generation and analysis of data in non-linear situations," Ph. D. Thesis, University of Wisconsin, Madison, Wisconsin (1963).

Hunter, W. G., and R. Mezaki. "A model-building technique for chemical engineering kinetics," *Am. Inst. Chem. Engrs. J.*, **10**, 3, 315–322 (1964).

Hunter, W. G., and A. M. Reiner. "Designs for discriminating between two rival models," Technical Report no. 32, University of Wisconsin, Department of Statistics, Madison, Wisconsin (1964). *Technometrics*, **7**, 307–323 (1965).

Levenberg, K. "A method for the solution of certain non-linear problems in least squares," *Quart. Appl. Math.*, **2**, 164–168 (1944).

Lipton, S., and C. A. McGilchrist. "The derivation of methods for fitting exponential curves," *Biometrika*, **51**, 504–508 (1964).

*Marquardt, D. W. "An algorithm for least squares estimation of non-linear parameters," *J. Soc. Ind. Appl. Math.*, **2**, 431–441 (1963).

*Marquardt, D. W. "Least squares estimation of non-linear parameters," A computer program in FORTRAN IV language; I.B.M. Share Library, Distribution no. 309401 (August, 1966) (Successor to Distribution no. 1428 and 3094).

Marquardt, D. W. "Solution of non-linear chemical engineering models," *Chem. Eng. Progr.*, **55**, 65–70 (1959).

Marquardt, D. W., R. G. Bennett, and E. J. Burrell. "Least-squares analysis of electron paramagnetic resonance spectra," *J. Mol. Spectr.*, **7**, 269–279 (1961).

*Marquardt, D. W., and R. M. Stanley. "NLIN 2—Least squares estimation of non-linear parameters." Supplement to SDA 3094 (NLIN), mimeo manuscript, 11/20/64.

Meeter, D. A. "Program GAUSHAUS," Numerical Analysis Laboratory, University of Wisconsin, Madison, Wisconsin, 1964 (Revised, 1966).

Meeter, D. A. "Problems in the analysis of non-linear models by least squares," Ph. D. Thesis, University of Wisconsin, Madison, Wisconsin, 1964 (see also Guttman and Meeter).

Moore, R. H. "On the least squares estimation of parameters in non-linear models," Ph. D. Thesis, Oklahoma State University, Stillwater, Oklahoma, 1962.

Morrison, D. D. "Methods for non-linear least squares problems and convergence proofs," *Proc. Jet Propulsion Lab. Sem.*; Tracking Problems and Orbit Determination, 1–9 (1960), (Available from Space Technology Laboratory, Redondo Beach, California).

Nelder, J. A., and R. Mead. "A simplex method for function minimization," *Computer J.*, **7**, 308–313 (1965).

Patterson, H. D. "The use of autoregression in fitting an exponential curve," *Biometrika*, **45**, 389–400 (1958).

Patterson, H. D., and S. Lipton. "An investigation of Hartley's method for fitting an exponential curve," *Biometrika*, **46**, 281–292 (1959).

Powell, M. J. D. "An efficient method for finding the minimum of a function of several variables without calculating derivatives," *Computer J.*, **7**, 2, 155–162 (1964).

Powell, M. J. D. "A method for minimizing a sum of squares of non-linear functions without calculating derivatives," *Computer J.*, **7**, 303–307 (1965).

Rosen, J. B. "The gradient projection method for non-linear programming. Part I: Linear constraints," *J. Soc. Ind. Appl. Math.*, **8**, 181–217 (1960).

Rosen, J. B. "The gradient projection method for non-linear programming. Part II: Nonlinear constraints," *J. Soc. Ind. Appl. Math.*, **9**, 514–532 (1961).

Rosenbrock, H. H. "An automatic method for finding the greatest or least value of a function," *Computer J.*, **3**, 175–184 (1960).

Rubin, D. I. "Non-linear least squares parameter estimation and its application to chemical kinetics," *Chem. Eng. Progr. Symp. Ser.*, **59**, 42, 90–94 (1963).

Saaty, T. L., and J. Bram. *Nonlinear Mathematics*, McGraw-Hill Book Co., New York, 1964.

Shah, B. V., R. J. Buehler, and O. Kempthorne. "Some algorithms for minimizing a function of several variables," *J. Soc. Ind. Appl. Math.*, **12**, 74–92 (1964).

Smith, H., and S. D. Dubey. "Some reliability problems in the chemical industry," *IQC*, **22**, 2, 64–70 (1964).

Smith, N. H. "Transient operation of continuous, stirred tank reactors," Ph. D. Thesis, University of Wisconsin, Madison, Wisconsin, 1963.

*Spang, H. A. "A review of minimization techniques for non-linear functions," *Soc. Ind. Appl. Math. Rev.*, **4**, 343–365 (1962).

Spendley, W., G. R. Hext, and F. R. Himsworth. "Sequential application of simplex designs in optimization and evolutionary operation," *Technometrics*, **4**, 441–461 (1962).

Turner, M. E. "The single process law: a study in non-linear regression," Dissertation no. 59-6571. University Microfilms, Ann Arbor, Michigan.

Turner, M. E., and E. D. Campbell. "A biometric theory of middle and long distance track records," *Biometrics*, **17**, 275–282 (1961).

Turner, M. E., R. J. Monroe, and H. L. Lucas. "Generalized asymptotic regression and non-linear path analysis," *Biometrics*, **17**, 120–143 (1961).

Wilk, M. B. "An identity of use in non-linear least squares," Abstract no. 24, *Annals Math. Statist.*, **29**, 618 (1958).

Williams, E. J. "Exact fiducial distributions in non-linear estimation," *J. Roy. Statist. Soc.*, **B-24**, 125–139 (1962).

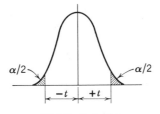

Distribution of *t*

"Probability = Area in Two Tails of Distribution Outside ±*t*-Value in Table "

Degrees of Freedom	Probability									
	0.9	0.7	0.5	0.3	0.2	0.1	0.05	0.02	0.01	0.001
1	0.158	0.510	1.000	1.963	3.078	6.314	12.706	31.821	63.657	636.619
2	0.142	0.445	0.816	1.386	1.886	2.920	4.303	6.965	9.925	31.598
3	0.137	0.424	0.765	1.250	1.638	2.353	3.182	4.541	5.841	12.924
4	0.134	0.414	0.741	1.190	1.533	2.132	2.776	3.747	4.604	8.610
5	0.132	0.408	0.727	1.156	1.476	2.015	2.571	3.365	4.032	6.869
6	0.131	0.404	0.718	1.134	1.440	1.943	2.447	3.143	3.707	5.959
7	0.130	0.402	0.711	1.119	1.415	1.895	2.365	2.998	3.499	5.408
8	0.130	0.399	0.706	1.108	1.397	1.860	2.306	2.896	3.355	5.041
9	0.129	0.398	0.703	1.100	1.383	1.833	2.262	2.821	3.250	4.781
10	0.129	0.397	0.700	1.093	1.372	1.812	2.228	2.764	3.169	4.587
11	0.129	0.396	0.697	1.088	1.363	1.796	2.201	2.718	3.106	4.437
12	0.128	0.395	0.695	1.083	1.356	1.782	2.179	2.681	3.055	4.318
13	0.128	0.394	0.694	1.079	1.350	1.771	2.160	2.650	3.012	4.221
14	0.128	0.393	0.692	1.076	1.345	1.761	2.145	2.624	2.977	3.140
15	0.128	0.393	0.691	1.074	1.341	1.753	2.131	2.602	2.947	4.073
16	0.128	0.392	0.690	1.071	1.337	1.746	2.120	2.583	2.921	4.015
17	0.128	0.392	0.689	1.069	1.333	1.740	2.110	2.567	2.898	3.965
18	0.127	0.392	0.688	1.067	1.330	1.734	2.101	2.552	2.878	3.922
19	0.127	0.391	0.688	1.066	1.328	1.729	2.093	2.539	2.861	3.883
20	0.127	0.391	0.687	1.064	1.325	1.725	2.086	2.528	2.845	3.850
21	0.127	0.391	0.686	1.063	1.323	1.721	2.080	2.518	2.831	3.819
22	0.127	0.390	0.686	1.061	1.321	1.717	2.074	2.508	2.819	3.792
23	0.127	0.390	0.685	1.060	1.319	1.714	2.069	2.500	2.807	3.767
24	0.127	0.390	0.685	1.059	1.318	1.711	2.064	2.492	2.797	3.745
25	0.127	0.390	0.684	1.058	1.316	1.708	2.060	2.485	2.787	3.725
26	0.127	0.390	0.684	1.058	1.315	1.706	2.056	2.479	2.779	3.707
27	0.127	0.389	0.684	1.057	1.314	1.703	2.052	2.473	2.771	3.690
28	0.127	0.389	0.683	1.056	1.313	1.701	2.048	2.467	2.763	3.674
29	0.127	0.389	0.683	1.055	1.311	1.699	2.045	2.462	2.756	3.659
30	0.127	0.389	0.683	1.055	1.310	1.697	2.042	2.457	2.750	3.646
40	0.126	0.388	0.681	1.050	1.303	1.684	2.021	2.423	2.704	3.551
60	0.126	0.387	0.679	1.046	1.296	1.671	2.000	2.390	2.660	3.460
120	0.126	0.386	0.677	1.041	1.289	1.658	1.980	2.358	2.617	3.373
∞	0.126	0.385	0.674	1.036	1.282	1.645	1.960	2.326	2.576	3.291

Abridged from Table III of Fisher and Yates *Statistical Tables for Biological, Agricultural and Medical Research* published by Oliver and Boyd, Ltd., Edinburgh, by permission of the authors and publishers.

F-Distribution, Upper 5% Points $[F(v_1, v_2, 0.95)]$

Degrees of Freedom for Numerator

v_2	v_1=1	2	3	4	5	6	7	8	9	10	12	15	20	24	30	40	60	120	∞
1	161.4	199.5	215.7	224.6	230.2	234.0	236.8	238.9	240.5	241.9	243.9	245.9	248.0	249.1	250.1	251.1	252.2	253.3	254.3
2	18.51	19.00	19.16	19.25	19.30	19.33	19.35	19.37	19.38	19.40	19.41	19.43	19.45	19.45	19.46	19.47	19.48	19.49	19.50
3	10.13	9.55	9.28	9.12	9.01	8.94	8.89	8.85	8.81	8.79	8.74	8.70	8.66	8.64	8.62	8.59	8.57	8.55	8.53
4	7.71	6.94	6.59	6.39	6.26	6.16	6.09	6.04	6.00	5.96	5.91	5.86	5.80	5.77	5.75	5.72	5.69	5.66	5.63
5	6.61	5.79	5.41	5.19	5.05	4.95	4.88	4.82	4.77	4.74	4.68	4.62	4.56	4.53	4.50	4.46	4.43	4.40	4.36
6	5.99	5.14	4.76	4.53	4.39	4.28	4.21	4.15	4.10	4.06	4.00	3.94	3.87	3.84	3.81	3.77	3.74	3.70	3.67
7	5.59	4.74	4.35	4.12	3.97	3.87	3.79	3.73	3.68	3.64	3.57	3.51	3.44	3.41	3.38	3.34	3.30	3.27	3.23
8	5.32	4.46	4.07	3.84	3.69	3.58	3.50	3.44	3.39	3.35	3.28	3.22	3.15	3.12	3.08	3.04	3.01	2.97	2.93
9	5.12	4.26	3.86	3.63	3.48	3.37	3.29	3.23	3.18	3.14	3.07	3.01	2.94	2.90	2.86	2.83	2.79	2.75	2.71
10	4.96	4.10	3.71	3.48	3.33	3.22	3.14	3.07	3.02	2.98	2.91	2.85	2.77	2.74	2.70	2.66	2.62	2.58	2.54
11	4.84	3.98	3.59	3.36	3.20	3.09	3.01	2.95	2.90	2.85	2.79	2.72	2.65	2.61	2.57	2.53	2.49	2.45	2.40
12	4.75	3.89	3.49	3.26	3.11	3.00	2.91	2.85	2.80	2.75	2.69	2.62	2.54	2.51	2.47	2.43	2.38	2.34	2.30
13	4.67	3.81	3.41	3.18	3.03	2.92	2.83	2.77	2.71	2.67	2.60	2.53	2.46	2.42	2.38	2.34	2.30	2.25	2.21
14	4.60	3.74	3.34	3.11	2.96	2.85	2.76	2.70	2.65	2.60	2.53	2.46	2.39	2.35	2.31	2.27	2.22	2.18	2.13
15	4.54	3.68	3.29	3.06	2.90	2.79	2.71	2.64	2.59	2.54	2.48	2.40	2.33	2.29	2.25	2.20	2.16	2.11	2.07
16	4.49	3.63	3.24	3.01	2.85	2.74	2.66	2.59	2.54	2.49	2.42	2.35	2.28	2.24	2.19	2.15	2.11	2.06	2.01
17	4.45	3.59	3.20	2.96	2.81	2.70	2.61	2.55	2.49	2.45	2.38	2.31	2.23	2.19	2.15	2.10	2.06	2.01	1.96
18	4.41	3.55	3.16	2.93	2.77	2.66	2.58	2.51	2.46	2.41	2.34	2.27	2.19	2.15	2.11	2.06	2.02	1.97	1.92
19	4.38	3.52	3.13	2.90	2.74	2.63	2.54	2.48	2.42	2.38	2.31	2.23	2.16	2.11	2.07	2.03	1.98	1.93	1.88
20	4.35	3.49	3.10	2.87	2.71	2.60	2.51	2.45	2.39	2.35	2.28	2.20	2.12	2.08	2.04	1.99	1.95	1.90	1.84
21	4.32	3.47	3.07	2.84	2.68	2.57	2.49	2.42	2.37	2.32	2.25	2.18	2.10	2.05	2.01	1.96	1.92	1.87	1.81
22	4.30	3.44	3.05	2.82	2.66	2.55	2.46	2.40	2.34	2.30	2.23	2.15	2.07	2.03	1.98	1.94	1.89	1.84	1.78
23	4.28	3.42	3.03	2.80	2.64	2.53	2.44	2.37	2.32	2.27	2.20	2.13	2.05	2.01	1.96	1.91	1.86	1.81	1.76
24	4.26	3.40	3.01	2.78	2.62	2.51	2.42	2.36	2.30	2.25	2.18	2.11	2.03	1.98	1.94	1.89	1.84	1.79	1.73
25	4.24	3.39	2.99	2.76	2.60	2.49	2.40	2.34	2.28	2.24	2.16	2.09	2.01	1.96	1.92	1.87	1.82	1.77	1.71
26	4.23	3.37	2.98	2.74	2.59	2.47	2.39	2.32	2.27	2.22	2.15	2.07	1.99	1.95	1.90	1.85	1.80	1.75	1.69
27	4.21	3.35	2.96	2.73	2.57	2.46	2.37	2.31	2.25	2.20	2.13	2.06	1.97	1.93	1.88	1.84	1.79	1.73	1.67
28	4.20	3.34	2.95	2.71	2.56	2.45	2.36	2.29	2.24	2.19	2.12	2.04	1.96	1.91	1.87	1.82	1.77	1.71	1.65
29	4.18	3.33	2.93	2.70	2.55	2.43	2.35	2.28	2.22	2.18	2.10	2.03	1.94	1.90	1.85	1.81	1.75	1.70	1.64
30	4.17	3.32	2.92	2.69	2.53	2.42	2.33	2.27	2.21	2.16	2.09	2.01	1.93	1.89	1.84	1.79	1.74	1.68	1.62
40	4.08	3.23	2.84	2.61	2.45	2.34	2.25	2.18	2.12	2.08	2.00	1.92	1.84	1.79	1.74	1.69	1.64	1.58	1.51
60	4.00	3.15	2.76	2.53	2.37	2.25	2.17	2.10	2.04	1.99	1.92	1.84	1.75	1.70	1.65	1.59	1.53	1.47	1.39
120	3.92	3.07	2.68	2.45	2.29	2.17	2.09	2.02	1.96	1.91	1.83	1.75	1.66	1.61	1.55	1.50	1.43	1.35	1.25
∞	3.84	3.00	2.60	2.37	2.21	2.10	2.01	1.94	1.88	1.83	1.75	1.67	1.57	1.52	1.46	1.39	1.32	1.22	1.00

Degrees of Freedom for Denominator

Reproduced with permission from E. S. Pearson and H. O. Hartley, *Biometrika Tables for Statisticians*, Vol. 1, Cambridge University Press, New York, 1954.

F-Distribution, Upper 1% Points [$F(v_1, v_2, 0.99)$]

Degrees of Freedom for Numerator

v_2 \ v_1	1	2	3	4	5	6	7	8	9	10	12	15	20	24	30	40	60	120	∞
1	4052	4999.5	5403	5625	5764	5859	5928	5982	6022	6056	6106	6157	6209	6235	6261	6287	6313	6339	6366
2	98.50	99.00	99.17	99.25	99.30	99.33	99.36	99.37	99.39	99.40	99.42	99.43	99.45	99.46	99.47	99.47	99.48	99.49	99.50
3	34.12	30.82	29.46	28.71	28.24	27.91	27.67	27.49	27.35	27.23	27.05	26.87	26.69	26.60	26.50	26.41	26.32	26.22	26.13
4	21.20	18.00	16.69	15.98	15.52	15.21	14.98	14.80	14.66	14.55	14.37	14.20	14.02	13.93	13.84	13.75	13.65	13.56	13.46
5	16.26	13.27	12.06	11.39	10.97	10.67	10.46	10.29	10.16	10.05	9.89	9.72	9.55	9.47	9.38	9.29	9.20	9.11	9.02
6	13.75	10.92	9.78	9.15	8.75	8.47	8.26	8.10	7.98	7.87	7.72	7.56	7.40	7.31	7.23	7.14	7.06	6.97	6.88
7	12.25	9.55	8.45	7.85	7.46	7.19	6.99	6.84	6.72	6.62	6.47	6.31	6.16	6.07	5.99	5.91	5.82	5.74	5.65
8	11.26	8.65	7.59	7.01	6.63	6.37	6.18	6.03	5.91	5.81	5.67	5.52	5.36	5.28	5.20	5.12	5.03	4.95	4.86
9	10.56	8.02	6.99	6.42	6.06	5.80	5.61	5.47	5.35	5.26	5.11	4.96	4.81	4.73	4.65	4.57	4.48	4.40	4.31
10	10.04	7.56	6.55	5.99	5.64	5.39	5.20	5.06	4.94	4.85	4.71	4.56	4.41	4.33	4.25	4.17	4.08	4.00	3.91
11	9.65	7.21	6.22	5.67	5.32	5.07	4.89	4.74	4.63	4.54	4.40	4.25	4.10	4.02	3.94	3.86	3.78	3.69	3.60
12	9.33	6.93	5.95	5.41	5.06	4.82	4.64	4.50	4.39	4.30	4.16	4.01	3.86	3.78	3.70	3.62	3.54	3.45	3.36
13	9.07	6.70	5.74	5.21	4.86	4.62	4.44	4.30	4.19	4.10	3.96	3.82	3.66	3.59	3.51	3.43	3.34	3.25	3.17
14	8.86	6.51	5.56	5.04	4.69	4.46	4.28	4.14	4.03	3.94	3.80	3.66	3.51	3.43	3.35	3.27	3.18	3.09	3.00
15	8.68	6.36	5.42	4.89	4.56	4.32	4.14	4.00	3.89	3.80	3.67	3.52	3.37	3.29	3.21	3.13	3.05	2.96	2.87
16	8.53	6.23	5.29	4.77	4.44	4.20	4.03	3.89	3.78	3.69	3.55	3.41	3.26	3.18	3.10	3.02	2.93	2.84	2.75
17	8.40	6.11	5.18	4.67	4.34	4.10	3.93	3.79	3.68	3.59	3.46	3.31	3.16	3.08	3.00	2.92	2.83	2.75	2.65
18	8.29	6.01	5.09	4.58	4.25	4.01	3.84	3.71	3.60	3.51	3.37	3.23	3.08	3.00	2.92	2.84	2.75	2.66	2.57
19	8.18	5.93	5.01	4.50	4.17	3.94	3.77	3.63	3.52	3.43	3.30	3.15	3.00	2.92	2.84	2.76	2.67	2.58	2.49
20	8.10	5.85	4.94	4.43	4.10	3.87	3.70	3.56	3.46	3.37	3.23	3.09	2.94	2.86	2.78	2.69	2.61	2.52	2.42
21	8.02	5.78	4.87	4.37	4.04	3.81	3.64	3.51	3.40	3.31	3.17	3.03	2.88	2.80	2.72	2.64	2.55	2.46	2.36
22	7.95	5.72	4.82	4.31	3.99	3.76	3.59	3.45	3.35	3.26	3.12	2.98	2.83	2.75	2.67	2.58	2.50	2.40	2.31
23	7.88	5.66	4.76	4.26	3.94	3.71	3.54	3.41	3.30	3.21	3.07	2.93	2.78	2.70	2.62	2.54	2.45	2.35	2.26
24	7.82	5.61	4.72	4.22	3.90	3.67	3.50	3.36	3.26	3.17	3.03	2.89	2.74	2.66	2.58	2.49	2.40	2.31	2.21
25	7.77	5.57	4.68	4.18	3.85	3.63	3.46	3.32	3.22	3.13	2.99	2.85	2.70	2.62	2.54	2.45	2.36	2.27	2.17
26	7.72	5.53	4.64	4.14	3.82	3.59	3.42	3.29	3.18	3.09	2.96	2.81	2.66	2.58	2.50	2.42	2.33	2.23	2.13
27	7.68	5.49	4.60	4.11	3.78	3.56	3.39	3.26	3.15	3.06	2.93	2.78	2.63	2.55	2.47	2.38	2.29	2.20	2.10
28	7.64	5.45	4.57	4.07	3.75	3.53	3.36	3.23	3.12	3.03	2.90	2.75	2.60	2.52	2.44	2.35	2.26	2.17	2.06
29	7.60	5.42	4.54	4.04	3.73	3.50	3.33	3.20	3.09	3.00	2.87	2.73	2.57	2.49	2.41	2.33	2.23	2.14	2.03
30	7.56	5.39	4.51	4.02	3.70	3.47	3.30	3.17	3.07	2.98	2.84	2.70	2.55	2.47	2.39	2.30	2.21	2.11	2.01
40	7.31	5.18	4.31	3.83	3.51	3.29	3.12	2.99	2.89	2.80	2.66	2.52	2.37	2.29	2.20	2.11	2.02	1.92	1.80
60	7.08	4.98	4.13	3.65	3.34	3.12	2.95	2.82	2.72	2.63	2.50	2.35	2.20	2.12	2.03	1.94	1.84	1.73	1.60
120	6.85	4.79	3.95	3.48	3.17	2.96	2.79	2.66	2.56	2.47	2.34	2.19	2.03	1.95	1.86	1.76	1.66	1.53	1.38
∞	6.63	4.61	3.78	3.32	3.02	2.80	2.64	2.51	2.41	2.32	2.18	2.04	1.88	1.79	1.70	1.59	1.47	1.32	1.00

Degrees of Freedom for Denominator

Reproduced with permission from E. S. Pearson and H. O. Hartley, *Biometrika Tables for Statisticians*, Vol. 1, Cambridge University Press, New York, 1954.

BIBLIOGRAPHY

This is a selection and not a complete listing of material on regression.

(Note. *JASA* = *J. Am. Statist. Assoc.*; *JRSS* = *J. Roy. Statist. Soc.*)

Acton, F. S. *Analysis of Straight-Line Data*, John Wiley and Sons, New York, 1959.
  A broad treatment of regression, emphasizing the functional relationship point of view. The author's insights on data handling are excellent.

Anderson, R. L., and T. A. Bancroft. *Statistical Theory in Research*, McGraw-Hill Book Co., New York, 1952.
  Basic regression analysis is covered in Chapters 13, 14, 15, and 16. Chapter 15 discusses "hand"-computation procedures. This is an applied exposition, well worth reading and studying.

Anderson, T. W. "Least squares and best unbiased estimates," *Ann. Math. Statist.*, **33**, 266–272 (1962).
  An investigation of estimates found by least squares which are not necessarily linear functions of the observations but which are unbiased and have smallest variances under certain conditions.

Anscombe, F. J. "Examination of residuals," *Proc. Fourth Berkeley Symp. Math. Statist. Prob.*, **I**, 1–36 (1961).
  A well written and extensive discussion of various tests and plots which can be used to examine residuals.

Anscombe, F. J., and J. W. Tukey. "The analysis of residuals," unpublished notes given at the ASQC Chemical Division Meeting, Buffalo, October 2, 1958.
  An article expounding clever ways of examining the residuals from a regression analysis.

Anscombe, F. J., and J. W. Tukey. "The examination and analysis of residuals," *Technometrics*, **5**, 141–160 (1963).
  Primarily a discussion of residuals in two-way analysis of variance but the ideas have more general application.

Arrow, K. J., and M. Hoffenberg. *A Time Series Analysis of Interindustry Demands*, North Holland Publishing Co., Amsterdam, 1959.
  The method of fitting a multiple linear regression function when the sum of the absolute values of deviations is to be minimized.

Askovitz, S. E. "A short-cut graphic method for fitting the best straight line to a series of points according to the criterion of least squares," *JASA*, **52**, 13–17 (1957).
  A simple graphical procedure for obtaining the least squares regression line when there is a single independent variable whose levels are equally spaced.

Bartlett, M. S. "The use of transformations," *Biometrics*, **3**, 39–52 (1947).

Bartlett, M. S. "Fitting a straight line when both variables are subject to error," *Biometrics*, **5**, 207–212 (1949).
  Provides a simple method of fitting a straight line when both X and Y are subject to error, together with a numerical example; clearly written. (See also articles by Dudzinski, *Biometrics*, **16**, 399–407 (1960), Halperin, *JASA*, **56**, 657–669 (1961), Jessop, *Appl. Statist.*, **1**, 131–137 (1952), and Wald, *Ann. Math. Statist.*, **11**, 284–300 (1940).)

Berkson, J. "Are there two regressions?" *JASA*, **45**, 164–180 (1950).

This paper discusses the two cases: (1) when both X and Y are subject to error (i.e., each pair (X_i, Y_i) is considered a sample from a bivariate distribution), and (2) when X_i is subject to error but is a controlled variable.

Box, G. E. P. "Multifactor designs of first order," *Biometrika*, **39**, 49–57 (1952).

An excellent discussion of designs suitable for fitting linear first order models in k variables.

Box, G. E. P. "The effects of errors in the factor levels and experimental design," *Technometrics*, **5**, 247–262 (1963).

This paper examines the way in which errors in the X-variables affect a regression analysis using a polynomial model.

Box, G. E. P. "The exploration and exploitation of response surfaces: Some general considerations and examples," *Biometrics*, **10**, 16–60 (1954).

The use of regression techniques with designed experiments. This paper is a must for applied statisticians. (See also papers by the same author and co-authors, in *JRSS* **B-13** (1951); *Biometrics*, **11** (1955); *Ann. Math. Statist.*, **28** (1957); *JASA*, **54** (1959) for associated work.)

Box, G. E. P., and D. R. Cox. "An analysis of transformations," *JRSS* **B-26**, 211–243, discussion 244–252 (1964).

Discusses transformations on the dependent variable to attain the basic requirements of normality, additivity, and equality of variance.

Box, G. E. P., and P. W. Tidwell. "Transformation of the independent variables," *Technometrics*, **4**, 531–550 (1962).

Bright, J. W., and G. S. Dawkins. "Some aspects of curve fitting using orthogonal polynomials," *Ind. Eng. Chem. Fund.*, **4**, 93–97 (1965).

Cochran, W. G. "The omission or addition of an independent variate in multiple linear regression," *JRSS*, **B-5**, 171–176 (1938).

Useful when calculations are done "by hand" and it is desired to add or remove a variable without redoing the complete least squares calculation.

Corlett, T. "Ballade of multiple regression," *Appl. Statist.*, **12**, 145 (1963).

Cowden, Dudley J. "Procedure for computing regression coefficients," *JASA*, **53**, 144–150 (1958).

A method for calculating regression coefficients, practically feasible for four or less independent variables, which makes use of regression coefficients obtained from subsets of variables.

Cox, D. R. "Regression analysis when there is prior information about supplementary variables," *JRSS*, **B-22**, 172–176 (1960).

Discusses, with an example, how information on intermediate variables can provide increased precision for the estimates of regression coefficients.

Crouse, C. F. "On a point arising in multiple regression fitting," *Biometrika*, **51**, 501–503 (1964).

Daniel, Cuthbert "Factor screening in process development," *Ind. Eng. Chem.*, **55**, 5, 45–48 (1963).

An excellent paper which discusses a method of organizing both known and unknown effects of many factors, a systematic scheme for analyzing all data available, and entirely new experimental designs to supplement existing data.

Daniel, C., J. W. Gorman, and R. J. Toman. "Fitting collections of multifactor data," unpublished paper presented at the Gordon Research Conference on Statistics in Chemistry and Chemical Engineering, 1962.

A practical paper pointing out the use of dummy variables, the analysis of the disposition of data points in factor space, the interpretation of clustered data points, and various other methods of handling multifactor data.

Davies, M. "Multiple linear regression analysis with adjustment for class differences," *JASA*, **56**, 729–735 (1961).

An iterative procedure for solving the normal equations in multiple regression when the data can be subdivided into "*p*" mutually exclusive classes.

Davies, O. L. *Statistical Methods in Research and Production* (3rd ed.), Hafner Publishing Co., New York, 1957.

Chapters 7 and 8 of this well written book should be read by all students of regression analysis.

DeBaun, R. M., and V. Chew. "Optimum allocation in regression experiments with two components of error," *Biometrics*, **16**, 451–463 (1960).

Optimum experimental designs for regression analysis under cost constraints are derived. The case of "split-plotting" is considered.

DeLury, D. B. *Values and Integrals of the Orthogonal Polynomials up to $N = 26$*, University of Toronto Press, Ontario, 1960.

Dolby, J. L. "A quick method for choosing a transformation," *Technometrics*, **5**, 317–325 (1963).

Dolby, J. L. "Graphical procedure for fitting the best line to a set of points," *Technometrics*, **2**, 477–481 (1960).

The use of mapping in the dual space (coordinate system) for determining the slope and intercept of a straight line of best fit to a set of points in two dimensions.

Durbin, J. "A note on regression when there is extraneous information about one of the coefficients," *JASA*, **48**, 799–808 (1953).

Provides methods of obtaining better (i.e., smaller variance) unbiased estimates of regression coefficients (compared with the usual least squares estimates) when a separate unbiased estimate of one coefficient is available.

Dwyer, Paul S. *Linear Computations*, John Wiley and Sons, New York, 1951.

An excellent book on methods of solving simultaneous linear equations ranging from simple elimination to matrix inversion methods. Some problems of concern to statisticians are discussed.

Edgett, G. L. "Multiple regression with missing observations among the independent variables," *JASA*, **51**, 122–131 (1956).

The maximum likelihood estimators for the parameters of a trivariate normal population are derived for the case when some of the sample observations for one of the variates are missing.

Ehrenberg, A. S. C. "Bivariate regression is useless," *App. Statist.*, **12**, 161–179 (1963).

Eisenhart, C. "Interpretation of certain regression methods and their use in biological and industrial research," *Ann. Math. Statist.*, **10**, 162–186 (1939).

This practical, well written, expository article is highly recommended.

Eisenhart, C. "The assumptions underlying the analysis of variance," *Biometrics*, **3**, 1–21 (1947).

An excellent exposition of analysis of variance methods, highly recommended.

Eisenpress, H. "Regression techniques applied to seasonal corrections and adjustments for calendar shifts," *JASA*, **51**, 615–620 (1956).

This paper presents a general method for handling calendar shifts in conjunction with seasonal adjustments. The method is illustrated with the seasonal correction of bank deposits outside New York City for 1946–1954 and of tobacco manufacturers for 1947–1953.

Efroymson, M. A. (*See* Ralston, A., and Wilf, H. S.)

Faddeeva, V. N. *Computational Methods of Linear Algebra*, Dover Publications, New York, 1959.

> This is an excellent introduction to matrices and to the various computational methods needed in practical numerical work.

Farrar, D. E., and R. R. Glauber. "Multicollinearity in regression analysis: The problem revisited," Working paper 105–164, The Alfred P. Sloan School of Management, MIT, 1964.

> A well-written paper on the problem of highly-intercorrelated independent variables. The authors propose procedures for gaining insight into the location, pattern and severity of interdependence through standard statistical procedures.

Fisher, R. A., and F. Yates. *Statistical Tables for Biological, Agricultural, and Medical Research* (6th ed.), Hafner Publishing Co., New York, 1964.

Fisher, Walter D. "A note on curve fitting with minimum deviations by linear programming," *JASA*, **56**, 359–362 (1961).

> The fitting of a linear multiple regression with the criterion of minimizing the sum of absolute deviations by use of linear programming.

Foote, Richard J. "A modified Doolittle approach for multiple and partial correlation and regression," *JASA*, **53**, 133–144 (1958).

> Simplified desk calculator procedures for obtaining multiple and partial correlation and regression coefficients are discussed.

Freund, R. J. "A warning of round-off errors in regression," *Am. Statist.*, **17**, 13–15 (1963).

> This paper provides an example to illustrate that round-off errors can invalidate the results of regression calculations.

Freund, R. J., R. W. Vail, and C. W. Clunies-Ross. "Residual analysis," *JASA*, **56**, 98–104 (1961).

> This paper examines the bias of estimates in the second stage of a stagewise regression procedure (see Goldberger, 1961, below).

Garside, M. J. "A short list of multiple regression programs available in Britain," *Appl. Statist.*, **12**, 151–160 (1963).

Gibson, W. M., and G. H. Jowett. "Three-group regression analysis. Part I: Simple regression analysis," *Appl. Statist.*, **6**, 114–122 (1957).

> A quick, simple, approximate method for constructing a regression line for one or two independent variables.

Goldberger, A. S. "Best linear unbiased prediction in the generalized linear regression model," *JASA*, **57**, 369–375 (1962).

> This paper demonstrates the use of patterns in residuals in getting a better unbiased predictor.

Goldberger, A. S. "Stepwise least-squares: Residual analysis and specification," *JASA*, **56**, 998–1000 (1961).

> More on stagewise regression. Goldberger discusses the bias of estimates in the first stage of a two-stage, stagewise procedure. In spite of their titles, this paper, the next one, and the paper by Wallace (see page 315) involve *stagewise* regression and *not* the Efroymson stepwise procedure.

Goldberger, A. S., and D. S. Jochems. "Note on stepwise least squares," *JASA*, **56**, 105–110 (1961).

> An article similar to the Freund, Vail, Clunies-Ross paper.

Gustafson, R. L. "Partial correlations in regression computations," *JASA*, **56**, 363–367 (1961).

> An exposition of the calculation of a partial correlation coefficient between the

dependent variable and the kth independent variable using the kth regression coefficient and its standard error.

Hald, A. *Statistical Theory with Engineering Applications*, John Wiley and Sons, New York, 1952.

Hamaker, H. C. "On multiple regression analysis," *Statist. Neerlandica*, **16**, 31–56 (1962).

An excellent article on the difficulties of the procedures used in selecting a subset k from a larger set of "p" variables in regression. The article is clearly written and is excellent supplemental reading to this text.

Hoerl, A. E. "Fitting curves to data," *Chemical Business Handbook*, McGraw-Hill, New York, 1954.

Discusses ways to determine transformations.

Huddleston, H. F. "An inverted matrix approach for determining crop-weather regression equations," *Biometrics*, **11**, 231–236 (1955).

This paper demonstrates how the use of the inverse matrix solution of a set of normal equations reduces the amount of computation when the same set of independent variables is used repeatedly.

Karst, Otto J. "Linear curve fitting using least deviations," *JASA*, **53**, 118–132 (1958).

The development of a method for finding a best fitting straight line to a set of two-dimensional points such that the sum of the absolute values of the vertical deviations of the points from the line is a minimum. An iterative procedure for the general case is outlined.

Kempthorne, O. "Answer to a query," *Biometrics*, **9**, 528–533 (1953).

A clear exposition of the difficulties encountered in doing multiple regression problems; recommended reading.

Kendall, M. G. "Regression, structure, and functional relationships, Part I," *Biometrika*, **38**, 11–25 (1951).

A discussion of the implications of the various assumptions that can be made about the dependent and independent variable in regression situations.

Kendall, M. G. "Regression, structure, and functional relationships, Part II," *Biometrika*, **39**, 96–108 (1952).

A continuation of Part I, and a general discussion on the problems of structural analysis.

Lindley, D. V. "Estimation of a functional relationship," *Biometrika*, **40**, 47–49 (1953).

Discusses a controversy arising from the papers of Berkson (*JASA*, **45**) and Kendall (*Biometrika*, **39**).

Lindley, D. V. "Regression lines and the linear functional relationship," *JRSS*, **B-9**, 218–244 (1947).

An examination of the conditions under which a linear regression remains linear when there are errors in the independent variables, and a discussion of the use of regression lines and functional relationships in statistical methodology.

Mandansky, Albert. "The fitting of straight lines when both variables are subject to error," *JASA*, **54**, 173–205 (1959).

This paper is a fairly complete exposition of this problem with an excellent set of references. It includes a discussion of grouping, variance components, cumulants, and their relationship to the estimation problem.

Mandel, John. "Fitting a straight line to certain types of cumulative data," *JASA*, **52**, 552–566 (1957).

An examination of the consequences of performing a standard regression analysis when the errors are cumulative rather than independent. Examples of such situations are discussed.

Myers, R. H. "Methods for estimating the composition of a three-component liquid mixture," *Technometrics*, **6**, 343–356 (1964).

An illustration of an application of regression techniques.

Pearson, E. S., and H. O. Hartley. *Biometrika Tables for Statisticians*, Vol. 1, Cambridge University Press, 1958.

Plackett, R. L. *Regression Analysis*, Clarendon Press, Oxford, 1960.

The emphasis of this book is on theory. An excellent mathematical exposition of assumptions in regression, properties of estimates and properties of test criteria.

Quandt, Richard E. "The estimation of the parameter of a linear regression system obeying two separate regimes," *JASA*, **53**, 873–880 (1958).

An interesting paper on a method of determining at what point in time a switch from one regime to another occurs. This is particularly useful in marketing work. See also papers by E. S. Page (*Biometrika*, **42** and *Biometrika*, **44**).

Quandt, Richard E. "Tests of hypotheses that a linear regression system obeys two separate regimes," *JASA*, **55**, 324–330 (1960).

Several approaches are explored for testing the hypothesis that no switch has occurred in the true values of the parameters of a linear regression system.

Ralston, A., and H. S. Wilf. *Mathematical Methods for Digital Computers*, John Wiley and Sons, New York, 1962.

Article 17, "Multiple Regression Analysis," covers the stepwise regression problem. It was written by M. A. Efroymson, Esso Research and Engineering Company.

Rand Corporation. *A Million Random Digits with 100,000 Normal Deviates*. The Free Press, Glencoe, Illinois, 1955.

An extremely useful volume.

Read, D. R. "The design of chemical experiments," *Biometrics*, **10**, 1–15 (1954).

The use of multiple regression in chemical experiments with a particular emphasis on response surface designs.

Reinfeld, N. V., and W. R. Vogel. *Mathematical Programming*, Prentice-Hall, Englewood Cliffs, New Jersey, 1958.

Explains the various methods of solving problems by linear programming and describes how the methods may be set up for computers.

Robson, D. S. "A sample method for constructing orthogonal polynomials when the independent variable is unequally spaced," *Biometrics*, **15**, 187–191 (1959).

Rubin, H., and R. B. Leipnik. "Measuring the equation systems of dynamic economics" in the book *Statistical Inference in Dynamic Economic Models, Cowles Commission Monograph 10* (T. C. Koopmans, editor), pp. 93–110, John Wiley and Sons, New York, 1950.

A difficult article concerned with the construction of (and estimation of parameters in) equation systems useful in economics.

Sargan, J. D. "Three-stage least squares and full maximum likelihood estimates," *Econometrica*, **32**, 77–81 (1964).

Proves that under certain mild restrictions the three-stage least squares estimates of regression parameters differ asymptotically from maximum likelihood estimates by order $1/T$ where T is the number of time periods.

Scheffé, Henry. "Fitting straight lines when one variable is controlled," *JASA*, **53**, 106–118 (1958).

A generalization of the problem considered by Berkson (1950 *JASA*).

Schipper, Lewis. "Consumer discretionary behavior, a comparative study in alternative methods of empirical research," *Contributions to Economic Analysis*, North-Holland Publishing Co., Amsterdam, 1964.

This is an interesting article which discusses covariance regression. An introduction by Suits discusses the relationship between the use of dummy variables and covariance regression.

Shah, B. K., and C. G. Khatri. "A method of fitting the regression curve $E(y) = \alpha + \delta x + \beta \rho^x$," *Technometrics*, **7**, 59–65 (1965).

Sorenson, Fred A. "Methods of performing multiple regression analysis," (90.10-100G) (1) United States Steel Technical Report, Applied Research Laboratory, Monroeville, Pennsylvania.

A report on theory and the steps to be followed in analyzing multifactor data. This paper includes some of the screening procedures mentioned in this book.

Spurrell, D. J. "Some metallurgical applications of principal components," *Appl. Statist.*, **12**, 180–188 (1963).

Suits, Daniel B. "Use of dummy variables in regression equations," *JASA*, **52**, 548–551 (1957).

Two procedures for handling dummy variables in regression are covered: (1) set the constant term equal to zero, and (2) omit one of the dummy variables from the equation. The article emphasizes the advantages of dummy variables in the presence of curvature.

Swed, Frieda S., and C. Eisenhart, "Tables for testing randomness of grouping in a sequence of alternatives," *Ann. Math. Statist.*, **14**, 66–87 (1943).

Telser, L. G. "Iterative estimation of a set of linear regression equations," *JASA*, **59**, 845–862 (1964).

Concerns the iterative calculation of estimates of regression coefficients for sets of regression equations when correlation between corresponding errors of various sets exists.

Thomson, Louis M. "Weather and technology in the production of corn and soybeans," CAED Report 17, Iowa State University, Ames, Iowa, 1963.

An interesting application of multiple regression techniques. It is particularly useful for students as it contains all the actual data for study.

Thonstad, T., and D. B. Jochems. "The influence of entrepreneurial appraisals and expectations on production planning," *Intern. Econ. Rev.*, **2**, 135–152 (1961).

Regression analysis applied to qualitative type data.

Tukey, J. W. "Components in regression," *Biometrics*, **7**, 33–70 (1951).

This is an excellent introduction to the simple linear regression problem when both variables are subject to error.

Tukey, J. W. "On the comparative anatomy of transformations," *Ann. Math. Statist.*, **28**, 602–632 (1957).

Discusses the family of transformations, $T(x) = (c + x)^p$, which for various values of c and p can represent a number of standard transformations.

Tukey, J. W. "Where do we go from here?" *JASA*, **55**, 80–93 (1960).

An excellent article in which a balance of bias and variability in regression situation is proposed.

Turner, M. E. "Straight-line regression through the origin," *Biometrics*, **16**, 483–485 (1960).

When the model is a straight line through the origin and Y is positive, normal error assumptions are impossible. This note shows that the assumption of other error distributions sometimes provides the same estimates.

Wagner, Harvey M. "Linear programming techniques for regression analysis," *JASA*, **54**, 206–212 (1959). *See also JASA*, **57**, 1962, 572–578.

This paper demonstrates that using the criterion, "least absolute deviations" leads to a p equation linear programming model with bounded variables, and using the criterion, "least maximum deviations," leads to a standard form $p + 1$ equation linear programming model.

Wallace, T. D. "Efficiencies for stepwise regressions," *JASA*, **59**, 1179–1182 (1964).

This paper deals with what we call "stagewise regression." It points out that under certain conditions, the stagewise procedure yields lower mean square errors than does least squares though the parameter estimates are biased.

Wegner, P. "Relations between multivariate statistics and mathematical programming," *Appl. Statist.*, **12**, 146–150 (1963).

Whiston, W. B. "Sequential selection of variables in multiple regression," unpublished Master's degree thesis, University of Cincinnati, 1964.

A mathematical explanation of the Efroymson procedure with illustrations.

Williams, E. J. *Regression Analysis*, John Wiley and Sons, New York, 1959.

An intermediate text covering a wide range of topics in regression analysis. The problem of constructing models using regression as a tool is covered very briefly.

Wishart, J., and T. Metakides. "Orthogonal polynomial fitting," *Biometrika*, **40**, 361–369 (1953).

A systematic procedure for determining the coefficients of orthogonal polynomials when the data are unequally spaced or should be weighted.

Wold, Herman O. "Econometric model building: Essays on the casual chain approach," *Contributions to Economic Analysis*, North-Holland Publishing Co., Amsterdam, 1964.

This is a collection of lectures on stochastic economic models. The article on the Janus Quotient as a measure of the accuracy of prediction, and the article on the standard errors of regression coefficients when the residuals are autocorrelated are excellent.

Zellner, A., and H. Theil. "Three-stage least squares: Simultaneous estimation of simultaneous equations," *Econometrica*, **30**, 54–78 (1962).

ANSWERS TO EXERCISES

Chapter 1

A. (a) $b_1 = \frac{158}{110} = 1.44$; $b_0 = \frac{102}{11} = 9.27$; $\hat{Y} = 9.27 + 1.44X$.

(b)

Analysis of Variance

Source of Variation	Degrees of Freedom	Sum of Squares	Mean Square	F
Total (corrected)	10	248.18		
Regression	1	$\dfrac{(158)^2}{110}$	226.95⌐	96.17*
Residual	9	21.23	2.36⌐	

The hypothesis, H_0: $\beta_1 = 0$ is tested with $\alpha = 0.05$ by comparing the computed $F(1, 9)$ statistic with the critical $F(1, 9)$ for $\alpha = 0.05$. From the $\alpha = 0.05$ F table, page 306, $F(1, 9, 0.95) = 5.12$. Since 96.17 is greater than 5.12, reject the hypothesis $\beta_1 = 0$.

(c) The 95% confidence limits for β_1 are

$$1.11 \leq \beta_1 \leq 1.77.$$

(d) The 95% confidence limits for the true average value of Y at $X_0 = 3$ are

$$12.15 \leq \text{true average } Y \text{ at } X_0 = 3 \leq 15.03.$$

(e) The 95% confidence limits for the difference between the true average value of Y at $X_1 = 3$ and the true average value of Y at $X_2 = -2$. First, determine the algebraic difference between $\hat{Y}_1$ and $\hat{Y}_2$:

$$\hat{Y}_1 = b_0 + b_1(3) \qquad \hat{Y}_2 = b_0 + b_1(-2)$$

Thus

$$\hat{Y}_1 - \hat{Y}_2 = b_1(3 + 2) = 5b_1 = 5(1.44) = 7.20$$

$$s^2_{(\hat{Y}_1 - \hat{Y}_2)} = 25s^2_{b_1} = 25\left(\frac{2.36}{110}\right) = 0.53635$$

$$s_{(\hat{Y}_1 - \hat{Y}_2)} = \sqrt{0.53635} = 0.732$$

$$ts_{(\hat{Y}_1 - \hat{Y}_2)} = (2.262)(0.732) = 1.656.$$

Thus, the 95% confidence band on the true difference is

$$7.20 - 1.66 \leq \text{true difference} \leq 7.20 + 1.66,$$

$$5.54 \leq \text{true difference} \leq 8.86.$$

(f) Calculate the residuals and look for patterns.

X	Y	$\hat{Y}$	$Y - \hat{Y}$
−5	1	2.07	−1.07
−4	5	3.51	1.49
−3	4	4.95	−0.95
−2	7	6.39	0.61
−1	10	7.83	2.17
0	8	9.27	−1.27
1	9	10.71	−1.71
2	13	12.15	0.85
3	14	13.59	0.41
4	13	15.03	−2.03
5	18	16.47	1.53

There is no obvious alternative to the model.

(g) If the tentative assumption of a first-order model is reasonable, there is little point in using eleven different experimental levels. Of course, we need at least two levels to estimate the parameters in the model, and at least one more to detect curvature in the true model, if curvature exists. By taking repeat observations at some or all levels, we can obtain a pure error estimate of σ^2 to use in checking lack of fit. Thus for an experiment of about the same size, one possibility would be to choose three widely spaced levels—the extremes of the X range and the center, for example— and to take four observations at each of these levels. This would lead to an analysis of variance table of the form below.

ANOVA ($n = 12$)

Source of Variation	df
Total (corrected)	11
Regression	1
Residual	10
lack of fit	1
pure error	9

Since we now have only 1 degree of freedom for lack of fit, this is not entirely satisfactory. Slightly better would be the choice of three runs at each of four levels:

ANOVA ($n = 12$)

Source of Variation	df
Total (corrected)	11
Regression	1
Residual	10
lack of fit	2
pure error	8

There are many other possibilities that the student can consider.

B.   **1.**   Randomized order.

2(a) $\hat{Y} = 0.5 + 0.5X$.

2(b)

Source	SS	df	MS
Corrected total $\sum (Y_i - \bar{Y})^2$	83.2	19	
Due to regression $\dfrac{[\sum (X_i - \bar{X})(Y_i - \bar{Y})]^2}{\sum (X_i - \bar{X})^2}$	40.0	1	40.0 *
Residual	43.2	18	2.4

2(c)   (1) $\hat{Y} = 3.0 \pm 0.73 = 2.27$ to 3.73.

     (2) $\hat{Y} = 5.0 \pm 1.26 = 3.74$ to 6.26.

3(a)

Residual	43.2	18	2.4
Lack of fit	1.2	3	0.4 NS
Pure error	42.0	15	2.8

No significant lack of fit.

3(b)   Yes.

4(a)   Same ANOVA and conclusion as in 3(a).

4(b) Confidence limits not applicable. Error variance is dependent on level of X.

4(c) First-order model suitable

5(a)

Residual	43.2	18	2.4
Lack of fit	20.0	3	6.67⎤*
Pure error	23.2	15	1.55⎦

Significant lack of fit term indicates inadequacy of model.

5(b) Since the model is incorrect, confidence intervals will be invalid.

5(c) A second-order model is suggested.

C.   Best fitting straight line is

$$\hat{Y} = b_0 + b_1 X = 129.7872 - 24.0199 X.$$

ANOVA

Source	df	SS	MS	F
Total	12	3396.62		
Regression	1	3293.77	3293.77	
Residual	11	102.85	9.35	
lack of fit	5	91.08	18.22	9.30*
pure error	6	11.77	1.96	

The model is inadequate.

E.   Models: $Y_{iu} - \bar{Y}_i = \beta_i(X_{iu} - \bar{X}_i) \cdot i = 1, 2, \ldots, m$

$$u = 1, 2, \ldots, n$$

1.

X_1	Y_1	$(X_{iu} - \bar{X}_i)$	$(Y_{iu} - \bar{Y}_i)$
3.5	24	-1.529	-17.286
4.1	32	-0.929	-9.286
4.4	37	-0.629	-4.286
5.0	40	-0.029	-1.286
5.5	43	0.471	1.714
6.1	51	1.071	9.714
6.6	62	1.571	20.714

$$\bar{X}_1 = 5.029 \qquad \bar{Y}_1 = 41.286 \qquad n_1 = 7.$$

$$b_1 = \left\{ \sum_{u=1}^{7} (X_{iu} - \bar{X}_i)(Y_{iu} - \bar{Y}_i) \right\} \bigg/ \left\{ \sum_{u=1}^{7} (X_{iu} - \bar{X}_i)^2 \right\}$$

$$= 81.542858/7.434 = 10.969.$$

$$SS(b_1) = b_1^2 \left\{ \sum_{u=1}^{7} (X_{iu} - \bar{X}_i)^2 \right\}$$

$$= (120.318961)(7.434) = 894.451.$$

2.

X_2	Y_2	$\bar{X}_2 = 5.533$	$(X_{iu} - \bar{X}_i)$	$(Y_{iu} - \bar{Y}_i)$
3.2	22	$\bar{Y}_2 = 41.333$	-2.333	-19.333
3.9	33		-1.633	-8.333
4.9	39	$n_2 = 6$	-0.633	-2.333
6.1	44		0.567	2.667
7.0	53		1.467	11.667
8.1	57		2.567	15.667

$$b_2 = \left\{ \sum_{u=1}^{6} (X_{iu} - \bar{X}_i)(Y_{iu} - \bar{Y}_i) \right\} \Big/ \left\{ \sum_{u=1}^{6} (X_{iu} - \bar{X}_i)^2 \right\}$$

$$= 119.033334/17.573334 = 6.773520$$

$$SS(b_2) = b_2^2 \left\{ \sum_{u=1}^{6} (\bar{X}_{iu} - \bar{X}_i)^2 \right\}$$

$$= (45.880573)(17.573334) = 806.274633.$$

3.

X_3	Y_3	$(X_{iu} - \bar{X}_i)$	$(Y_{iu} - \bar{Y}_i)$
3.0	32	-2.775	-18.750
4.0	36	-1.775	-14.750
5.0	47	-0.775	-3.750
6.0	49	0.225	-1.750
6.5	55	0.725	4.250
7.0	59	1.225	8.250
7.3	64	1.525	13.250
7.4	64	1.625	13.250

$$\bar{X}_3 = 5.775 \qquad \bar{Y}_3 = 50.750.$$

$$n_3 = 8$$

$$b_3 = \left\{ \sum_{u=1}^{8} (X_{iu} - \bar{X}_i)(Y_{iu} - \bar{Y}_i) \right\} \Big/ \left\{ \sum_{u=1}^{8} (X_{iu} - \bar{X}_i)^2 \right\}$$

$$= (135.650000)/(18.495) = 7.334415.$$

$$SS(b_3) = (53.793643)(18.495) = 994.913427.$$

$$b = \left\{ \sum_{i=1}^{m=3} \sum_{u=1}^{n_i} (X_{iu} - \bar{X}_i)(Y_{iu} - \bar{Y}_i) \right\} \Big/ \left\{ \sum_{i=1}^{m=3} \sum_{u=1}^{n_i} (X_{iu} - \bar{X}_i)^2 \right\}$$

$$= (81.542858 + 119.033334 + 135.650)$$

$$\div (7.434 + 17.573334 + 18.495)$$

$$= 336.226192/43.502334 = 7.728923.$$

$$SS(b) = b^2 \left\{ \sum_{i=1}^{3} \sum_{u=1}^{n_i} (X_{iu} - \bar{X}_i)^2 \right\}$$

$$= (59.736251)(43.502334) = 2598.666299.$$

$$SS \text{ due to all } b_i \mid b = \sum_{i=1}^{3} SS(b_i) - SS(b)$$

$$= 894.451000 + 806.274633 + 994.913427$$

$$- 2598.666343 = 96.972717.$$

$$\text{Residual} = \text{Total } SS - SS(b) - SS \text{ due to all } b_i \mid b$$

$$= 2792.261906 - 2598.666343 - 96.972717 = 96.622846.$$

ANOVA

Source	df	SS	MS	F
b	1	2598.666343	2598.666343	403.42
All $b_i \mid b$	2	96.972717	48.486359	7.53
Residual	15	96.622846	6.441523	
Total	18	2792.261906		

$H_0: \beta_i = \beta \qquad F_2 = 7.53 > F(2, 15, 0.95) = 3.68,$
$\therefore H_0$ is rejected.

F. 1. $\hat{Y} = -21.33 + 5X$
   2. $2.984 \leq \beta_1 \leq 7.016$
   3.

ANOVA

Source	df	SS	MS	F
Corrected total	11	69.67		
Due to regression	1	52.50	52.50	
Residual	10	17.17	1.72	
Lack of fit	4	5.50	1.375	0.706 (not significant
Pure error	6	11.67	1.945	at $\alpha = 0.05$)

The model seems to be adequate.

G. 1. $\hat{Y} = 323.628 + 131.717X.$

ANOVA

Source	df	SS	MS	F
Corrected Total	16	2,305,042		
Due to regression	1	1,099,641.1	1,099,641.10	13.68 significant
Residual	15	1,205,400.9	80,360.06	at $\alpha = 0.05$

2.

ANOVA

Source	df	SS	MS	F
Corrected total	16	2,305,042		
Due to regression	1	1,099,641.1		
Residual	15	1,205,400.9		
Lack of fit	5	520,648.6	104,129.72	1.52 not significant
Pure error	10	684,752.3	68,475.23	at $\alpha = 0.05$

A. straight line relationship seems reasonable.

H.　　Prediction Equation: $\hat{Y} = 1.222 + 0.723X$

ANOVA

Source	df	SS	MS	F	$F_{0.95}$
Corrected total	13	2.777			
Due to regression	1	1.251	1.251⎤	9.850	4.75
Residual	12	1.526	0.127⎦		

$9.850 > F(1, 12, 0.95) = 4.75$; $\therefore$ reject $H_0: \beta_1 = 0$

ANOVA

Source	df	SS	MS	F	$F_{0.95}$
Corrected total	13	2.777			
Due to regression	1	1.251			
Residual	12	1.526			
Lack of fit	7	0.819	0.117⎤	0.830	4.88
Pure error	5	0.707	0.141⎦		

$0.830 < F(7, 5, 0.95) = 4.88$, $\therefore$ lack of fit is not significant.
Conclusion: Use the prediction equation

Cup loss(%) $= 1.222 + (0.723)[$bottle loss(%)$]$.

I. Prediction equation: $\hat{Y} = 17.146 + 11.836X$.

ANOVA

Source	df	SS	MS	F	$F_{0.95}$
Corrected total	12	22,126.308			
Due to regression	1	6,034.379	6,034.379	4.125	4.840
Residual	11	16,091.929	1,462.903		

$4.125 < F(1, 11, 0.95) = 4.840$; $\therefore$ do not reject $H_0: \beta_1 = 0$. The regression is not significant.

$$R^2 = \frac{\text{SS due to regression}}{\text{Corrected total SS}} = \frac{6034.379}{22,126.308} = 27.27\%.$$

Conclusions: (1) The model is inadequate. (2) Further investigation of alternative variables will be necessary.

J. 1. $\hat{Y} = 2.5372000 - 0.004718X$

2.

ANOVA

Source	df	SS	MS	F	$F_{0.95}$
Corrected total	14	0.209333			
Due to regression	1	0.110395	0.110395	14.50	4.67
Residual	13	0.098938	0.007611		

$14.50 > F(1, 13, 0.95) = 4.67$; $\therefore$ reject $H_0: \beta_1 = 0$. The regression is significant.

ANOVA

Source	df	SS	MS	F	$F_{0.95}$
Residual	13	0.098938			
Lack of fit	5	0.018938	0.003788	0.38	3.69
Pure error	8	0.080000	0.010000		

$0.38 < F(5, 8, 0.95) = 3.69$; $\therefore$ Lack of fit not significant.

3. 95% confidence interval on the true mean value of Y, calculated at four points: $X = 0$, $X = \bar{X}$, $X = 400$, $X = 460$:

at $X = 0$	$\hat{Y} \pm (2.160)(0.527) = \hat{Y} \pm 1.138$
at $X = \bar{X}$	$\hat{Y} \pm (2.160)(0.022) = \hat{Y} \pm 0.048$
at $X = 400$	$\hat{Y} \pm (2.160)(0.039) = \hat{Y} \pm 0.084$
at $X = 460$	$\hat{Y} \pm (2.160)(0.048) = \hat{Y} \pm 0.104$

Chapter 3

A.   The plot of the residuals against $\hat{Y}_i$ indicates the assumption of equal variances is incorrect.

B.   1. $\hat{Y} = -2.679 + 9.5X$

2.   ANOVA

Source of Variation	df	SS	MS	F
Total	6	651.714		
Regression: $b_1 \mid b_0$	1	631.750	631.750	158.33
Residual	5	19.964	3.99	

Since $158.33 > 6.61$, the regression is significant, $\alpha = 0.05$.

3.   No evidence to suggest a more complicated model needed.

C.   1. $b_1 = 9.13$; $\hat{Y} = 9.13X$

Obs. No.	X	Y	$\hat{Y}$	Residuals
1	3.5	24.4	31.955	−7.555
2	4.0	32.1	36.520	−4.420
3	4.5	37.1	41.085	−3.985
4	5.0	40.4	45.650	−5.250
5	5.5	43.3	50.215	−6.915
6	6.0	51.4	54.780	−3.380
7	6.5	61.9	59.345	2.555
8	7.0	66.1	63.910	2.190
9	7.5	77.2	68.475	8.725
10	8.0	79.2	73.040	6.160

3.   The plot of the residuals against $\hat{Y}$ indicates the omission of the β_0 term in the model.

4.   The model $Y = \beta_0 + \beta_1 X_1 + \epsilon$ is recommended, restricting its use to within the X_1 range, 3.5 to 8.0, shown by the data.

    If the true model really has $\beta_0 = 0$, then data will have to be obtained nearer to the zero response before any insight into the right model for the range of X from 0 to a large value of X can be obtained.

D. **1.** $\hat{Y} = -252.298 + 8.529X$

2.

ANOVA

Source of Variation	df	SS	MS	F
Total	9	219270.5000		
$b_1 \mid b_0$	1	200772.3188		
Residual	8	18498.1812		
Lack of fit	4	18454.6812	4613.6703	424.2455
Pure error	4	43.5000	10.8750	

The lack of fit test indicates that the model is an inadequate. The plot of the residuals against $\hat{Y}$ indicates a definite trend from negative residuals to positive residuals as the value of $\hat{Y}$ increases.

There is also evidence of an outlier; namely, $Y = 415$ when $X = 90$. This point should be investigated further.

E. **1.** $\hat{Y} = 7.950 - 0.0179T$.

2.

ANOVA

Source of Variation	df	SS	MS	Calc. F	$F_{0.95}$
Total	8	6.260			
Regression	1	1.452	1.452	2.114	5.59
Residual	7	4.808	0.687		
Lack of fit	3	1.763	0.588	<1	
Pure error	4	3.045	0.761		

The regression is nonsignificant. $R^2 = 23.2\%$.

3. $s_{b_1} = 0.01228$. $-0.04689 \le \beta_1 \le 0.01119$.

4.

Batch No.	Y_i	$\hat{Y}_i$	$Y - \hat{Y}_i$
1	2.10	2.95	−0.85
2	3.00	3.49	−0.49
3	3.20	2.59	0.61
4	1.40	2.24	−0.84
5	2.60	2.42	0.18
6	3.90	2.95	0.95
7	1.30	2.24	−0.94
8	3.40	2.59	0.81
9	2.80	2.24	0.56

No discernible pattern in the residuals.

5. $0.193 \le \hat{Y}_k \le 4.453$.

6. No. The slope of the line fitted is not significant and, in any case, 360 is well beyond the temperature range, making the use of the fitted equation even more dangerous.

F. 1. The negative residuals occur at low levels of concentration and the positive residuals occur at high levels of concentration.

2. $\hat{Y} = 2.693374 - 0.277361 X_1 + 0.365028 X_2$.

3.

ANOVA

Source of Variation	df	SS	MS	Calc. F	$F_{0.95}$
Total (corrected)	8	6.2600			
Regression	2	4.7381	2.3690	9.34	5.14
$b_1 \mid b_0$	1	1.4521	1.4521	5.73	5.99
$b_2 \mid b_0, b_1$	1	3.2860	3.2860	12.96	5.99
Residual	6	1.5219	0.2536		

a. Since there are no replicates, the lack of fit test cannot be done.

b. A model $\hat{Y} = \bar{Y}$ explains $62.41/68.67 = 90.88\%$ of the crude variation in the data measured from $Y = 0$. Of the remaining variation, the model $\hat{Y} = b_0 + b_1 X_1 + b_2 X_2$ explains 75.69% of it, or a total of 97.78% of the crude variation.

c. The addition of β_2 to the model improves the fit as shown by R^2 going from 23.20% to 75.69%.

4. $R^2 = 75.69\%$.

5. sd $\tilde{b}_1 = 0.00795$
    sd $\tilde{b}_2 = 0.10141$.

6.

Batch	Y	$\hat{Y}$	$Y - \hat{Y}$
1	2.100	2.518	−0.418
2	3.000	3.350	−0.350
3	3.200	2.693	0.507
4	1.400	1.774	−0.374
5	2.600	2.781	−0.181
6	3.900	3.248	0.652
7	1.300	1.044	0.256
8	3.400	3.058	0.342
9	2.800	3.234	−0.434

7. Var $\hat{Y}$ (coded) $= 0.044871$.

Chapter 4

A. 1. $b_0 = 14$ $b_1 = -2$ $b_2 = -\frac{1}{2}$.

2.   Analysis of Variance

Source of Variation	df	SS	MS	F
Total (corrected)	10	190		
Due to regression	2	122	61.0	7.17*
Residual	8	68	8.5	

3. Test of significance:

Compare $F = \dfrac{\text{MS regression}}{\text{MS residual}}$ with $F(2, 8, 0.95) = 4.46$

Since 7.17 is greater than the critical F, we reject the hypothesis of no planar fit and use the fitted equation

$$\hat{Y} = 14 - 2x_1 - \tfrac{1}{2}x_2.$$

4. $R^2 = \dfrac{122}{190} = 64.21\%$.

5(a) Variance of $b_1 = 1.4365$.
  (b) Variance of $b_2 = 0.3587$.
  (c) Variance of $\hat{Y} = 1.95075$.

6.   Analysis of Variance

Source of Variation	df	SS	MS	F
Total (corrected)	10	190.00		
Regression	2	122.00	61.00	7.18
due to b_1	1	116.08	116.08	13.64*
b_2 given b_1	1	5.92	5.92	<1
Residual	8	68.00	8.50	

7.   Analysis of Variance

Source of Variation	df	SS	MS	F
Total (corrected)	10	190.00		
Regression	2	122.00	61.00	7.18
due to b_2	1	98.33	98.33	11.57*
b_1 given b_2	1	23.67	23.67	2.78
Residual	8	68.00	8.50	

8. *Conclusions*

(a) While the regression equation, $\hat{Y} = 14 - 2X_1 - \frac{1}{2}X_2$, is statistically significant, the mean square error is larger than that obtained when $\hat{Y} = 9.162 - 1.027X_1$ is used.

(b) Independent estimates of β_1 and β_2 are not obtainable from these data. If independent estimates of β_1 and β_2 are desired, a balanced experiment in X_1 and X_2 should be done.

(c) When problems arise as to a choice of a model, more experimental work of a balanced nature is usually necessary.

B.

$$\hat{X} = 1.0607 + 0.0056\,Y - 0.0013Z.$$

Analysis of Variance

ANOVA

Source of Variation	df	SS	MS	F
Total (corrected)	11	0.294867		
Regression	2	0.236409	0.118204	18.20
due to Y	1	0.236275	0.236275	36.38*
due to $Z \mid Y$	1	0.000134	0.000134	<1
due to Z	1	0.236006	0.236006	36.34*
due to $Y \mid Z$	1	0.000403	0.000403	<1
Residual	9	0.058458	0.006495	

Conclusion: The inclusion of both Y and Z in the model is not useful. This is further demonstrated by the correlation coefficient, $r_{yz} = -0.9978$.

C.

ANOVA

Source of Variation	df	SS	MS	F
Total (corrected)	14	85386.0000		
Regression	2	49791.1751	24895.58755	8.39*
due to X_1 (alone)	1	48186.1482	48186.1482	16.24*
due to $X_2 \mid X_1$	1	1605.0269	1605.0269	<1
Residual	12	35594.8249	2966.2354	

Conclusions

1. The predictive model

$$\hat{Y} = 124.063977 + 3.512038X_1 + 0.834632X_2$$

explains only 58.31 % of the total variability in G.C.E. examination scores. While it proves to be statistically significant for an α-risk of 0.011, the standard deviation of the residuals is **54.46** and expressed as a per cent of the mean exam score, 9.725%. This shows that there is a great deal of unexplained variation, and thus the equation will not be very useful for prediction.

2. The addition of X_2, the previous performance in S. C. English Language, adds nothing to the predictability of a candidate's total mark in the G.C.E. examination. A simple model $Y = \beta_0 + \beta_1 X_1 + \epsilon$ would do as well.

D. 1. $\hat{Y} = -94.552026 + 2.801551 X_1 + 1.072683 X_2$.

2.
ANOVA

Source of Variation	df	SS	MS	F
Total (corrected)	7	2662.14		
Regression	2	2618.98	1309.49	151.74*
Residual	5	43.16	8.63	

Since $F(2, 5, 0.95) = 5.79$, the overall regression is statistically significant, i.e., $151.74 > 5.79$.

3. $R^2 = \dfrac{\text{SS Regression}}{\text{Total SS}} = \dfrac{2618.98}{2662.14} = 98.38\%$.

E. 1. $\hat{Y} = 67.234527 + 0.906089(X_1 - 164) - 0.064122(X_2 - 213)$.

2.
ANOVA

Source of Variation	df	SS	MS	F
Total (corrected)	15	8429.14444		
Regression	2	6796.77105	3398.385525	26.90*
Residual	13	1632.37339	126.336415	

Multiple $R^2 = 80.5\%$.
Standard deviation of residuals = 11.239947.
The fitted model is statistically significant. However 20% of the variability remains unexplained; more work needs to be done on this problem.

3.

ANOVA

Source of Variation	df	SS	MS	F
Total (corrected)	15	8429.14444		
Regression	2	6796.77105	3398.385525	26.90*
X_1	1	6777.72877	6777.72877	53.65*
$X_2 \mid X_1$	1	19.04228	19.04228	NS
X_2	1	25.10057	25.10057	NS
$X_1 \mid X_2$	1	6771.67048	6771.67048	53.60*
Residual	13	1632.37339	126.33645	

X_1 is the more important variable.

4.   *Conclusions*

(a) In the region of this machine's operation as indicated by the levels of Plate Clearance and Plate Temperature, the Plate Clearance has a pronounced effect on the per cent properly sealed.

(b) There is little evidence present that the Plate Temperature has an additive effect on the per cent properly sealed.

3.   There are some indications in the data that a different model should be fitted. It is helpful to examine the observations after they have been rearranged into the format below.

Sealer Plate Temperature

		176–208	210–220	225–240
	130–148	35 42.5 43.5	34.5	56.7 51.7
Sealer Plate Clearance	156–178	81.7 94.3	44.3 91.4	52.7
	186–194	82.0	98.3 83.3	95.4 84.4

One can see a definite interaction occurring between the clearance and the temperature. Thus a second-order model would be more appropriate.

Chapter 5

A. 1. $b_0 = -\frac{1}{2}$, intercept of line #1.
   $b_1 = 2$, slope of line #1.
   $b_2 = 1$, slope of line #2.

  2. $b_3 = 0.2$ indicates that $\hat{Y}$ predicted at the fifth observation by line #2 is +0.2 of a unit higher than $\hat{Y}$ predicted at the fifth observation by line #1. Thus, the point of intersection of the two lines is to the right of the fifth observation.

  3.

ANOVA

Source of Variation	df	SS	MS	F
Total (corrected)	8	146.50		
Regression	3	145.20	48.40	186.15*
$b_1 \mid b_0$	1	132.22	132.22	508.54*
$b_2 \mid b_0, b_1$	1	12.83	12.83	49.35*
$b_3 \mid b_0, b_1, b_2$	1	0.15	0.15	<1 NS
Residual	5	1.30	0.26	

B.   Final matrix is

$$\mathbf{Z} = \begin{bmatrix} 1 & -\frac{17}{5} & \frac{215}{186} & 0 \\ 1 & -\frac{12}{5} & -\frac{822}{186} & 0 \\ 1 & -\frac{2}{5} & \frac{824}{186} & 0 \\ 1 & \frac{13}{5} & \frac{131}{186} & 0 \\ 1 & \frac{18}{5} & -\frac{348}{186} & 0 \end{bmatrix}$$

C. 1.   The model contains ten parameters. Examination of the data reveals only eight different data points. Thus, it is impossible to fit the model as stated.

  2. $s^2 = 4.325$ with 10 degrees of freedom.

D.   The basic reason for the computer's failure to provide estimates is that the experimental design and model used provide a singular $(\mathbf{X'X})$ matrix which cannot be inverted. Here, $X_1^4 = 10X_1^2 - 9$ and $X_2^4 = 10X_2^2 - 9$, for each row of $\mathbf{X}$.
   Both models have the same singularity problems.

E.   It is simpler to work in terms of transformed observations $Y_i - 0.56$ instead of the original Y_i. If a first-order model is fitted, the residual plots indicate the need for higher order terms. The analysis of variance table below shows that only the first and fifth order coefficients are statistically significant. If we retain all terms up to and including fifth order we obtain the rather unwieldy fitted equation below, which could be rearranged to collect terms involving the same powers of $(Z - 1957)$.

$$\hat{Y} = 1.060 + 0.070143(Z - 1957) - 0.000386[3(Z - 1957)^2 - 56]$$

$$- \frac{0.000824}{6}[5(Z - 1957)^3 - 167(Z - 1957)]$$

$$- \frac{0.000058}{12}[35(Z - 1957)^4 - 1655(Z - 1957)^2 + 9072]$$

$$- 0.000102\left(\frac{21}{20}\right)\left[(Z - 1957)^5 - \frac{545}{9}(Z - 1957)^3\right.$$

$$\left. + \frac{708,032}{1008}(Z - 1957)\right]$$

Source of Variation	df	SS	MS
a_0	1	3.750000	3.750000
a_1	1	1.377606	1.377606*
a_2	1	0.005539	0.005539
a_3	1	0.027012	0.027012
a_4	1	0.022108	0.022108
a_5	1	0.109029	0.109029*
Residual	9	0.071906	0.007990
Total	15	5.363200	

F.   Let Z = week number. The fitted equation is

$$\hat{Y} = 136.227 + 2.687Z + 0.167Z^2.$$

Source of Variation	df	SS	MS	F
Total	20	74,628.00		
a_0	1	48,609.80		
a_1	1	25,438.75	25,438.75	4,558.92*
a_2	1	489.00	489.00	87.63*
a_3	1	1.15	1.15	0.21
Residual	16	89.30	5.58	

G. 1.

$$
\begin{array}{cccc}
& X_0 & X_1 & X_2 & X_3 \\
\end{array}
$$

$$
\mathbf{X} = \begin{bmatrix}
1 & 1 & 0 & -1 \\
1 & 1 & 0 & 0 \\
1 & 1 & 0 & 1 \\
1 & 0 & 1 & -1 \\
1 & 0 & 1 & 0 \\
1 & 0 & 1 & 1 \\
1 & -1 & -1 & -1 \\
1 & -1 & -1 & 0 \\
1 & -1 & -1 & 1
\end{bmatrix}
$$

$$\hat{Y} = 248 + 2X_1 - 10X_2 - 7.33X_3$$

2.

ANOVA

Source of Variation	df	SS	MS	F
Total (corrected)	8	1466.0		
Regression	3	826.7	257.57	NS
$b_1 \mid b_0$	1	54.0 ⎫ 504.0	252.00	NS
$b_2 \mid b_0, b_1$	1	450.0 ⎭		
$b_3 \mid b_0, b_1, b_2$	1	322.7	322.70	NS
Residual	5	639.3	127.9	

Operator differences are not statistically significant; i.e. $\dfrac{252.0}{127.9} = 1.97$ is less than $F(2, 5, 0.95) = 5.79$.

Operator #1: $\hat{Y} = 248 + 2(1) = 250$
Operator #2: $\hat{Y} = 248 - 10(1) = 238$
Operator #3: $\hat{Y} = 248 + 2(-1) - 10(-1) = 256$

3. There is not sufficient evidence to say line speed affects bar appearance with an α risk of 0.05.

4. Residual plots indicate a second-order model with line speed a better choice.

H. 1. $\hat{Y} = -.0037 - 2.8008t + .2314t^2$

2. Residual Analysis: There is no evidence for increasing the degree of the polynomial in t.

I. 1. $\hat{Y} = 22.561235 + 1.668017X - 0.067958X^2$.

2.

ANOVA

Source of Variation	df	SS	MS	F
Total (corrected)	18	204.481053		
Regression	2	201.994394	100.997197	649.8*
Residual	16	2.486659	0.155416	

The regression is statistically significant at $\alpha = 0.05$.

3. Test for lack of fit (breakdown of residual term in the above ANOVA table)

Source	df	SS	MS	F
Residual	16	2.486659		
Lack of fit	8	1.733325	0.216666	2.30
Pure error	8	0.753334	0.094168	

The lack of fit is nonsignificant since $2.30 < F(8, 8, 0.95) = 3.44$. Thus, the quadratic model is sufficient for predictive purposes.

4. $\hat{Y} = 23.346374 + 1.045463X$.

ANOVA

Source of Variation	df	SS	MS	F
Total	18	204.481053		
Regression	1	195.242967	195.242967	359.29*
Residual	17	9.238086	0.543417	
Lack of fit	9	8.484752	0.942750	10.01*
Pure error	8	0.753334	0.094168	

The lack of fit is statistically significant since $10.01 > F(9, 8, 0.95) = 3.39$.

Residuals from the fitted equation also demonstrate the inadequacy of the model.

By plotting the residuals against the values of X, the independent variable, one sees a definite curvature in the residuals which indicates the need for a second-order term in X.

5.   *Conclusions*

The cloud point can be predicted by a second-order function of the per cent I — 8 in the base stock,

$$\hat{Y} = 22.561235 + 1.668017X - 0.067958X^2.$$

There is no indication that a more complicated model is needed.

Chapter 6

A. 2. $b_4 \pm t(14, 0.95) \cdot \text{s.e.} (b_4)$

$-21.567372 \pm (2.145)(8.9558802)$

$-40.777735 \le \beta_{4Y \cdot 2} \le 2.357009$

3.

$$R^{-1} = \begin{bmatrix} 1.024102384 & -0.335666758 & 0.392338720 \\ -0.335666766 & 6.504097791 & -6.001426988 \\ 0.392338726 & -6.001426959 & 6.544384320 \end{bmatrix}$$

B. 1. $\hat{Y} = \hat{X}_6 = f(X_1, X_2, X_3, X_4)$

or $\hat{Y} = 6360.3385 + 13.868864X_1 + 0.211703X_2$
$\qquad - 126.690360X_3 - 21.817974X_4.$

2.   The above least squares equation shows an R^2 of 76.702710 which is the highest one in the given regression information. The overall $F = 9.8770256$ is statistically significant. The partial F-values are also all statistically significant. None of the 95% confidence limits on the β coefficients includes zero. The standard deviation as per cent of response mean = 7.536% which is lower than any other in the given information.

3.   The random vector X_5 does not contribute significantly to the explanation of variation. Actually, it contributes less than 1% and it increases the standard deviation as per cent of the response from 7.536 to 7.759. The partial F-test also shows that this variable is not statistically significant.

C.   The regression model $Y = f(X_2, X_4)$ explains only 57.4% of the total variation. The standard deviation is 9.4% of the response mean. All F's are statistically significant. However, if one plots the residuals versus the observations, there is a definite indication that another variable is needed; there is a trend in the residuals.

The regression model $Y = f(X_1, X_2, X_3, X_4)$ produces an $R^2 = 76.7\%$; the standard deviation is reduced to 7.5% of the mean response; all variables are statistically significant. However, there is still evidence of problems when the water usage is high, even though the model works pretty well.

D. 1. $Y = f(X_1)$ because X_1 has the highest correlation with Y.
    2. $s^2 = 13.444444$.
    3.

ANOVA

Source of Variation	df
Total	21
Mean (b_0)	1
Corrected total	20
Due to regression	3
$b_1 \mid b_0$	1
$b_2 \mid b_0, b_1$	1
$b_3 \mid b_0, b_1, b_2$	1
Residuals	17
Lack of fit	11
Pure error	6

4. The F-test for lack of fit, because replicates are available for an estimate of pure error.
5. Yes, while 6 df for pure error is not many, it should be sufficient.
6. $\hat{Y} = -50.358836 + 0.6711545X_1 + 1.295351X_2$.
7. Plots do not suggest that an obvious variable is missing.
8. The residuals on page 211 show that the fitted equation is least satisfactory at $(X_1, X_2, X_3) = (70, 20, 91)$; thus one would be reluctant to use the equation in that neighborhood. Future runs should be chosen to provide more balanced coverage of the X-space.

E. 2. *Step* 1 (Put in X_1)

ANOVA

Source of Variation	df	SS	MS	F	Partial F
Total (corrected)	8	1279.20010			
$b_1 \mid b_0$	1	822.27852	822.27852	12.597	12.597
Residual	7	456.92166	65.27452		

Step 2 (Put in X_2)

ANOVA

Source of Variation	df	SS	MS	F	Partial F
Total (corrected)	8	1279.20010			
Regression	2	938.29314	469.14657	8.257	
$b_1 \mid b_0$	1	822.27852	822.27852	14.472	3.236
$b_2 \mid b_0, b_1$	1	116.01462	116.01462	2.042	2.042
Residual	6	340.90696	56.81783		

Step 3 (Put in X_3)

ANOVA

Source of Variation	df	SS	MS	F	Partial F
Total (corrected)	8	1279.20010			
Regression	3	1081.34820	360.44940	9.109	
$b_1 \mid b_0$	1	822.27852	822.27852	20.780	0.056
$b_2 \mid b_0, b_1$	1	116.01462	116.01462	2.932	5.599
$b_3 \mid b_0, b_1, b_2$	1	143.05506	143.05506	3.615	3.615
Residual	5	197.85190	39.57038		

Step 4 (Reject X_1)

ANOVA

Source of Variation	df	SS	MS	F	Partial F
Total (corrected)	8	1279.20010			
Regression	2	1079.12600	539.56300	16.181	
$b_2 \mid b_0$	1	754.40445	754.40445	22.624	17.197
$b_3 \mid b_0, b_2$	1	324.72155	324.72155	9.738	9.738
Residual	6	200.07410	33.34568		

3. Final equation: $\hat{Y} = 63.021 + 11.517X_2 - 0.816X_3$

4. Residual plots reveal no problems.

Chapter 7

A. $\hat{Y} = 0.4368012 + 0.0001139X_1 - 0.0051897X_3$
   $- 0.0018887X_4 + 0.0044263X_5.$
   The plot of the residuals versus $\hat{Y}$ indicates that the variance is not homogeneous. One should try weighted least squares, or perhaps a transformation on the Y_i.

This model explains only 76.9% of the total variation, and the confidence limits on $\beta_{1Y\cdot345}$ and $\beta_{4Y\cdot135}$ include zero. The standard deviation of the residuals is 3.3% of the mean response. Thus, the model predicts well, but is not as good as one would like. If one can get rid of the large variance for large Y's, the model will be much better.

B.   Model $Y = \alpha X_1{}^\beta X_2{}^\gamma X_3{}^\delta \cdot \epsilon$.

By taking logarithms to the base e we can convert the model into the linear form,

$$\ln Y = \ln \alpha + \beta \ln X_1 + \gamma \ln X_2 + \delta \ln X_3 + \ln \epsilon$$

or $\ln Y = 8.5495297 + 0.1684244 \ln X_1$
$$- 0.537137 \ln X_2 - 0.0144135 \ln X_3.$$

X_2 or Peripheral wheel velocity.

All except $X_3 =$ Feed viscosity, which provides an F-ratio of

$$2.15 < F(1, 31, 0.95) = 4.16.$$

With 95.52% variation explained and a small standard deviation of 1.563% of the response mean, this looks like a good prediction equation. Plots of residuals reveal no peculiarities.

C.   1   The model chosen is

$$\hat{Y} = 120.627 + 490.412X_2 - 5.716X_3 - 1107.847X_2{}^2.$$

A plot of the residuals reveals runs of $+$ and $-$ signs indicating the presence of unconsidered X-variables.

Adding second-order terms in X_2 and X_3 is of only marginal help. This equation has $R^2 = 90.27\%$, with a standard deviation of 6.2233.

The model in Anderson and Bancroft is

$$\hat{Y} - 84.204 = 2.463(X_1 - 1.86) - 75.369(X_2 - 0.188)$$
$$+ 1.584(X_3 - 7.64)$$
$$- 1.380(X_1 X_2 - 0.3507).$$

This model is not as good a fit. The residuals have a definite pattern and $R^2 = 75.49\%$.

Warning: The example in Anderson and Bancroft was used to illustrate regression calculations; there was no intention of building a best model.

D.   Model: $\hat{Y} = b_0 + b_2 X_2 + b_8 X_8$

or $\hat{Y} = 9.4742224 + 0.7616482X_2 - 0.0797608X_8$

$R^2 = 86.0\%$

Standard deviation as per cent of response mean $= 6.761\%$,

E.   1. $\hat{Y} = 87.158859 + 0.8519104X_1 + 0.5988662X_2$
$+ 2.3613018X_6 - 0.9755309X_9.$

where X_1 = year,

X_2 = Preseason precipitation in inches,

X_6 = Rainfall in July in inches,

X_9 = August temperature.

2.   The most important variable is X_1, which accounts for the upward trend in corn yield. Of all the other variables, only preseason precipitation, July rainfall, and August temperature contribute significantly to the regression.

3.   With an R^2 of 72.06% and standard deviation as per cent of response mean of 14.903%, this prediction equation needs to be improved. New variables should be found to bring R^2 up, and to decrease the standard deviation of residuals. Investigation of the residuals may yield some insight into this problem.

F.   1.   There is a lot of replication in the data. Thus, an independent estimate of pure error can be obtained. The analysis of variance can be written down as

ANOVA

Source of Variation	df
Total	47
Regression	5
Residuals	42
Lack of fit	2
Pure error	40

2. $\hat{Y} = 134.258 + 0.050X_1 - 0.012X_2$
$+ 0.834X_3 - 0.154X_4 - 3.804X_5.$

3.   The model is not adequate since the lack of fit test is statistically significant at $\alpha = 0.05$.

ANOVA

Source of Variation	df	SS	MS	F
Total	47	2850.3107		
Regression	5	1817.1055		
Residual	42	1033.2052		
Lack of fit	2	383.7052	191.8526⎤	11.82*
Pure error	40	649.5000	16.2375⎦	

4. This model explains only 63.75% of the variation, and it is not a good one.
    The residuals show definite nonrandom patterns.
5. This experiment is poorly designed; there are too many replicates and not enough different design points.

G. 1. The *prediction* equation obtained by the stepwise procedure using a critical F of 2.00 for acceptance and rejection is

$$\hat{Y} = 250.1875 - 2.3124998\left(\frac{X_1 - 146}{3}\right)$$

$$- 14.687499\left(\frac{X_2 - 69.5}{3.5}\right) - 2.8124997\left(\frac{X_6 - 289.5}{93.5}\right)$$

The optimum rate using this prediction equation will be at the point $\hat{Y} = 270$; $X_1 = 143$, $X_2 = 66$, $X_6 = 196$; and the other variables held at their mean levels, namely, $X_3 = -10$, $X_4 = 132.5$, $X_5 = 91.5$.

H. $\hat{Y}_1 = -2.80512 + 0.15176X_1 + 3.60191X_3$
    $\hat{Y}_2 = -2.84492 + 0.11344X_1 + 3.67343X_3$.

Chapter 9

A. a. Both methods of analysis will yield the following analysis of variance table:

ANOVA

Source of Variation	df	SS	MS	F	$F_{0.95}$
Total SS	18	24,403.750			
Mean	1	22,352.027			
Corr. total	17	2,051.723			
Steam pressure	2	963.721	481.861	11.728*	4.26
Blowing time	2	37.481	18.741	0.456	4.26
Interaction	4	680.756	170.189	4.142*	3.63
Pure error	9	369.765	41.085		

Thus, both steam pressure and the interaction of steam pressure and blowing time are statistically significant.

The regression model obtained for this problem is

$$\hat{Y} = 35.239 + 4.794X_1 - 10.339X_2 - 0.206X_3 + 1.861X_4$$
$$+ 5.773X_1X_3 - 2.394X_1X_4 + 6.506X_2X_3 - 3.361X_2X_4$$

where X_1, X_2 are dummy variables for steam pressure defined as follows:

X_1	X_2	
1	0	= 10# steam pressure
0	1	= 20# steam pressure
-1	-1	= 30# steam pressure

and X_3, X_4 are dummy variables for blowing time defined as

X_3	X_4	
1	0	= blowing time 1
0	1	= blowing time 2
-1	-1	= blowing time 3

Residual analysis indicates that the experiments have a much smaller variance at the low level of steam pressure. Since only two repeat runs are available at each set of conditions, the analysis is not necessarily invalid, but further investigation is clearly indicated. To appreciate the interaction effect, a table of mean values could be examined, or the mean responses could be plotted against blowing time for each level of steam pressure.

ANOVA

b.

Source of	df	SS	MS	F	$F_{0.95}$
Total	18	417.000			
Mean	1	34.722			
Corrected total	17	382.278			
Premix speed	2	24.111	12.055	1.080	4.26
Finished mix speed	2	7.444	3.722	0.333	4.26
Interaction	4	250.223	62.556	5.602*	3.63
Pure error	9	100.500	11.167		

Thus, only the interaction term is statistically significant in this experiment.

The regression model obtained for this problem is

$$\hat{Y} = 1.39 - 1.39X_1 + 1.44X_2 - 0.89X_3 + 0.61X_4 + 1.39X_1X_2$$
$$+ 1.89X_1X_4 + 4.56X_2X_3 - 1.44X_2X_4,$$

where X_1, X_2 are dummy variables for premix speeds, and X_3, X_4 are dummy variables for finished mix speeds as follows:

X_1	X_2		X_3	X_4	
1	0	= premix speed 1	1	0	= finished mix speed 1
0	1	= premix speed 2	0	1	= finished mix speed 2
−1	−1	= premix speed 3	−1	−1	= finished mix speed 3

Residual analysis indicates a much larger variability in the observations at the lowest level of finished mix speed. This should be investigated.

The significant interaction is most easily seen in the following table:

Table of Mean Responses

		X_2		
		−1	0	+1
	−1	0.5	2.5	−3
X_1	0	6.5	2.0	0
	+1	−5.5	1.5	8.0

B. 1(a)

ANOVA

Source	df	SS	MS	F
Total	16	921.0000		
Regression	4	881.2500	220.3125	66.51*
b_0	1	798.0625	798.0625	
b_1	1	18.0625	18.0625	5.45*
b_2	1	60.0625	60.0625	18.13*
b_{12}	1	5.0625	5.0625	1.53 NS
Residual	12	39.7500	3.3125	

(b) (1) Regression equation is significant.
   (2) All except b_{12}.
(c) $R^2 = 95.68\%$.

2(a)

$$b_0 = \frac{798.0625}{113} = 7.0625$$

$$b_1 = \frac{18.0625}{17} = 1.0625$$

$$b_2 = \frac{60.0625}{31} = 1.9375$$

$$b_{12} = \frac{5.0625}{-9} = -0.5625$$

$$\therefore \hat{Y} = 7.0625 + 1.0625X_1 + 1.9375X_2 - 0.5625X_1X_2.$$

(b) $s^2(\mathbf{X'CX}) = 0.6875$

$(3.3125)(\mathbf{X'CX}) = 0.6875$

$$\mathbf{X'CX} = \frac{0.6875}{3.3125} = 0.207547.$$

Variance of a single observation

$$= s^2(1 + \mathbf{X'CX}) = (3.3125)(1 + 0.207547)$$
$$= 4.0000.$$

(c) $\hat{Y}$ is 54 at $X_1 = 70$ and $X_2 = 150$

$V(\hat{Y}) = 0.6875$

$\therefore$ Confidence limits for the true mean value of Y are

$$\hat{Y} \pm t(11, 0.95)s(\hat{Y}) = 54 \pm (2.201)\sqrt{0.6875}$$
$$= 54 \pm (2.201)(0.8292)$$
$$= 54 \pm 1.8251.$$

3(a) The prediction equation determined by this analysis is

$$\hat{Y} = 7.0625 + 1.0625X_1 + 1.9375X_2 - 0.5625X_1X_2.$$

(b) The interaction term, X_1X_2, is not statistically significant at an α level of 0.05. Thus, there is some doubt as to the validity of the assumed model. However, this doubt is based on a small number of observations, $n = 16$, and the original model was based on the knowledge of the chemist. Before considering dropping the X_1X_2 term, more experimental work should be done. This is an example of the statement, "Even though a variable is nonsignificant statistically, it should *not* be considered to have a zero effect on the result of the experiment."

Chapter 10

A. $\hat{\theta} = 0.20345$, $S(\hat{\theta}) = 0.00030$; $0.179 \le \theta \le 0.231$.

B. $\hat{\theta} = 0.20691$, $S(\hat{\theta}) = 0.01202$; $0.190 \le \theta \le 0.225$.

C. $(\hat{\alpha}, \hat{\beta}) = (0.38073, 0.07949)$, $S(\hat{\alpha}, \hat{\beta}) = 0.00005$;
      $S(\alpha, \beta) = 0.001$ (or 0.0009 to one more decimal place)

D. $(\hat{\alpha}, \hat{\beta}, \hat{\rho}) = (72.4326, 28.2519, 0.5968)$, $S(\hat{\alpha}, \hat{\beta}, \hat{\rho}) = 3.5688$;
      $S(\alpha, \beta, \rho) = 106.14$.

E. $(\hat{\alpha}, \hat{\beta}, \hat{\gamma}) = (5.2673, 8.5651, 294.9931)$,
      $S(\hat{\alpha}, \hat{\beta}, \hat{\gamma}) = 1718.2108$; $S(\alpha, \beta, \gamma) = 3400$.

F.   Write the model as

$$Y = \theta + \alpha(X_1 X_3 + \gamma X_1) + \beta(X_2 X_3 + \gamma X_2) + \epsilon.$$

Fix γ, solve for $\hat{\theta}$, $\hat{\alpha}$, $\hat{\beta}$. Repeat for other values of γ, iterating on γ until minimum $S(\hat{\theta}, \hat{\alpha}, \hat{\beta}, \hat{\gamma})$ is obtained.

APPENDIX A

Explanation of Data and Symbols
in the Appendices

Identification of the Variables

In each appendix the first page gives a statement of the regression problem followed

by the identification of the variables used in the problem.

Original and/or Transformed Data

The input data is then listed. Each row represents a simultaneous set of observa-

tions on each independent variable and dependent variable. The number of such

observations is identified by the row number at the left of the input data matrix.

Means of Transformed Variables

This is a row of the average values of each column.

Std. Deviations of Transformed Variables

Each item in this row is the standard deviation of a column of the input data.

Correlation Matrix

This is a matrix of calculated correlation coefficients, r_{ij}. The diagonal elements should be exactly one. The failure of this to be true in a computer printout is due to a machine rounding error.

Control Information

Listed under control information are:

a) No. of observations--This is the number of observations on the response in the problem.

b) Response variable is no. --This identifies which column of the input data matrix is to be considered the response. For example, the no. 4 would indicate that the data in column 4 is to be considered the response or Y variable.

c) Risk level for β confidence interval--This line specifies the probability, α, of making a type I error. It is used to place $1-\alpha$ confidence intervals on the regression coefficients.

d) List of excluded variables--These numbers identify which vectors are not to be considered in fitting the regression model.

e) Variable entering--This identifies which independent variable has been selected to enter regression at this stage.

f) Sequential F-test--This is the F-test for testing whether the last variable in regression has contributed significantly in reducing the amount of unexplained variation. This test is discussed in detail in Chapters II and IV.

g) Percent variation explained-R-SQ--This is the square of the multiple correlation coefficient, R^2. An explanation of this statistic is given in Chapters I, II and IV.

h) Standard deviation of residuals--This is the square root of the mean square error in the Analysis of Variance Table.

) Mean of the response--This is the arithmetic average of all the observed values of the response in the data.

) Std. dev. as % of response mean--This is a measure of the size of the standard deviation of the residuals relative to the average response and is calculated as

$$\frac{\text{Standard deviation of residuals}}{\text{Average response}}$$

) Degrees of freedom--This is the number of degrees of freedom used to calculate the standard deviation of the residuals.

) Determinant value--This is the value of the determinant of the correlation matrix of all variables in the regression on each printout.

ANOVA

This is the Analysis of Variance Table for the regression problem.

Source - Under the heading, source, are listed all the sources of variation in the problem.

d.f. - This column shows the degrees of freedom in each of the sources of variation.

Sums sqs. - This column headed sums of squares indicates the sums of squares for each source of variation.

Mean sq. - The mean square column is obtained by dividing each sum of squares by its corresponding degrees of freedom.

Overall F - This is the F-statistic for determining the statistical significance of the regression model under consideration at each step. It is calculated as,

$$F = \frac{\text{mean square due to regression}}{\text{mean square due to residual}}$$

Coefficients and Confidence Limits

Var No. - This indicates the number of each independent variable in the regression model.

Mean - The average value of all the observations on the independent variable.

Decoded B Coefficient - If any coding was performed on the input after its submission to the computer, the program decodes the b-coefficient into original units and prints it out here.

Limits Upper/Lower - These are the 95% confidence limits on the true regression coefficient, β. The calculation formulas for these limits are shown in Chapters I and IV.

Standard error - This is the standard error of the "b" coefficient. The formula for its calculation is shown in Chapters I and IV.

Partial F-test - This is the F-test for each variable as if it were the last variable entering the regression. A discussion of this statistic is found in Chapters V and VI.

Constant Term in Prediction Equation - This is the least squares estimate of β_0.

Squares of Partial Correlation Coefficients of Variables Not in Regression

The partial correlation coefficient of each variable not in regression with the response is calculated and squared. A discussion of this statistical indicator is found in Chapters V and VI.

Residual Analysis

The data recorded here are the results of the fitted model to exactly fit all the data.

Observed Y - the observation on the response, Y, for each data point shown in the input data matrix.

Predicted Y - the value of the predicted Y, or $\hat{Y}$, using the fitted model.

Residual - the difference between the observed Y and the predicted Y, or $Y-\hat{Y}$.

Normal deviate - the difference, $Y-\hat{Y}$, divided by the standard deviation of the residuals, s, or $\dfrac{Y-\hat{Y}}{s}$.

HFS STEAM TARGET REGRESSION

Machine Printouts for the
Explanation of a 3-Variable
Multiple Regression Problem

The Original Data

Col. 1   Response vector--Pounds of Steam Used Monthly

Col. 2   Pounds of Real Fatty Acid in storage per month

Col. 3   Pounds of Crude Glycerine Made

Col. 4   Average wind velocity in miles per hour

Col. 5   Calendar days per month

Col. 6   Operating days per month

Col. 7   Days below 32°F.

Col. 8   Average Atmospheric Temperature, degrees Fahrenheit

Col. 9   (Average wind velocity)^2

Col. 10   Number of startups

Original and/ or Transformed Data

	X_1	X_2	X_3	X_4	X_5	X_6	X_7	X_8	X_9	X
1	10.980	5.2000	.61000	7.4000	31.000	20.000	22.000	35.300	54.800	4.0
2	11.130	5.1200	.64000	8.0000	29.000	20.000	25.000	29.700	64.000	5.0
3	12.510	6.1900	.78000	7.4000	31.000	23.000	17.000	30.800	54.800	4.0
4	8.400	3.8900	.49000	7.5000	30.000	20.000	22.000	58.800	56.300	4.0
5	9.270	6.2800	.84000	5.5000	31.000	21.000	.000	61.400	30.300	5.0
6	8.730	5.7600	.74000	8.9000	30.000	22.000	.000	71.300	79.200	4.0
7	6.360	3.4500	.42000	4.1000	31.000	11.000	.000	74.400	16.800	2.0
8	8.500	6.5700	.87000	4.1000	31.000	23.000	.000	76.700	16.800	5.0
9	7.820	5.6900	.75000	4.1000	30.000	21.000	.000	70.700	16.800	4.0
10	9.140	6.1400	.76000	4.5000	31.000	20.000	.000	57.500	20.300	5.0
11	8.240	4.8400	.65000	10.3000	30.000	20.000	11.000	46.400	106.100	4.0
12	12.190	4.8800	.62000	6.9000	31.000	21.000	12.000	28.900	47.600	4.0
13	11.880	6.0300	.79000	6.6000	31.000	21.000	25.000	28.100	43.600	5.0
14	9.570	4.5500	.60000	7.3000	28.000	19.000	18.000	39.100	53.300	5.0
15	10.940	5.7100	.70000	8.1000	31.000	23.000	5.000	46.800	65.600	4.0
16	9.580	5.6700	.74000	8.4000	30.000	20.000	7.000	48.500	70.600	4.0
17	10.090	6.7200	.85000	6.1000	31.000	22.000	.000	59.300	37.200	6.0
18	8.110	4.9500	.67000	4.9000	30.000	22.000	.000	70.000	24.000	4.0
19	6.830	4.6200	.45000	4.6000	31.000	11.000	.000	70.000	21.200	3.0
20	8.880	6.0000	.95000	3.7000	31.000	23.000	.000	74.500	13.700	4.0
21	7.680	5.0100	.64000	4.7000	30.000	20.000	.000	72.100	22.100	4.0
22	8.470	5.6800	.75000	5.3000	31.000	21.000	1.000	58.100	28.100	6.0
23	8.860	5.2800	.70000	6.2000	30.000	20.000	14.000	44.600	38.400	4.C
24	10.360	5.3600	.67000	6.8000	31.000	20.000	22.000	33.400	46.200	4.0
25	11.080	5.8700	.70000	7.5000	31.000	22.000	28.000	28.600	56.300	5.0

Means of Transformed Variables

	X_1	X_2	X_3	X_4	X_5	X_6	X_7	X_8	X_9	X
1	9.424	5.4424	.69520	6.3560	30.480	20.240	9.160	52.600	43.364	4.3

Std. Deviations of Transformed Variables

	X_1	X_2	X_3	X_4	X_5	X_6	X_7	X_8	X_9	X
1	1.630	.8169	.12586	1.7540	.770	3.017	10.282	17.265	23.198	.8

Correlation Matrix

1	1.00000	.38318	.30555	.47431	.13674	.53612	.64065	-.84524	.39454	.3
2	.38318	1.00000	.94364	-.12609	.38213	.68509	-.19112	-.00188	-.13135	.6
3	.30555	.94364	1.00000	-.14367	.24823	.76446	-.22636	.06774	-.13419	.6
4	.47431	-.12609	-.14367	1.00000	-.31677	.23114	.55810	-.61634	.98996	.0
5	.13674	.38213	.24823	-.31677	1.00000	.02008	-.20475	.07738	-.32100	-.0
6	.53612	.68509	.76446	.23114	.02008	1.00000	.11688	-.20976	.21248	.6
7	.64065	-.19112	-.22636	.55810	-.20475	.11688	1.00000	-.85761	.49151	.1
8	-.84524	-.00188	.06774	-.61634	.07738	-.20976	-.85761	1.00000	-.54142	-.2
9	.39454	-.13135	-.13419	.98996	-.32100	.21248	.49151	-.54142	1.00000	.0
10	.38212	.61630	.60130	.07390	-.05330	.60059	.11751	-.23695	.02842	1.C

Simple linear regression, $\hat{X}_1 = f(X_8)$

This corresponds to $\hat{Y} = f(X_8)$ in the notes or the amount of steam

as a linear function of average atmospheric temperature in $^{\circ}F$.

Control Information

No. of observations	25
Response variable is no.	1
Risk level for B conf. interval	5%
List of excluded variables	2, 3, 4, 5, 6, 7, 9, 10

Variable entering	8
Sequential F-test	57.5427930
Percent variation explained R-SQ	71.4437600
Standard deviation of residuals	.8901244
Mean of the Response	9.4240000
Std. dev. as % of response mean	9.445%
Degrees of freedom	23
Determinant value	.9999998

ANOVA

Source	d.f.	Sums sqs.	Mean sq.	Overall F
Total	24	63.8158000		
Regression	1	45.5924060	45.5924060	57.5428050
Residual	23	18.2233950	.7923215	

B Coefficients and Confidence Limits

Var No.	Mean	Decoded B Coefficient	Limits Upper/Lower	Standard Error	Partial F-test
8	52.6000000	-.0798287	-.0580554 -.1016020	.0105236	57.5427970

Constant Term in Prediction Equation   13.6229890

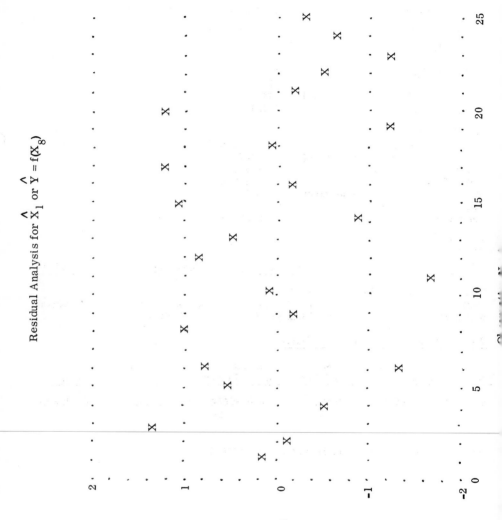

Residual Analysis for $\hat{X}_1$ or $\hat{Y} = f(X_8)$

Residual Analysis for $\hat{X}_1$ or $\hat{Y} = f(X_8)$

Obs. No.	Observed Y	Predicted Y	Residual	Normal Deviate
1	10.9800000	10.8050370	.1749630	.1965602
2	11.1300000	11.2520770	-.1220770	-.1371460
3	12.5100000	11.1642660	1.3457340	1.5118492
4	8.4000000	8.9290620	-.5290620	-.5943686
5	9.2700000	8.7215080	.5484920	.6161970
6	8.7300000	7.9312040	.7987960	.8973981
7	6.3600000	7.6837350	-1.3237350	-1.4871347
8	8.5000000	7.5001290	.9998710	1.1232935
9	7.8200000	7.9791010	-.1591010	-.1787402
10	9.1400000	9.0328400	.1071600	.1203877
11	8.2400000	9.9189380	-1.6789380	-1.8861834
12	12.1900000	11.3159400	.8740600	.9819526
13	11.8800000	11.3798030	.5001970	.5619405
14	9.5700000	10.5016880	-.9316880	-1.0466941
15	10.9400000	9.8870070	1.0529930	1.1829727
16	9.5800000	9.7512980	-.1712980	-.1924428
17	10.0900000	8.8891480	1.2008520	1.3490832
18	8.1100000	8.0349810	.0750190	.0842792
19	6.8300000	8.0349810	-1.2049810	-1.3537219
20	8.8800000	7.6757520	1.2042480	1.3528984
21	7.6800000	7.8673410	-.1873410	-.2104661
22	8.4700000	8.9849420	-.5149420	-.5785056
23	8.8600000	10.0626300	-1.2026300	-1.3510807
24	10.3600000	10.9567110	-.5967110	-.6703681
25	11.0800000	11.3398890	-.2598890	-.2919693

Simple linear regression, $\hat{X}_6 = f(X_8)$

or the operating days per month written as a linear

function of the average atmospheric temperature in ^{O}F.

Control Information

No. of observations	25
Response variable is no.	6
Risk level for B conf. interval	5%
List of excluded variables	1, 2, 3, 4, 5, 7, 9, 10

Variable entering	8
Sequential F-test	1.0585742
Percent variation explained R-SQ	4.4000000
Standard deviation of residuals	3.0140493
Mean of the response	20.2400000
Std. dev. as % of response mean	14.892%
Degrees of freedom	23
Determinant value	.9999998

ANOVA

Source	d.f.	Sums sqs.	Mean sq.	Overall F
Total	24	218.5599900		
Regression	1	9.6166395	9.6166395	1.0585773
Residual	23	208.9433500	9.0844934	

B Coefficients and Confidence Limits

Var. No.	Mean	Decoded B Coefficient	Limits Upper/Lower	Standard Error	Partial F-test
8	52.6000000	-.0366626	.0370639	.0356339	1.0585740
			-.1103892		

Constant Term in Prediction Equation   22.1684550

$$\text{Residual Analysis for } \hat{X}_6 = f(X_8)$$

Obs. No.	Observed Y	Predicted Y	Residual	Normal Deviate
1	20.0000000	20.8742640	-.8742640	-.2900629
2	20.0000000	21.0795750	-1.0795750	-.3581809
3	23.0000000	21.0392460	1.9607540	.6505381
4	20.0000000	20.0126920	-.0126920	-.0042109
5	21.0000000	19.9173690	1.0826310	.3591949
6	22.0000000	19.5544090	2.4455910	.8113971
7	11.0000000	19.4407550	-8.4407550	-2.8004701
8	23.0000000	19.3564310	3.6435690	1.2088617
9	21.0000000	19.5764060	1.4235940	.4723194
10	20.0000000	20.0603530	-.0603530	-.0200239
11	20.0000000	20.4673090	-.4673090	-.1550436
12	21.0000000	21.1089050	-.1089050	-.0361325
13	21.0000000	21.1382350	-.1382350	-.0458635
14	19.0000000	20.7349460	-1.7349460	-.5756196
15	23.0000000	20.4526440	2.5473560	.8451607
16	20.0000000	20.3903170	-.3903170	-.1294992
17	22.0000000	19.9943610	2.0056390	.6654301
18	22.0000000	19.6020700	2.3979300	.7955842
19	11.0000000	19.6020700	-8.6020700	-2.8539911
20	23.0000000	19.4370880	3.5629120	1.1821014
21	20.0000000	19.5250790	.4749210	.1575691
22	21.0000000	20.0383560	.9616440	.3190538
23	20.0000000	20.5333010	-.5333010	-.1769384
24	20.0000000	20.9439230	-.9439230	-.3131744
25	22.0000000	21.1199040	.8800960	.2919979

Residual Analysis I

The residuals of $X_1 - \hat{X}_1$ where $\hat{X}_1 = f(X_8)$ are plotted against $X_6 - \hat{X}_6$ where $\hat{X}_6 = f(X_8)$. Then the residuals of $X_1 - \hat{X}_1$ are regressed against $X_6 - \hat{X}_6$. Finally, the residuals of this fit are calculated.

Analysis for $X_1 - \hat{X}_1 = f(X_6 - \hat{X}_6)$

where $\hat{X}_1 = f(X_8)$ and $\hat{X}_6 = f(X_8)$

Residuals for

$\hat{X}_1 = f(X_8)$

-1.679

-1.079

-.479

.121

.721

1.321

-8.602   -4.602   -.602   3.398   7.398   11.398

Residuals for $\hat{X}_6 = f(X_8)$

Control Information

No. of observations	25
Response variable is	$X_1 - \hat{X}_1$
Risk level for B conf. interval	5%

Variable entering	$X_6 - \hat{X}_6$
Sequential F-test	20.5300700
Percent variation explained R-SQ	47.1629600
Standard deviation of residuals	.6470233
Mean of the response	.0000006
Std. dev. as % of response mean	999.999%
Degrees of freedom	23
Determinant value	1.0000001

ANOVA

Source	d.f.	Sums sqs.	Mean sq.	Overall F
Total	24	18.2233910		
Regression	1	8.5946906	8.5946906	20.5300660
Residual	23	9.6287007	.4186392	

B Coefficients and Confidence Limits

Var No.	Mean	Decoded B Coefficient	Limits Upper/Lower	Standard Error	Partial F-test
2	-.00000044	.2028154	.2954271 .1102036	.0447616	20.5300700

Constant Term in Prediction Equation   .0000006892387

Obs. No.	Observed Y	Predicted Y	Residual	Normal Deviate
1	.1749640	-.1773135	.3522775	.5444587
2	-.1220760	-.2189537	.0968777	.1497283
3	1.3457350	.3976717	.9480633	1.4652691
4	-.5290620	-.0025734	-.5264886	-.8137088
5	.5484930	.2195749	.3289181	.5083559
6	.7987970	.4960041	.3027929	.4679783
7	-1.3237340	-1.7119142	.3881802	.5999478
8	.9998720	.7389725	.2608995	.4032305
9	-.1591000	.2887272	-.4478272	-.6921346
10	.1071610	-.0122398	.1194008	.1845387
11	-1.6789370	-.0947768	-1.5841603	-2.4483821
12	.8740610	-.0220869	.8961479	1.3850318
13	.5001980	-.0280355	.5282335	.8164057
14	-.9316870	-.3518730	-.5798140	-.8961253
15	1.0529940	.5166436	.5363504	.8289507
16	-.1712970	-.0791616	-.0921354	-.1423989
17	1.2008530	.4067751	.7940780	1.2272788
18	.0750200	.4863377	-.4113177	-.6357078
19	-1.2049800	-1.7446314	.5396514	.8340525
20	1.2042490	.7226138	.4816353	.7443863
21	-.1873400	.0963220	-.2836620	-.4384107
22	-.5149420	.1950369	-.7099789	-1.0973002
23	-1.2026290	-.1081612	-1.0944679	-1.6915432
24	-.5967100	-.1914414	-.4052686	-.6263586
25	-.2598880	.1784977	-.4383857	-.6775423

The bivariate regression of $\hat{X}_1$ or $\hat{Y} = f(X_6, X_8)$

The residuals are calculated and plotted by observation number.

Control Information

No. of observations	25
Response variable is no.	1
Risk level for B conf. interval	5%
List of excluded variables	2, 3, 4, 5, 7, 9, 10

Variables entering	8, 6
Sequential F-test	19.6374510
Percent variation explained R-SQ	84.9117300
Standard deviation of residuals	.6615651
Mean of the response	9.4240000
Std. dev. as % of response mean	7.020%
Degrees of freedom	22
Determinant value	.9559998

ANOVA

Source	d.f.	Sums sqs.	Mean sq.	Overall F
Total	24	63.8158000		
Regression	2	54.1870990	27.0935490	61.9042930
Residual	22	9.6287033	.4376683	

B Coefficients and Confidence Limits

Var No.	Mean	Decoded B Coefficient	Limits Upper/Lower	Standard Error	Partial F-test
8	52.6000000	-.0723929	-.0558022 -.0889837	.0079994	81.8992020
6	20.2400000	.2028154	.2977374 .1078933	.0457676	19.6374520

Constant Term in Prediction Equation 9.1268861

Residual Analysis for $\hat{Y} = f(X_6, X_8)$

Obs. No.	Observed Y	Predicted Y	Residual	Normal Deviate
1	10.9800000	10.6277220	.3522780	.5324918
2	11.1300000	11.0331220	.0968780	.1464376
3	12.5100000	11.5619360	.9480640	1.4330623
4	8.4000000	8.9264882	-.5264882	-.7958223
5	9.2700000	8.9410818	.3289182	.4971819
6	8.7300000	8.4272071	.3027929	.4576918
7	6.3600000	5.9718199	.3881801	.5867603
8	8.5000000	8.2391005	.2608995	.3943671
9	7.8200000	8.2678274	-.4478274	-.6769212
10	9.1400000	9.0205990	.1194010	.1804826
11	8.2400000	9.8241607	-1.5841607	-2.3945652
12	12.1900000	11.2938520	.8961480	1.3545878
13	11.8800000	11.3517660	.5282340	.7984612
14	9.5700000	10.1498130	-.5798130	-.8764263
15	10.9400000	10.4036490	.5363510	.8107305
16	9.5800000	9.6721355	-.0921355	-.1392690
17	10.0900000	9.2959224	.7940780	1.2003022
18	8.1100000	8.5213179	-.4113179	-.6217346
19	6.8300000	6.2903489	.5396511	.8157189
20	8.8800000	8.3983649	.4816351	.7280238
21	7.6800000	7.9636620	-.2836620	-.4287742
22	8.4700000	9.1799785	-.7099785	-1.0731801
23	8.8600000	9.9544680	-1.0944680	-1.6543618
24	10.3600000	10.7652690	-.4052690	-.6125913
25	11.0800000	11.5183850	-.4383850	-.6626484

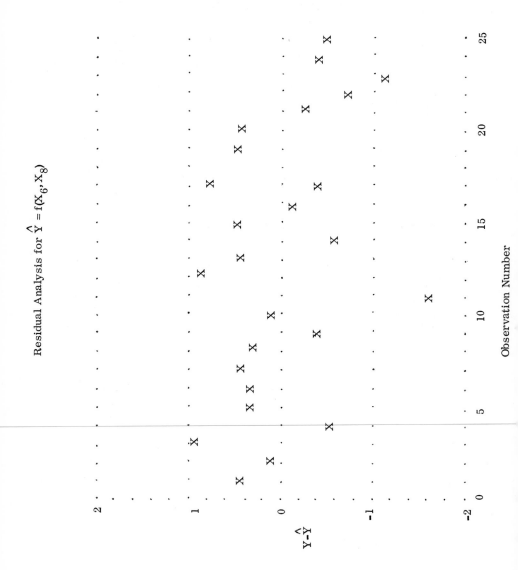

Residual Analysis for $\hat{Y} = f(X_6, X_8)$

Observation Number

APPENDIX B

Experiment from Hald's Statistical Theory with
Engineering Applications, page 647

Data Code

X_1 = am't of tricalcium aluminate, $3\ CaO \cdot Al_2O_3$

X_2 = am't of tricalcium silicate, $3\ CaO \cdot SiO_2$

X_3 = am't of tetracalcium alumino ferrite, $4\ CaO \cdot Al_2O_3 \cdot Fe_2O_3$

X_4 = am't of dicalcium silicate, $2\ CaO \cdot SiO_2$

(Response)$Y = X_5$ = heat evolved in calories per gram of cement

X_1, X_2, X_3, and X_4 are measured as percent of the weight of the

clinkers from which the cement was made.

This data was first printed in an article in Industrial and

Engineering Chemistry, 24, 1932, 1207-14, Table I, by H. Woods,

H. H. Steinour, and H. R. Starke: "Effect of Composition of

Portland Cement on Heat Evolved during Hardening."

Original and/ or Transformed Data

	X_1	X_2	X_3	X_4	X_5
1	7.00000000	26.00000000	6.00000000	60.00000000	78.5000000
2	1.00000000	29.00000000	15.00000000	52.00000000	74.3000000
3	11.00000000	56.00000000	8.00000000	20.00000000	104.3000000
4	11.00000000	31.00000000	8.00000000	47.00000000	87.6000000
5	7.00000000	52.00000000	6.00000000	33.00000000	95.9000000
6	11.00000000	55.00000000	9.00000000	22.00000000	109.2000000
7	3.00000000	71.00000000	17.00000000	6.00000000	102.7000000
8	1.00000000	31.00000000	22.00000000	44.00000000	72.5000000
9	2.00000000	54.00000000	18.00000000	22.00000000	93.1000000
10	21.00000000	47.00000000	4.00000000	26.00000000	115.9000000
11	1.00000000	40.00000000	23.00000000	34.00000000	83.8000000
12	11.00000000	66.00000000	9.00000000	12.00000000	113.3000000
13	10.00000000	68.00000000	8.00000000	12.00000000	109.4000000

Means of Transformed Variables

1	7.46153830	48.15384500	11.76923000	29.99999900	95.4230750

Std. Deviations of Transformed Variables

1	5.88239440	15.56087900	6.40512590	16.73817800	15.0437240

Correlation Matrix

1	.99999991	.22857948	-.82413372	-.24544512	.7307174
2	.22857948	1.00000010	-.13924238	-.97295516	.8162526
3	-.82413372	-.13924238	.99999991	.02953700	-.5346706
4	-.24544512	-.97295516	.02953700	1.00000010	-.8213051
5	.73071745	.81625268	-.53467065	-.82130513	.9999999

Control Information

No. of observations	13
Response variable is no.	5
Risk level for B conf. interval	5%
List of excluded variables	2, 3, 4

Variable entering	1
Sequential F-test	12.6025160
Percent variation explained R-SQ	53.3948000
Standard deviation of residuals	10.7267170
Mean of the response	95.4230750
Std. dev. as % of response mean	11.241%
Degrees of freedom	11
Determinant value	.9999999

ANOVA

Source	d.f.	Sums sqs.	Mean sq.	Overall F
Total	12	2715.7635000		
Regression	1	1450.0764000	1450.0764000	12.6025160
Residual	11	1265.6870000	115.0624500	

B Coefficients and Confidence Limits

Var No.	Mean	Decoded B Coefficient	Limits Upper/Lower	Standard Error	Partial F-test
1	7.4615383	1.8687477	3.0273705 .7101249	.5264075	12.6025150

Constant Term in Prediction Equation   81.4793430

Squares of Partial Correlation Coefficients of Variables Not in Regression

Variables	Square of Partials
2	.95425
3	.03051
4	.94093
5	1.00000

Residual Analysis for $\hat{X}_5 = f(X_1)$

Obs. No.	Observed Y	Predicted Y	Residual	Normal Deviate
1	78.5000000	94.5605760	-16.0605760	-1.4972499
2	74.3000000	83.3480900	-9.0480900	-.8435097
3	104.3000000	102.0355600	2.2644400	.2111028
4	87.6000000	102.0355600	-14.4355600	-1.3457575
5	95.9000000	94.5605760	1.3394240	.1248680
6	109.2000000	102.0355600	7.1644400	.6679061
7	102.7000000	87.0855860	15.6144200	1.4556569
8	72.5000000	83.3480900	-10.8480900	-1.0113150
9	93.1000000	85.2168380	7.8831620	.7349091
10	115.9000000	120.7230400	-4.8230400	-.4496287
11	83.8000000	83.3480900	.4519100	.0421294
12	113.3000000	102.0355600	11.2644400	1.0501293
13	109.4000000	100.1668200	9.2331800	.8607648

Control Information

No. of observations	13
Response variable is no.	5
Risk level for B conf. interval	5%
List of excluded variables	1, 3, 4

Variable entering	2
Sequential F-test	21.9606150
Percent variation explained R-SQ	66.6268400
Standard deviation of residuals	9.0771249
Mean of the response	95.4230750
Std. dev. as % of response mean	9.513%
Degrees of freedom	11
Determinant value	1.0000000

ANOVA

Source	d.f.	Sums sqs.	Mean sq.	Overall F
Total	12	2715.7635000		
Regression	1	1809.4274000	1809.4274000	21.9606160
Residual	11	906.3361700	82.3941970	

B Coefficients and Confidence Limits

Var No.	Mean	Decoded B Coefficient	Limits Upper/ Lower	Standard Error	Partial F-test
2	48.1538450	.7891250	1.1597575 .4184924	.1683928	21.9606140

Constant Term in Prediction Equation   57.4236730

Squares of Partial Correlation Coefficients of Variables Not in Regression

Variables	Square of Partials
1	.93611
3	.54162
4	.04133
5	1.00000

Residual Analysis for $\hat{X}_5 = f(X_2)$

Obs. No.	Observed Y	Predicted Y	Residual	Normal Devia
1	78.500000	77.9409220	.5590780	.0615920
2	74.300000	80.3082970	-6.0082970	-.6619163
3	104.300000	101.6146700	2.6853300	.2958349
4	87.600000	81.8865470	5.7134530	.6294342
5	95.900000	98.4581720	-2.5581720	-.2818262
6	109.200000	100.8255400	8.3744600	.9225895
7	102.700000	113.4515400	-10.7515400	-1.1844653
8	72.500000	81.8865470	-9.3865470	-1.0340881
9	93.100000	100.0364200	-6.9364200	-.7641649
10	115.900000	94.5125470	21.3874600	2.3561932
11	83.800000	88.9886720	-5.1886720	-.5716206
12	113.300000	109.5059200	3.7940800	.4179826
13	109.400000	111.0841700	-1.6841700	-.1855400

Control Information

No. of observations	13
Response variable is no.	5
Risk level for B conf. interval	5%
List of excluded variables	1, 2, 4

Variable entering	3
Sequential F-test	4.4034159
Percent variation explained R-SQ	28.5872800
Standard deviation of residuals	13.2781460
Mean of the response	95.4230750
Std. dev. as % of response mean	13.915%
Degrees of freedom	11
Determinant value	.9999998

ANOVA

Source	d.f.	Sums sqs.	Mean sq.	Overall F
Total	12	2715.7635000		
Regression	1	776.3629100	776.3629100	4.4034177
Residual	11	1939.4008000	176.3091600	

B Coefficients and Confidence Limits

Var no.	Mean	Decoded B Coefficient	Limits Upper/ Lower	Standard Error	Partial F-test
3	11.7692300	-1.2557812	.0613807 -2.5729431	.5984380	4.4034159

Constant Term in Prediction Equation   110.2026500

Squares of Partial Correlation Coefficients of Variables Not in Regression

Variables	Square of Partials
1	.36729
2	.78579
4	.90939
5	1.00000

Residual Analysis for $\hat{X}_5 = f(X_3)$

Obs. No.	Observed Y	Predicted Y	Residual	Normal Deviate
1	78.5000000	102.6679700	-24.1679700	-1.8201313
2	74.3000000	91.3659400	-17.0659400	-1.2852652
3	104.3000000	100.1564100	4.1435900	.3120609
4	87.6000000	100.1564100	-12.5564100	-.9456448
5	95.9000000	102.6679700	-6.7679700	-.5097075
6	109.2000000	98.9006200	10.2993800	.7756640
7	102.7000000	88.8543700	13.8456300	1.0427381
8	72.5000000	82.5754700	-10.0754700	-.7588010
9	93.1000000	87.5985900	5.5014100	.4143206
10	115.9000000	105.1795300	10.7204700	.8073770
11	83.8000000	81.3196900	2.4803100	.1867964
12	113.3000000	98.9006200	14.3993800	1.0844420
13	109.4000000	100.1564100	9.2435900	.6961507

Control Information

No. of observations	13
Response variable is no.	5
Risk level for B conf. interval	5%
List of excluded variables	1, 2, 3

Variable entering	4
Sequential F-test	22.7985280
Percent variation explained R-SQ	67.4542100
Standard deviation of residuals	8.9639014
Mean of the response	95.4230750
Std. dev. as % of response mean	9.394%
Degrees of freedom	11
Determinant value	.9999999

ANOVA

Source	d.f.	Sums sqs.	Mean sq.	Overall F
Total	12	2715.7635000		
Regression	1	1831.8968000	1831.8968000	22.7985300
Residual	11	883.8668200	80.3515290	

B Coefficients and Confidence Limits

Var No.	Mean	Decoded B Coefficient	Limits Upper/Lower	Standard Error	Partial F-test
4	29.9999990	-.7381619	-.3978962 -1.0784277	.1545960	22.7985270

Constant Term in Prediction Equation   117.5679300

Squares of Partial Correlation Coefficients of Variables Not in Regression

Variables	Square of Partials
1	.91541
2	.01696
3	.80117
5	1.00000

Residual Analysis for $\hat{X}_5 = f(X_4)$

Obs. No.	Observed Y	Predicted Y	Residual	Normal Deviate
1	78.5000000	73.2782200	5.2217800	.5825343
2	74.3000000	79.1835100	-4.8835100	-.5447974
3	104.3000000	102.8047000	1.4953000	.1668135
4	87.6000000	82.8743200	4.7256800	.5271901
5	95.9000000	93.2085900	2.6914100	.3002498
6	109.2000000	101.3283700	7.8716300	.8781478
7	102.7000000	113.1389600	-10.4389600	-1.1645554
8	72.5000000	85.0888100	-12.5888100	-1.4043896
9	93.1000000	101.3283700	-8.2283700	-.9179452
10	115.9000000	98.3757200	17.5242800	1.9549835
11	83.8000000	92.4704300	-8.6704300	-.9672608
12	113.3000000	108.7099900	4.5900100	.5120549
13	109.4000000	108.7099900	.6900100	.0769765

Control Information

No. of observations	13
Response variable is no.	5
Risk level for B conf. interval	5%
List of excluded variables	3, 4

Variable entering	1
Sequential F-test	146.5229400
Percent variation explained R-SQ	97.8678500
Standard deviation of residuals	2.4063327
Mean of the response	95.4230750
Std. dev. as % of response mean	2.522%
Degrees of freedom	10
Determinant value	.9477514

ANOVA

Source	d.f.	Sums sqs.	Mean sq.	Overall F
Total	12	2715.7635000		
Regression	2	2657.8593000	1328.9296000	229.5042100
Residual	10	57.9043680	5.7904368	

B Coefficients and Confidence Limits

Var No.	Mean	Decoded B Coefficient	Limits Upper/Lower	Standard Error	Partial F-test
	48.1538450	.6622507	.7644149 .5600865	.0458547	208.5823200
	7.4615383	1.4683057	1.7385638 1.1980476	.1213008	146.5229400

Constant Term in Prediction Equation   52.5773400

Squares of Partial Correlation Coefficients of Variables Not in Regression

Variables	Square of Partials
3	.16914
4	.17152
5	1.00000

Residual Analysis for $\hat{X}_5 = f(X_2, X_1)$

Obs. No.	Observed Y	Predicted Y	Residual	Normal Deviate
1	78.5000000	80.0739960	-1.5739960	-.6541057
2	74.3000000	73.2509140	1.0490860	.4359688
3	104.3000000	105.8147300	-1.5147300	-.6294765
4	87.6000000	89.2584720	-1.6584720	-.6892114
5	95.9000000	97.2925130	-1.3925130	-.5786868
6	109.2000000	105.1524800	4.0475200	1.6820284
7	102.7000000	104.0020500	-1.3020500	-.5410931
8	72.5000000	74.5754150	-2.0754150	-.8624805
9	93.1000000	91.2754870	1.8245130	.7582131
10	115.9000000	114.5375400	1.3624600	.5661977
11	83.8000000	80.5356710	3.2643290	1.3565576
12	113.3000000	112.4372400	.8627600	.3585373
13	109.4000000	112.2934400	-2.8934400	-1.2024272

Control Information

No. of observations	13
Response variable is no.	5
Risk level for B conf. interval	5%
List of excluded variables	2, 4

Variable entering	3
Sequential F-test	.3146887
Percent variation explained R-SQ	54.8166700
Standard deviation of residuals	11.0773310
Mean of the response	95.4230750
Std. dev. as % of response mean	11.609%
Degrees of freedom	10
Determinant value	.3208036

ANOVA

Source	d.f.	Sums sqs.	Mean sq.	Overall F
Total	12	2715.7635000		
Regression	2	1488.6911000	744.3455500	6.0660270
Residual	10	1227.0726000	122.7072600	

B Coefficients and Confidence Limits

ar o.	Mean	Decoded B Coefficient	Limits Upper/Lower	Standard Error	Partial F-test
1	7.4615383	2.3124675	4.4508523 .1740827	.9597778	5.8051026
3	11.7692300	4944674	2.4583356 -1.4694008	.8814489	.3146887

Constant Term in Prediction Equation   72.3490110

Squares of Partial Correlation Coefficients of Variables Not in Regression

Variables	Square of Partials
2	.96079
4	.95857
5	1.00000

Residual Analysis for $\hat{X}_5 = f(X_1, X_3)$

Obs. No.	Observed Y	Predicted Y	Residual	Normal Deviate
1	78.5000000	91.5030870	-13.0030870	-1.1738465
2	74.3000000	82.0784890	-7.7784890	-.7021988
3	104.3000000	101.7418900	2.5581100	.2309320
4	87.6000000	101.7418900	-14.1418900	-1.2766513
5	95.9000000	91.5030870	4.3969130	.3969289
6	109.2000000	102.2363500	6.9636500	.6286397
7	102.7000000	87.6923590	15.0076500	1.3548073
8	72.5000000	85.5397610	-13.0397610	-1.1771572
9	93.1000000	85.8743590	7.2256410	.6522908
10	115.9000000	122.8886900	-6.9886900	-.6309002
11	83.8000000	86.0342280	-2.2342280	-.2016937
12	113.3000000	102.2363500	11.0636500	.9987650
13	109.4000000	99.4294250	9.9705800	.9000887

Control Information

No. of observations	13
Response variable is no.	5
Risk level for B conf. interval	5%
List of excluded variables	2, 3

Variable entering	4
Sequential F-test	159.2951900
Percent variation explained R-SQ	97.2471100
Standard deviation of residuals	2.7342662
Mean of the response	95.4230750
Std. dev. as % of response mean	2.865%
Degrees of freedom	10
Determinant value	.9397566

ANOVA

Source	d.f.	Sums sqs.	Mean sq.	Overall F
Total	12	2715.7635000		
Regression	2	2641.0015000	1320.5007000	176.6269800
Residual	10	74.7621170	7.4762117	

B Coefficients and Confidence Limits

Var No.	Mean	Decoded B Coefficient	Limits Upper/Lower	Standard Error	Partial F-test
1	7.4615383	1.4399582	1.7483504 1.1315660	.1384166	108.2238900
4	29.9999990	-.6139537	-.5055737 -.7223338	.0486446	159.2952400

Constant Term in Prediction Equation   103.0973800

Squares of Partial Correlation Coefficients of Variables Not in Regression

Variables	Square of Partials
2	.35833
3	.32003
5	1.00000

Residual Analysis for $\hat{X}_5 = f(X_1, X_4)$

Obs. No.	Observed Y	Predicted Y	Residual	Normal Deviate
1	78.5000000	76.3398700	2.1601300	.7900218
2	74.3000000	72.6117500	1.6882500	.6174417
3	104.3000000	106.6578400	-2.3578400	-.8623301
4	87.6000000	90.0811000	-2.4811000	-.9074098
5	95.9000000	92.9166200	2.9833800	1.0911081
6	109.2000000	105.4299300	3.7700700	1.3788233
7	102.7000000	103.7335300	-1.0335300	-.3779917
8	72.5000000	77.5233800	-5.0233800	-1.8371949
9	93.1000000	92.4703200	.6296800	.2302921
10	115.9000000	117.3737000	-1.4737000	-.5389746
11	83.8000000	83.6629200	.1370800	.0501341
12	113.3000000	111.5694700	1.7305300	.6329047
13	109.4000000	110.1295100	-.7295100	-.2668028

Control Information

No. of observations	13
Response variable is no.	5
Risk level for B conf. interval	5%
List of excluded variables	1, 4

Variable entering	3
Sequential F-test	11.8161580
Percent variation explained R-SQ	84.7025600
Standard deviation of residuals	6.4454832
Mean of the response	95.4230750
Std. dev. as % of response mean	6.755%
Degrees of freedom	10
Determinant value	.9806113

ANOVA

Source	d.f.	Sums sqs.	Mean sq.	Overall F
Total	12	2715.7635000		
Regression	2	2300.3212000	1150.1606000	27.6851910
Residual	10	415.4425300	41.5442530	

B Coefficients and Confidence Limits

Var No.	Mean	Decoded B Coefficient	Limits Upper/Lower	Standard Error	Partial F-test
2	48.1538450	.7313298	1.0003577 .4623019	.1207486	36.6827680
3	11.7692300	-1.0083860	-.3547984 -1.6619736	.2933517	11.8161580

Constant Term in Prediction Equation   72.0746600

Squares of Partial Correlation Coefficients of Variables Not in Regression

Variables	Square of Partials
1	.88419
4	.82232
5	1.00000

Residual Analysis for $\hat{X}_5 = f(X_2, X_3)$

Obs. No.	Observed Y	Predicted Y	Residual	Normal Deviate
1	78.5000000	85.0389180	-6.5389180	-1.0144961
2	74.3000000	78.1574330	-3.8574330	-.5984707
3	104.3000000	104.9620400	-.6620400	-.1027138
4	87.6000000	86.6787950	.9212050	.1429226
5	95.9000000	104.0534900	-8.1534900	-1.2649928
6	109.2000000	103.2223200	5.9776800	.9274215
7	102.7000000	106.8565100	-4.1565100	-.6448717
8	72.5000000	72.5613910	-.0613910	-.0095247
9	93.1000000	93.4155200	-.3155200	-.0489521
10	115.9000000	102.4136100	13.4863900	2.0923784
11	83.8000000	78.1349730	5.6650270	.8789142
12	113.3000000	111.2669500	2.0330500	.3154224
13	109.4000000	113.7379900	-4.3379900	-.6730279

Control Information

No. of observations	13
Response variable is no.	5
Risk level for B conf. interval	5%
List of excluded variables	1, 3

Variable entering	4
Sequential F-test	.4310840
Percent variation explained R-SQ	68.0060600
Standard deviation of residuals	9.3213731
Mean of the response	95.4230750
Std. dev. as % of response mean	9.768%
Degrees of freedom	10
Determinant value	.0533585

ANOVA

Source	d.f.	Sums sqs.	Mean sq.	Overall F
Total	12	2715.7635000		
Regression	2	1846.8837000	923.4418500	10.6279560
Residual	10	868.8799600	86.8879960	

B Coefficients and Confidence Limits

Var No.	Mean	Decoded B Coefficient	Limits Upper/Lower	Standard Error	Partial F-test
2	48.1538450	.3109057	1.9787999 −1.3569887	.7486061	.1724847
4	29.9999990	−.4569411	1.0936398 −2.0075220	.6959520	.4310840

Constant Term in Prediction Equation   94.1600050

Squares of Partial Correlation Coefficients of Variables Not in Regression

Variables	Square of Partials
1	.94479
3	.91515
5	1.00000

Residual Analysis for $\hat{X}_5 = f(X_2, X_4)$

Obs. No.	Observed Y	Predicted Y	Residual	Normal Devia
1	78.5000000	74.8270850	3.6729150	.3940315
2	74.3000000	79.4153310	-5.1153310	-.5487744
3	104.3000000	102.4319000	1.8681000	.2004104
4	87.6000000	82.3218490	5.2781510	.5662418
5	95.9000000	95.2480440	.6519560	.0699421
6	109.2000000	101.2071100	7.9928900	.8574799
7	102.7000000	113.4926600	-10.7926600	-1.1578401
8	72.5000000	83.6926720	-11.1926720	-1.2007535
9	93.1000000	100.8962000	-7.7962000	-.8363789
10	115.9000000	96.8921020	19.0079000	2.0391738
11	83.8000000	91.0602340	-7.2602340	-.7788803
12	113.3000000	109.1964800	4.1035200	.4402270
13	109.4000000	109.8182900	-.4182900	-.0448743

Control Information

No. of observations	13
Response variable is no.	5
Risk level for B conf. interval	5%
List of excluded variables	1, 2

Variable entering	3
Sequential F-test	40.2945330
Percent variation explained R-SQ	93.5289700
Standard deviation of residuals	4.1921130
Mean of the response	95.4230750
Std. dev. as % of response mean	4.393%
Degrees of freedom	10
Determinant value	.9991272

ANOVA

Source	d.f.	Sums sqs.	Mean sq.	Overall F
Total	12	2715.7635000		
Regression	2	2540.0256000	1270.0128000	72.2673510
Residual	10	175.7381100	17.5738110	

B Coefficients and Confidence Limits

Var No.	Mean	Decoded B Coefficient	Limits Upper/Lower	Standard Error	Partial F-test
4	29.9999990	-.7246003	-.5634470 -.8857535	.0723309	100.3573400
3	11.7692300	-1.1998510	-.7787179 -1.6209841	.1890185	40.2945430

Constant Term in Prediction Equation   131.2824000

Squares of Partial Correlation Coefficients of Variables Not in Regression

Variables	Square of Partials
1	.71073
2	.57997
5	1.00000

Residual Analysis for $\hat{X}_5 = f(X_4, X_3)$

Obs. No.	Observed Y	Predicted Y	Residual	Normal Deviat
1	78.5000000	80.6072800	-2.1072800	-.5026773
2	74.3000000	75.6054300	-1.3054300	-.3114014
3	104.3000000	107.1915900	-2.8915900	-.6897691
4	87.6000000	87.6273800	-.0273800	-.0065313
5	95.9000000	100.1714900	-4.2714900	-1.0189348
6	109.2000000	104.5425400	4.6574600	1.1110053
7	102.7000000	106.5373400	-3.8373400	-.9153713
8	72.5000000	73.0032700	-.5032700	-.1200516
9	93.1000000	93.7438800	-.6438800	-.1535932
10	115.9000000	107.6433900	8.2566100	1.9695580
11	83.8000000	79.0494200	4.7505800	1.1332184
12	113.3000000	111.7885400	1.5114600	.3605485
13	109.4000000	112.9883900	-3.5883900	-.8559860

ontrol Information

No. of observations	13
Response variable is no.	5
Risk level for B conf. interval	5%
List of excluded variables	4

ariable entering	1
equential F-test	68.7166430
ercent variation explained R-SQ	98.2284800
tandard deviation of residuals	2.3120568
ean of the response	95.4230750
d. dev. as % of response mean	2.423%
egrees of freedom	9
eterminant value	.3016276

NOVA

Source	d.f.	Sums sqs.	Mean sq.	Overall F
otal	12	2715.7635000		
egression	3	2667.6532000	889.2177300	166.3455300
esidual	9	48.1104560	5.3456062	

Coefficients and Confidence Limits

r o.	Mean	Decoded B Coefficient	Limits Upper/Lower	Standard Error	Partial F-test
	11.7692300	.2500169	.6678322 -.1677985	.1847106	1.8321249
	48.1538450	.6569150	.7569728 .5568573	.0442342	220.5476100
	7.4615383	1.6958894	2.1586529 1.2331259	.2045816	68.7166370

Constant Term in Prediction Equation   43.1936420

Squares of Partial Correlation Coefficients of Variables Not in Regression

Variables	Square of Partials
4	.00513
5	1.00000

Residual Analysis for $\hat{X}_5 = f(X_3, X_2, X_1)$

Obs. No.	Observed Y	Predicted Y	Residual	Normal Devia
1	78.5000000	78.6447590	-.1447590	-.0626105
2	74.3000000	72.6903190	1.6096810	.6962117
3	104.3000000	105.6358000	-1.3358000	-.5777540
4	87.6000000	89.2129250	-1.6129250	-.6976148
5	95.9000000	95.7245500	.1754500	.0758848
6	109.2000000	105.2289000	3.9711000	1.7175616
7	102.7000000	104.1725600	-1.4725600	-.6369048
8	72.5000000	75.7542670	-3.2542670	-1.4075203
9	93.1000000	91.5591350	1.5408650	.6664477
10	115.9000000	115.6823900	.2176100	.0941197
11	83.8000000	81.9165190	1.8834810	.8146344
12	113.3000000	112.4549600	.8450400	.3654928
13	109.4000000	111.8228900	-2.4228900	-1.0479370

Control Information

No. of observations	13
Response variable is no.	5
Risk level for B conf. interval	5%
List of excluded variables	3

Variable entering	4
Sequential F-test	1.8632545
Percent variation explained R-SQ	98.2335600
Standard deviation of residuals	2.3087418
Mean of the response	95.4230750
Std. dev. as % of response mean	2.419%
Degrees of freedom	9
Determinant value	.0500394

ANOVA

Source	d.f.	Sums sqs.	Mean sq.	Overall F
Total	12	2715.7635000		
Regression	3	2667.7911000	889.2637000	166.8321800
Residual	9	47.9725980	5.3302886	

B Coefficients and Confidence Limits

Var no.	Mean	Decoded B Coefficient	Limits Upper/Lower	Standard Error	Partial F-test
2	48.1538450	.4161107	.8359611 -.0037398	.1856103	5.0258974
1	7.4615383	1.4519380	1.7165861 1.1872899	.1169974	154.0080400
4	29.9999990	-.2365395	.1554371 -.6285160	.1732876	1.8632548

Constant Term in Prediction Equation   71.6482410

Squares of Partial Correlation Coefficients of Variables Not in Regression

Variables	Square of Partials
3	.00227
5	1.00000

Residual Analysis for $\hat{X}_5 = f(X_2, X_1, X_4)$

Obs. No.	Observed Y	Predicted Y	Residual	Normal Deviate
1	78.5000000	78.4383160	.0616840	.0267176
2	74.3000000	72.8673360	1.4326640	.6205389
3	104.3000000	106.1909600	-1.8909600	-.8190435
4	87.6000000	89.4016340	-1.8016340	-.7803532
5	95.9000000	95.6437590	.2562410	.1109873
6	109.2000000	105.3017700	3.8982300	1.6884651
7	102.7000000	104.1286700	-1.4286700	-.6188089
8	72.5000000	75.5918720	-3.0918720	-1.3392021
9	93.1000000	91.8182250	1.2817750	.5551833
10	115.9000000	115.5461100	.3538900	.1532826
11	83.8000000	81.7022630	2.0977370	.9086062
12	113.3000000	112.2443900	1.0556100	.4572231
13	109.4000000	111.6246700	-2.2246700	-.9635854

Control Information

No. of observations	13
Response variable is no.	5
Risk level for B conf. interval	5%
List of excluded variables	2

Variable entering	3
Sequential F-test	4.2358482
Percent variation explained R-SQ	98.1281200
Standard deviation of residuals	2.3766478
Mean of the response	95.4230750
Std. dev. as % of response mean	2.491%
Degrees of freedom	9
Determinant value	.2716373

ANOVA

Source	d.f.	Sums sqs.	Mean sq.	Overall F
Total	12	2715.7635000		
Regression	3	2664.9276000	888.3092000	157.2658800
Residual	9	50.8360910	5.6484545	

B Coefficients and Confidence Limits

Var No.	Mean	Decoded B Coefficient	Limits Upper/Lower	Standard Error	Partial F-test
1	7.4615383	1.0518542	1.5578282 .5458802	.2236844	22.1126000
4	29.9999990	-.6427963	-.5420373 -.7435552	.0445442	203.2401700
3	11.7692300	-.4100433	.0406197 -.8607064	.1992321	4.2358519

Constant Term in Prediction Equation   111.6844000

Squares of Partial Correlation Coefficients of Variables Not in Regression

Variables	Square of Partials
2	.05847
5	1.00000

Residual Analysis for $\hat{X}_5 = f(X_1, X_4, X_3)$

Obs. No.	Observed Y	Predicted Y	Residual	Normal Deviat
1	78.5000000	78.0193500	.4806500	.2022386
2	74.3000000	73.1602000	1.1398000	.4795830
3	104.3000000	107.1185300	-2.8185300	-1.1859266
4	87.6000000	89.7630300	-2.1630300	-.9101180
5	95.9000000	95.3748500	.5251500	.2209625
6	109.2000000	105.4228900	3.7771100	1.5892594
7	102.7000000	104.0124500	-1.3124500	-.5522274
8	72.5000000	75.4322700	-2.9322700	-1.2337839
9	93.1000000	92.2658200	.8341800	.3509902
10	115.9000000	115.4204600	.4795400	.2017716
11	83.8000000	81.4501900	2.3498100	.9887077
12	113.3000000	111.8508500	1.4491500	.6097454
13	109.4000000	111.2090500	-1.8090500	-.7611772

Control Information

No. of observations	13	
Response variable is no.	5	
Risk level for B conf. interval	5%	
List of excluded variables	1	

Variable entering	2
Sequential F-test	12.4271010
Percent variation explained R-SQ	97.2819800
Standard deviation of residuals	2.8638569
Mean of the response	95.4230750
Std. dev. as % of response mean	3.001%
Degrees of freedom	9
Determinant value	.0411008

ANOVA

Source	d.f.	Sums sqs.	Mean sq.	Overall F
Total	12	2715.7635000		
Regression	3	2641.9485000	880.6495000	107.3743300
Residual	9	73.8150840	8.2016760	

B Coefficients and Confidence Limits

Var No.	Mean	Decoded B Coefficient	Limits Upper/Lower	Standard Error	Partial F-test
4	29.9999990	-1.5570434	-1.0113254 -2.1027614	.2412547	41.6533540
3	11.7692300	-1.4479704	-1.1153091 -1.7803317	.1470651	96.9393040
2	48.1538450	-.9234143	-.3308925 -1.5159360	.2619460	12.4270980

Constant Term in Prediction Equation   203.6418100

Squares of Partial Correlation Coefficients of Variables Not in Regression

Variables	Square of Partials
1	.35157
5	1.00000

Residual Analysis for $\hat{X}_5 = f(X_4, X_3, X_2)$

Obs. No.	Observed Y	Predicted Y	Residual	Normal Deviate
1	78.5000000	77.5226200	.9773800	.3412810
2	74.3000000	74.1769900	.1230100	.0429526
3	104.3000000	109.2059800	-4.9059800	-1.7130674
4	87.6000000	90.2511700	-2.6511700	-.9257341
5	95.9000000	95.5540200	.3459800	.1208091
6	109.2000000	105.5673400	3.6326600	1.2684502
7	102.7000000	104.1216500	-1.4216500	-.4964110
8	72.5000000	74.6507200	-2.1507200	-.7509872
9	93.1000000	93.4590200	-.3590200	-.1253624
10	115.9000000	113.9663400	1.9336600	.6751944
11	83.8000000	80.4624500	3.3375500	1.1654038
12	113.3000000	110.9802200	2.3197800	.8100195
13	109.4000000	110.5813600	-1.1813600	-.4125066

Control Information

No. of observations	13	
Response variable is no.	5	
Risk level for B conf. interval	5%	

Variable entering	1
Sequential F-test	4.3375998
Percent variation explained R-SQ	98.2375700
Standard deviation of residuals	2.4460044
Mean of the response	95.4230750
Std. dev. as % of response mean	2.563%
Degrees of freedom	8
Determinant value	.0010377

ANOVA

Source	d.f.	Sums sqs.	Mean sq.	Overall F
Total	12	2715.7635000		
Regression	4	2667.9000000	666.9750000	111.4795200
Residual	8	47.8634980	5.9829372	

B Coefficients and Confidence Limits

Var No.	Mean	Decoded B Coefficient	Limits Upper/ Lower	Standard Error	Partial F-test
4	29.9999990	-.1440588	1.4909970 -1.7791144	.7090441	.0412794
3	11.7692300	.1019111	1.8422494 -1.6384272	.7547001	.0182345
2	48.1538450	.5101700	2.1792063 -1.1588665	.7237799	.4968402
1	7.4615383	1.5511043	3.2685233 -.1663147	.7447611	4.3375858

Constant Term in Prediction Equation   62.4051530

Squares of Partial Correlation Coefficients of Variables Not in Regression

Variables	Square of Partials
5	1.00000

$$\text{Residual Analysis for } \hat{X}_5 = f(X_4, X_3, X_2, X_1)$$

Obs. No.	Observed Y	Predicted Y	Residual	Normal Deviate
1	78.5000000	78.4952410	.0047590	.0019456
2	74.3000000	72.7887950	1.5112050	.6178260
3	104.3000000	105.9709300	-1.6709300	-.6831263
4	87.6000000	89.3270940	-1.7270940	-.7060879
5	95.9000000	95.6492470	.2507530	.1025154
6	109.2000000	105.2745500	3.9254500	1.6048417
7	102.7000000	104.1486600	-1.4486600	-.5922557
8	72.5000000	75.6749840	-3.1749840	-1.2980287
9	93.1000000	91.7216450	1.3783550	.5635129
10	115.9000000	115.6184400	.2815600	.1151102
11	83.8000000	81.8090130	1.9909870	.8139752
12	113.3000000	112.3270100	.9729900	.3977875
13	109.4000000	111.6943300	-2.2943300	-.9379910

STEP-WISE SOLUTION FOR THE HALD DATA

Original and/ or Transformed Data

	X_1	X_2	X_3	X_4	X_5
1	7.00000000	26.00000000	6.00000000	60.00000000	78.50000000
2	1.00000000	29.00000000	15.00000000	52.00000000	74.30000000
3	11.00000000	56.00000000	8.00000000	20.00000000	104.30000000
4	11.00000000	31.00000000	8.00000000	47.00000000	87.60000000
5	7.00000000	52.00000000	6.00000000	33.00000000	95.90000000
6	11.00000000	55.00000000	9.00000000	22.00000000	109.20000000
7	3.00000000	71.00000000	17.00000000	6.00000000	102.70000000
8	1.00000000	31.00000000	22.00000000	44.00000000	72.50000000
9	2.00000000	54.00000000	18.00000000	22.00000000	93.10000000
10	21.00000000	47.00000000	4.00000000	26.00000000	115.90000000
11	1.00000000	40.00000000	23.00000000	34.00000000	83.80000000
12	11.00000000	66.00000000	9.00000000	12.00000000	113.30000000
13	10.00000000	68.00000000	8.00000000	12.00000000	109.40000000

Means of Transformed Variables

	X_1	X_2	X_3	X_4	X_5
1	7.46153830	48.15384500	11.76923000	29.99999900	95.42307500

Std. Deviations of Transformed Variables

	X_1	X_2	X_3	X_4	X_5
1	5.88239440	15.56087900	6.40512590	16.73817800	15.04372400

Correlation Matrix

	X_1	X_2	X_3	X_4	X_5
1	.99999991	.22857948	-.82413372	-.24544512	.73071745
2	.22857948	1.00000010	-.13924238	-.97295516	.81625268
3	-.82413372	-.13924238	.99999991	.02953701	-.53467065
4	-.24544512	-.97295516	.02953701	1.00000010	-.82130513
5	.73071745	.81625268	-.53467065	-.82130513	.99999999

Control Information

No. of observations	13
F level for entering a variable	3.28
F level for deleting a variable	3.28
Response variable is no.	5
Risk level for B conf. interval	5%

Step No. 1

Variable entering	4
Sequential F-test	22.7985280
Percent variation explained R-SQ	67.4542100
Standard error of Y	8.9639014
Mean of the response	95.4230750
Std. error as a % of mean response	9.394%
Degrees of freedom	11
Determinant value	1.0000001

ANOVA

Source	d.f.	Sums sqs.	Mean sq.	Overall F
Total	12	2715.7635000		
Regression	1	1831.8968000	1831.8968000	22.7985300
Residual	11	883.8668200	80.3515290	

B Coefficients and Confidence Limits

Var No.	Mean	Decoded B Coefficient	Limits Upper/Lower	Standard Error	Partial F-test
4	29.9999990	-.7381620	-.3978962 -1.0784277	.1545960	22.7985270

Constant Term in Prediction Equation   117.5679300

Squares of Partial Correlation Coefficients of Variables Not in Regression

Variables	Square of Partials
1	.91541
2	.01696
3	.80117
5	1.00000

Step No. 2

Variable entering	1
Sequential F-test	108.2240500
Percent variation explained R-SQ	97.2471100
Standard error of Y	2.7342642
Mean of the response	95.4230750
Std. error as a % of mean response	2.865%
Degrees of freedom	10
Determinant value	.9397567

ANOVA

Source	d.f.	Sums sqs.	Mean sq.	Overall F
Total	12	2715.7635000		
Regression	2	2641.0015000	1320.5007000	176.6272400
Residual	10	74.7620080	7.4762008	

B Coefficients and Confidence Limits

Var No.	Mean	Decoded B Coefficient	Limits Upper/Lower	Standard Error	Partial F-test
4	29.9999990	-.6139538	-.5055738 -.7223338	.0486445	159.2954900
1	7.4615383	1.4399582	1.7483502 1.1315662	.1384165	108.2240500

Constant Term in Prediction Equation   103.0973800

Squares of Partial Correlation Coefficients of Variables Not in Regression

Variables	Square of Partials
2	.35833
3	.32003
5	1.00000

Step No. 3

Variable entering	2
Sequential F-test	5.0258747
Percent variation explained R-SQ	98.2335500
Standard error of Y	2.3087426
Mean of the response	95.4230750
Std. error as a % of mean response	2.419%
Degrees of freedom	9
Determinant value	.0500394

ANOVA

Source	d.f.	Sums sqs.	Mean sq.	Overall F
Total	12	2715.7635000		
Regression	3	2667.7908000	889.2636000	166.8320500
Residual	9	47.9726310	5.3302923	

B Coefficients and Confidence Limits

Var No.	Mean	Decoded B Coefficient	Limits Upper/Lower	Standard Error	Partial F-test
4	29.9999990	-.2365401	.1554367 -.6285170	.1732877	1.8632619
1	7.4615383	1.4519379	1.7165861 1.1872897	.1169975	154.0079500
2	48.1538450	.4161100	.8359608 -.0037408	.1856104	5.0258730

Constant Term in Prediction Equation 71.6482910

Squares of Partial Correlation Coefficients of Variables Not in Regression

Variables	Square of Partials
3	.00227
5	1.00000

Step No. 4

Variable leaving is	4
Sequential F-test	1.8632611
Percent variation explained R-SQ	97.8678500
Standard error of Y	2.4063325
Mean of the response	95.4230750
Std. error as a % of mean response	2.522%
Degrees of freedom	10
Determinant value	.9477514

ANOVA

Source	d.f.	Sums sqs.	Mean sq.	Overall F
Total	12	2715.7635000		
Regression	2	2657.8593000	1328.9296000	229.5042500
Residual	10	57.9043570	5.7904357	

B Coefficients and Confidence Limits

Var No.	Mean	Decoded B Coefficient	Limits Upper/Lower	Standard Error	Partial F-test
1	7.4615383	1.4683057	1.7385638 1.1980476	.1213008	146.5229500
2	48.1538450	.6622507	.7644149 .5600864	.0458547	208.5821200

Constant Term in Prediction Equation   52.5773400

Squares of Partial Correlation Coefficients of Variables Not in Regression

Variables	Square of Partials
3	.16914
4	.17152
5	1.00000

Residual Analysis

Obs. No.	Observed Y	Predicted Y	Residual	Normal Deviate
1	78.5000000	80.0739960	-1.5739960	-.6541058
2	74.3000000	73.2509140	1.0490860	.4359688
3	104.3000000	105.8147300	-1.5147300	-.6294766
4	87.6000000	89.2584720	-1.6584720	-.6892115
5	95.9000000	97.2925130	-1.3925130	-.5786869
6	109.2000000	105.1524800	4.0475200	1.6820285
7	102.7000000	104.0020500	-1.3020500	-.5410931
8	72.5000000	74.5754150	-2.0754150	-.8624806
9	93.1000000	91.2754870	1.8245130	.7582132
10	115.9000000	114.5375400	1.3624600	.5661977
11	83.8000000	80.5356710	3.2643290	1.3565577
12	113.3000000	112.4372400	.8627600	.3585373
13	109.4000000	112.2934400	-2.8934400	-1.2024273

INDEX

Applied Probability and Statistics (*Continued*)

JOHNSON and KOTZ · Distributions in Statistics
Discrete Distributions
Continuous Univariate Distributions-1
Continuous Univariate Distributions-2
Continuous Multivariate Distributions

JOHNSON and LEONE · Statistics and Experimental Design: In Engineering and the Physical Sciences, Volumes I and II

KEENEY and RAIFFA · Decisions with Multiple Objectives

LANCASTER · The Chi Squared Distribution

LANCASTER · An Introduction to Medical Statistics

LEWIS · Stochastic Point Processes

MANN, SCHAFER and SINGPURWALLA · Methods for Statistical Analysis of Reliability and Life Data

MEYER · Data Analysis for Scientists and Engineers

MILTON · Rank Order Probabilities: Two-Sample Normal Shift Alternatives

OTNES and ENOCHSON · Digital Time Series Analysis

PRENTER · Splines and Variational Methods

RAO and MITRA · Generalized Inverse of Matrices and Its Applications

SARD and WEINTRAUB · A Book of Splines

SEAL · Stochastic Theory of a Risk Business

SEARLE · Linear Models

THOMAS · An Introduction to Applied Probability and Random Processes

WHITTLE · Optimization under Constraints

WONNACOTT and WONNACOTT · Econometrics

YOUDEN · Statistical Methods for Chemists

ZELLNER · An Introduction to Bayesian Inference in Econometrics

Tracts on Probability and Statistics

BHATTACHARYA and RAO · Normal Approximation and Asymptotic Expansions

BILLINGSLEY · Convergence of Probability Measures

CRAMER and LEADBETTER · Stationary and Related Stochastic Processes

JARDINE and SIBSON · Mathematical Taxonomy

KINGMAN · Regenerative Phenomena

RIORDAN · Combinatorial Identities